Risk Assessment for Biological Invasions

Risk Assessment for Biological Invasions

Guest Editor

Xubin Pan

MDPI

Basel • Beijing • Wuhan • Barcelona • Belgrade • Novi Sad • Cluj • Manchester

Guest Editor
Xubin Pan
Institute of Plant Inspection
and Quarantine
Chinese Academy of Quality
and Inspection & Testing
Beijing
China

Editorial Office
MDPI AG
Grosspeteranlage 5
4052 Basel, Switzerland

This is a reprint of the Special Issue, published open access by the journal *Biology* (ISSN 2079-7737), freely accessible at: https://www.mdpi.com/journal/biology/special_issues/biological_invasions_risk.

For citation purposes, cite each article independently as indicated on the article page online and as indicated below:

Lastname, A.A.; Lastname, B.B. Article Title. *Journal Name* **Year**, *Volume Number*, Page Range.

ISBN 978-3-7258-4497-5 (Hbk)
ISBN 978-3-7258-4498-2 (PDF)
https://doi.org/10.3390/books978-3-7258-4498-2

Cover image courtesy of Xubin Pan
Invasive alien plants

Contents

About the Editor

Xubin Pan

Xubin Pan is a pest risk analysis expert at the Chinese Academy of Quality and Inspection & Testing (formerly the Chinese Academy of Inspection and Quarantine). He reconstructed the Species—Area Theory using set theory and systematically summarized the progress of pest risk analysis. He is currently working on a National Key R&D Program of China (Grant No. 2023YFC2605204).

Preface

It is with great pleasure that I welcome you to this Special Issue on "Risk Assessment for Biological Invasions", published in the journal *Biology*. This collection of articles brings together a diverse range of research and insights into the complex challenges posed by invasive alien species and the critical role of risk assessment in managing these threats.

Biological invasions are a global concern, affecting ecosystems, economies, and societies worldwide. The introduction and spread of non-native species can lead to significant ecological and economic impacts, making it essential to prioritize and address these issues through robust risk assessment. This Special Issue aims to highlight the latest advancements in this field, providing a platform for researchers to share their findings and contribute to the broader understanding of biological invasions.

The contributions in this Special Issue come from various disciplines and geographical regions, reflecting the global nature of biological invasions and the collaborative efforts required to address them. The articles included in this issue cover a wide range of topics, from the distribution and ecological risks of specific invasive species to the development of risk assessment methodologies and the identification of data gaps in invasion management. We hope that this collection will not only enhance our understanding of the risks associated with invasive alien species, but also provide valuable insights for policymakers, conservationists, and researchers working to mitigate these risks.

I would like to express my sincere gratitude to all the authors who have contributed to this Special Issue, as well as the reviewers whose expertise and dedication have ensured the high quality of the published articles. I would also like to thank the editorial team at *Biology* for their support and assistance throughout the publication process. Additionally, this Special Issue also could not have been published without the support of the National Key R&D Program of China (Grant No. 2023YFC2605200).

I hope that this Special Issue will serve as a valuable resource for the scientific community and beyond, fostering further research and collaboration in the field of risk assessment for biological invasions.

Enjoy exploring the articles in this Special Issue, and we look forward to your continued interest in this important topic.

Xubin Pan
Guest Editor

biology

Editorial

Global Initiative for the Risk Assessment Working Group of Invasive Alien Species

Xubin Pan

Chinese Academy of Inspection and Quarantine, Beijing 100176, China; xubin.hu.pan@gmail.com

Biological invasion is not only an ecological phenomenon but also an administrative challenge. To manage invasive alien species (IAS), given our limited resources, we must be clear about which species, pathways, and habitats should be our top priority, and we can use risk assessment tools to identify and rank them. However, biological invasion risk assessment faces a plethora of dilemmas, which hinders our effective control of IAS. This Special Issue brings together contributions from researchers with different concerns on this topic, and it is hoped that such diverse concerns will attract more attention and motivate us to do more.

The spatial dispersal of IAS means that they cross the boundary of a geographical unit/ecoregion (source area) with these species to reach another geographical unit/ecoregion (sink area) without these species [1]. However, effective management of IAS cannot only consider the sink area, otherwise the source area will provide propagule pressure to the sink area again, resulting in the failure of the IAS control program. Conversely, sink areas can also provide propagule pressure to source areas, as well as other areas, causing more widespread damage. Thus, control measures of IAS must be taken globally.

In fact, international collaboration has become a consensus of IAS management. IAS is listed as an important issue in both the Convention on Biological Diversity agenda and the Sustainable Development Goals. SDSs 15.8 aims to "Prevent Invasive Alien Species on Land and in Water Ecosystems" [2]. However, current international cooperation remains limited to paper to a certain extent, and too much time and resources are wasted in endless meetings. Even Aichi Target 9 states that "By 2020, invasive alien species and pathways are identified and prioritized, priority species are controlled or eradicated, and measures are in place to manage pathways to prevent their introduction and establishment", which has been only partially achieved [3,4]. Therefore, it is necessary to both rethink and restructure our international efforts to produce concrete results.

Due to the prerequisites and operability of IAS risk assessment, it should become a priority development area for international cooperation [2,3,5]. Here, I propose to initiate a Risk Assessment Working Group of Invasive Alien Species (IAS-RAWG) on a global scale, which can pool the wisdom of global experts and provide imperative support for the targeted management of IAS in countries/areas with insufficient risk assessment capacities.

1. Risk Assessment Working Group of Invasive Alien Species

The members of IAS-RAWG should come from different countries with various academic backgrounds, and its operating mode can select an iterative mode, such as the two-step method [6]. The global working group uses global/regional data to generate spatial-implicit global reports for risk warning, and the national/local working group then uses the national/local data to generate spatial-explicit national/local reports for risk management (Figure 1). National/local IAS data can be extracted from global/regional data and further added by national/local working groups. At the same time, the global working group can also train national/local working groups to improve their capacities.

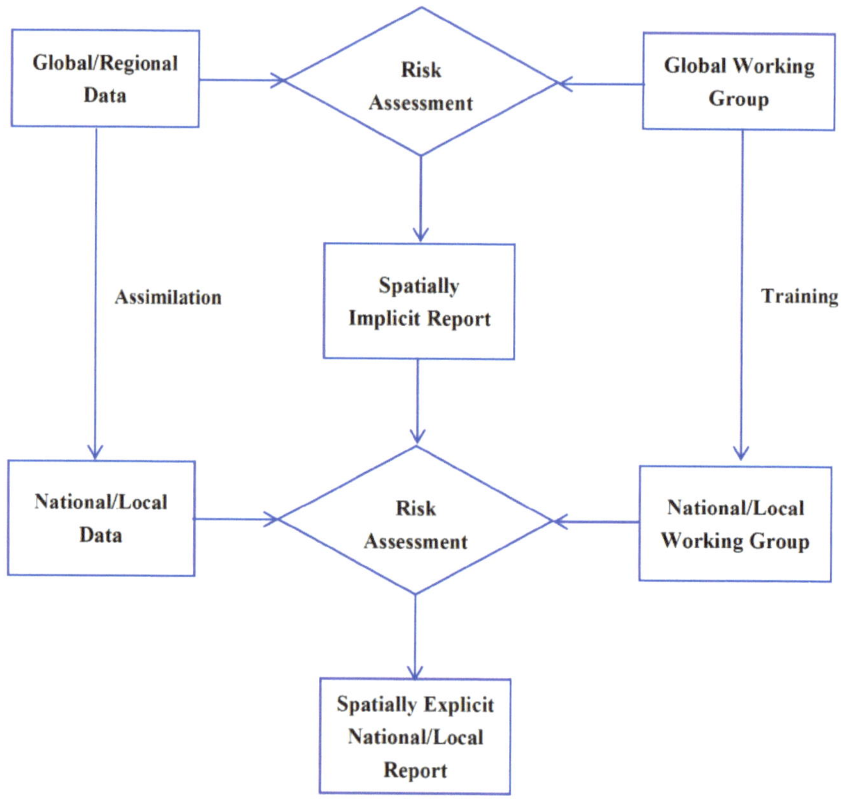

Figure 1. Working flowchart of risk assessment working group of invasive alien species.

2. Necessity Analysis of IAS-RAWG

Undoubtedly, risk assessment of IAS is an indispensable component of effective management actions of IAS, through which we can understand which alien species should be concerned. With limited financial resources and unignorable opportunity costs, it is best to give precedence to prevent and control specific species that are likely to cause the most damage.

However, there are still many countries/areas that lack adequate risk assessment capacities, although tremendous progress has been made over the past few decades in identifying IAS and prioritizing them based on the risk they pose, such as the Horizon Scanning Tool (HST) and Pest Risk Analysis (PRA) Tool developed by CABI. Nevertheless, only "invasive alien species are identified and prioritized" in Aichi Target 9 has been or is likely to be achieved, which is almost common sense for IAS, but "priority species are controlled or eradicated" has not been achieved yet, let alone the "measures" [4]. Therefore, it is time to go all out to focus on "priority species" at this stage.

There is a large body of literature focusing on the risk assessment of IAS, exploring not only the pathways and potential geographical distribution of various IAS, but also their impacts and economic losses [7,8]. However, the conclusions from most academic articles remain critically distant from policy and application, as do the recommendations for international cooperation in many places in the discussion section. As we all understand the importance of risk assessment and international cooperation, the global expert community should provide more concrete and practical support.

Risk assessment of IAS is not as simple as it may appear. Widely used methodologies such as SOM and Maxent for the establishment assessment of IAS are not easily learned by beginners through traditional training courses. Meanwhile, risk assessment is not only about techniques but also experiences, which means that learning by 'doing' is a better approach for countries or areas with insufficient risk assessment capacities. These countries/areas have little experience in this area, including a lack of experts and organizations, which means they need the fish first and fishing later.

It should be emphasized that the characteristics of risk assessment are particularly suitable for the establishment of a global working group. Global experts can bridge data and knowledge gaps for each other because no one is omnipotent. A reasonable team can have as much information as possible, including not only English, but also Arabic, Chinese, French, Russian, and Spanish. The same is true for assessment techniques, as one person or group cannot master all the methods required. More importantly, group communication can even spark new ideas. Furthermore, this small step in risk assessment with the help of the global group will improve the capacities of countries/areas that are not yet ready for IAS management. In other words, we already have enough "Why" knowledge about IAS, but we need more "How" guidance and even manuals for IAS management in specific countries/areas.

3. Feasibility Analysis of IAS-RAWG

As mentioned above, IAS risk assessment tasks require databases and assessment techniques. There are currently several open-access databases/datasets, including the Global Register of Introduced and Invasive Species, the Global Invasive Species Database, GBIF, CABI-ISC, EPPO Global Database, IPBES IAS Report, InvaCost, Global Naturalized Alien Flora, and so on. Assessment methods include Maxent and Biomod2, SOM, Risk Estimation Matrix [9], EICAT and SEICAT [10,11], and Pest Risk Analysis Standards developed by IPPC (ISPM 2, ISPM 11, and ISPM 21). At the same time, there are many significant findings in the academic literature and reports on the risk assessment of IAS at a global/regional scale, covering the spatial dynamics and the impacts of IAS. In other words, the main gap in the management of IAS exists between the global/regional and national/local scale.

In addition to the database teams and symposium committees, several global expert working groups are currently functioning well. The Global Health Work Group (also known as Epidemiology in Action) focuses on tuberculosis and HIV, and more recently, SARS-CoV-2. IUCN has the Species Survival Commission—Invasive Species Specialist Group. IPPC has established the Phytosanitary Measures Research Group and Beyond Compliance Facilitators. We have also established the International Pest Risk Research Group improving pest risk modeling methods, and another group of 36 experts identifying the drivers of potential alien species impacts [12]. The Mountain Invasion Research Network was established in 2005 to monitor non-native plant invasions in mountainous areas globally. However, it has to be said that compared with international demand and the damages IAS has caused, there are too few global expert working groups specialized in IAS risk assessment. Some countries, such as China and the United States, have indeed taken a similar approach by establishing task forces composed of senior experts from capable national departments and agencies [13].

For countries/areas without IAS management capacities, financial support is a major bottleneck, but for the global risk assessment expert group, this is not a principal problem. On the one hand, most experts should receive financial support from their agencies, international and national funding, or volunteer contributions from IAS assessment recipient countries/areas. If risk assessment groups demonstrate their capacities and contributions, they can apply for further international financial support for more countries/areas or species. On the other hand, the costs for global working groups are negligible, especially in the era of online documents and meetings, which will become the dominant way of working on risk assessment tasks. If working groups need face-to-face discussions on some key issues, they can acquire help from various international symposiums.

4. Opportunities and Challenges

IAS management is a systematic project, and risk assessment is an integral part of this work. Risk assessment alone will not prevent or control IAS, and similarly, without risk assessment, IAS management will be ineffective. The successful implementation of a global initiative, therefore, requires that we address the opportunities and challenges that come with it.

4.1. Opportunities

Now is the best time for the international community to reach an agreement on this global initiative. Firstly, the IPBES IAS report provides a baseline assessment of IAS, not only including information on status and trends, drivers, impacts, management, and governance and policy options, but also creating various databases. In other words, it is more than just a report; it builds an encyclopedic infrastructure for IAS management, especially risk assessment. Secondly, following Aichi Target 9, the Kunming-Montreal Global Biodiversity Framework provides institutional support in Target 6 to reduce IAS threats to biodiversity [14]. Meanwhile, the WTO/SPS and WHO are also concerned about the global spread of quarantine species. Thirdly, experts in the field of IAS continue to emerge, providing an adequate intelligence pool and academic support. Finally, and most importantly, the development of information technology has provided strong support for IAS management, not only in terms of databases and algorithms, but also through citizen science, online collaboration, and Artificial General Intelligence.

4.2. Challenges

However, the implementation of this global initiative faces several challenges. In the first instance, we should take effective actions against IAS as quickly as possible, which will reduce the total cost of damage by the IAS in a specific country/area or at the global/regional scale. In addition, globalization and environmental change will exacerbate the current invasion situation [8]. Next, we should guide national/local governments and experts to conduct a comprehensive IAS survey based on IAS risk assessments, and in turn the background list (IAS catalogue) can become the basis for future IAS risk assessments. Finally, we need to transform our operating mode of IAS management to adapt to the features of IAS management. For risk assessment, it is better to choose a centralized top-down rather than distributed bottom-up approach [15].

5. Conclusions

With the global spread and huge impact of IAS, it is time to take effective action as quickly as possible. As risk assessment is important for the management of IAS, I propose a global initiative to establish an international Risk Assessment Working Group of Invasive Alien Species (IAS-RAWG) to overcome the shortcomings of the current discrete assessment mode. The analysis of necessity and feasibility of IAS-RAWG indicates that IAS-RAWG is a useful attempt that is suitable for the development of current technology. Although this mode faces challenges, it also brings valuable opportunities for IAS management.

Funding: This work was jointly funded by the National Key R&D Program of China (Grant No. 2023YFC2605200) and the Basic Scientific Research Foundation of the Chinese Academy of Inspection and Quarantine (Grant No. 2021JK007).

Institutional Review Board Statement: Not applicable.

Informed Consent Statement: Not applicable.

Data Availability Statement: Not applicable.

Acknowledgments: Special thanks are due to the authors, the reviewers, and the editorial staff of Biology for their great contributions to this Special Issue.

Conflicts of Interest: The author declares no conflicts of interest.

References

1. Pan, X.; Zhang, J.; Xu, H.; Zhang, X.; Zhang, W.; Song, H.; Zhu, S. Spatial similarity in the distribution of invasive alien plants and animals in China. *Nat. Hazards* **2015**, *77*, 1751–1764. [CrossRef]
2. United Nations. Transforming Our World: The 2030 Agenda for Sustainable Development. 2015. Available online: https://sdgs.un.org/2030agenda (accessed on 1 October 2024).
3. Convention on Biological Diversity Aichi Biodiversity Target. 2010. Available online: https://www.cbd.int/sp/targets (accessed on 1 October 2024).
4. CBD. Global Biodiversity Outlook 5. 2020. Available online: https://www.cbd.int/gbo5 (accessed on 1 October 2024).
5. WTO. *The WTO Agreement on the Application of Sanitary and Phytosanitary Measures*; World Trade Organization: Geneva, Switzerland, 1994.
6. Pan, X. Two-step sampling can produce triphasic species-area relationship. *J. Plant Ecol.* **2021**, *14*, 673–678. [CrossRef]
7. Diagne, C.; Leroy, B.; Gozlan, R.E.; Vaissière, A.C.; Assailly, C.; Nuninger, L.; Roiz, D.; Jourdain, F.; Jarić, I.; Courchamp, F. InvaCost, a public database of the economic costs of biological invasions worldwide. *Sci. Data* **2020**, *7*, 277. [CrossRef] [PubMed]
8. IPBES. *Thematic Assessment Report on Invasive Alien Species and Their Control of the Intergovernmental Science-Policy Platform on Biodiversity and Ecosystem Services*; Roy, H.E., Pauchard, A., Stoett, P., Renard Truong, T., Eds.; IPBES Secretariat: Bonn, Germany, 2023.
9. Department of Agriculture and Water Resources. *Biosecurity Import Risk Analysis Guidelines 2016: Managing Biosecurity Risks for Imports into Australia*; Department of Agriculture and Water Resources: Canberra, Australia, 2016.
10. Blackburn, T.M.; Essl, F.; Evans, T.; Hulme, P.E.; Jeschke, J.M.; Kühn, I.; Kumschick, S.; Mrugała, A.; Marková, Z.; Nentwig, W.; et al. A unified classification of alien taxa based on the magnitude of their environmental impacts. *PLoS Biol.* **2014**, *12*, e1001850. [CrossRef] [PubMed]
11. Bacher, S.; Blackburn, T.M.; Essl, F.; Genovesi, P.; Heikkilä, J.; Jeschke, J.M.; Jones, G.; Keller, R.; Kenis, M.; Kueffer, C.M.; et al. Socio-economic impact classification of alien taxa (SEICAT). *Methods Ecol. Evol.* **2018**, *9*, 159–168. [CrossRef]
12. Essl, F.; Lenzner, B.; Bacher, S.; Bailey, S.; Capinha, C.; Daehler, C.; Dullinger, S.; Genovesi, P.; Hui, C.; Roura-Pascual, N.; et al. Drivers of future alien species impacts: An expert-based assessment. *Glob. Change Biol.* **2020**, *26*, 4880–4893. [CrossRef] [PubMed]
13. Herrick, C.N. A review of the U.S. invasive species policy mix: Questioning the prospect of an integrated regime. *Environ. Policy Gov.* **2019**, *29*, 262–278. [CrossRef]
14. CBD. 15/4. Kunming-Montreal Global Biodiversity Framework. Decision Adopted by the Conference of the Parties to the Convention on Biological Diversity. In Proceedings of the Fifteenth Meeting (Part II) of Conference of the Parties to the Convention on Biological Diversity, Montreal, QC, Canada, 7–19 December 2022.
15. Reed, M.S.; Vella, S.; Challies, E.; de Vente, J.; Frewer, L.; Hohenwallner-Ries, D.; Huber, T.; Neumann, R.K.; Oughton, E.A.; Sidoli del Ceno, J.; et al. A theory of participation: What makes stakeholder and public engagement in environmental management work? *Restor. Ecol.* **2018**, *26* (Suppl. 1), S7–S17. [CrossRef]

biology

MDPI

Article

Aliens on the Road: Surveying Wildlife Roadkill to Assess the Risk of Biological Invasion

Andrea Viviano [1], Marcello D'Amico [2,*] and Emiliano Mori [1,3]

1 Istituto di Ricerca sugli Ecosistemi Terrestri, Consiglio Nazionale delle Ricerche, 50019 Sesto Fiorentino, Italy
2 Department of Conservation Biology and Global Change, Doñana Biological Station, Spanish National Research Council (CSIC), 41092 Seville, Spain
3 National Biodiversity Future Center, 90133 Palermo, Italy
* Correspondence: damico@ebd.csic.es; Tel.: +34-954-232-340 (ext. 434113)

Simple Summary: Monitoring non-native species is important to assess their invasion risk. Here, we carried out a literature review on roadkill studies to investigate geographical patterns of biological invasions. We believe that roadkill data from published literature can be a valuable resource for people dealing with biological invasions, especially when field surveys cannot be performed. We retrieved 2314 studies published until January 2022, but only 40 of them (in addition to our own unpublished field data) included a full list of road-killed species (with a number of affected individuals for each species), so we were able to include them in our analyses. We classified all road-killed species from retrieved studies as native or introduced (domestic, introduced in historical times or recently released). We found that a higher number of introduced species would be recorded among roadkill in Mediterranean and Temperate regions with respect to Tropical and Desert areas. This result is in line with the current knowledge on non-native species distribution at the global scale, thus confirming that roadkill datasets can be used beyond the study of road impacts, such as for an assessment of different levels of biological invasions among different countries.

Abstract: Monitoring the presence and distribution of alien species is pivotal to assessing the risk of biological invasion. In our study, we carried out a worldwide review of roadkill data to investigate geographical patterns of biological invasions. We hypothesise that roadkill data from published literature can turn out to be a valuable resource for researchers and wildlife managers, especially when more focused surveys cannot be performed. We retrieved a total of 2314 works published until January 2022. Among those, only 41 (including our original data) fitted our requirements (i.e., including a total list of roadkilled terrestrial vertebrates, with a number of affected individuals for each species) and were included in our analysis. All roadkilled species from retrieved studies were classified as native or introduced (domestic, paleo-introduced, or recently released). We found that a higher number of introduced species would be recorded among roadkill in Mediterranean and Temperate areas with respect to Tropical and Desert biomes. This is definitely in line with the current knowledge on alien species distribution at the global scale, thus confirming that roadkill datasets can be used beyond the study of road impacts, such as for an assessment of different levels of biological invasions among different countries.

Keywords: alien species; allochthonous species; biological invasions; introduced species; invasive species; non-native species; road ecology; roadkill; vertebrates; wildlife–vehicle collisions

1. Introduction

In recent decades, the percentage of the planet covered by roads has increased worldwide [1–3], together with their impacts on biodiversity and ecosystems [4–6]. Fragmentation, light, noise and chemical pollution, as well as direct human disturbance, are among the most well-recognised road impacts [7–9]. Other impacts have also been described, such as roads promoting biological invasions, especially for plants [10–12], but also for some animal species [13–15]. However, wildlife-vehicle collisions are the most evident road impact, as well as the most investigated one in scientific literature [16–18].

Roadkill is so well studied because it can affect the persistence of animal populations [18–20], mainly by limiting dispersal possibilities and then disrupting metapopulations, thus triggering local extinction vortices [21–23]. Such extinctions are often a relevant issue also for economic reasons because, in some cases, they concern species providing valuable ecosystem services [24–26]. Furthermore, roadkill is so well studied because they represent a major and often overlooked issue for road safety, especially when involving large vertebrates [27–29].

Roadkill observations and even surveys are increasingly used to monitor and update species distribution, e.g., [30–34], by means of citizen science, e.g., [35–37]. Individual hobbyists and wildlife enthusiasts have greatly increased in the last 30 years, and the majority of them upload opportunistic records to online platforms [38–40], including projects on road mortality [35,41,42]. Similarly, the low cost of detecting and uploading roadkill to online databanks (e.g., iNaturalist, GBIF) has also been used to monitor biological invasions, particularly at early invasion stages [43–45]. Immediately after release or escape, invasive alien species are present in new areas at low densities, which may prevent their detection through standardised monitoring, therefore relying on accidental observations, such as roadkill as first detection [46–48]. For instance, in Central Italy, the monitoring of the invasive population of the Northern raccoon *Procyon lotor* started after the first detections of several roadkilled individuals [49]. Furthermore, given that often roadkilled individuals are juveniles in dispersal [50–52], the detection of invasive alien species may also provide evidence of the potential range of expansion of established populations [53–55].

The effectiveness of detecting invasive alien species or monitoring their distribution through roadkill has never been quantified, although field evidence of the successfulness of this method has been suggested [34,56,57]. Using roadkill to study alien species represents a very novel approach to biological invasions, and to the best of our knowledge, there are no studies on roadkill of alien species comparing different locations. With this work, we aimed to fill this gap, and we hypothesised that the percentage of roadkilled allochthonous species varies among regions, being higher in the Mediterranean biome (first prediction), especially in Italy (second prediction). In fact, the Mediterranean biome is one of the world's biodiversity hotspots, but it is also threatened by a high number of alien species [58–60], probably because it includes, at the same time, the highest biodiversity but also the highest levels of human pressure and trade [61–63]. In the Mediterranean biome, Italy is reported to be one of the richest countries in animal diversity [62,64,65], as well as one of the richest in alien species (nearly 17% of all mammals and over 57% of freshwater fish species are alien in Italy [66,67]). This is probably because of the intense trade pressure which has occurred in this area for millennia [68–70].

In this work, we carried out a roadkill survey focused on terrestrial vertebrates in our study area in a Mediterranean region of Central Italy (i.e., Tuscany), and then we compared our list of alien vs. native roadkilled species with all published and comparable lists from other regions of the world (including both Mediterranean and other biomes), which we obtained by a literature review. Into the framework of such comparisons, we tested if the percentage of roadkilled introduced species varied among biomes (and then countries), always considering several possible confounding factors at the same time, such as the latitude, Gross Domestic Product (GDP hereafter), and the number of neighbouring countries. High temperatures extend the available time for growth and reproduction, so low latitudes can comparatively host higher numbers of introduced species [71–73]. GDP has also been described as increasing the number of introduced species, probably because of the combined effects of trade, consumerism and urbanization [74–76]. Finally, the number of introduced species is usually higher in connected countries with established trade and tourism along shared infrastructures, such as neighbouring countries [74,77,78]. In order to investigate the reliability of roadkill data for investigating biological invasions at a macro-ecological scale, we also carried out an additional analysis comparing the whole lists of alien vs. native species inhabiting all the countries considered in the first analysis, always including the same approach and candidate predictors. This study represents a first example of how we can investigate patterns of biological invasions by analysing roadkill datasets, which can become a valuable resource for researchers and wildlife managers, especially for first assessments when more focused surveys cannot be performed.

2. Materials and Methods

2.1. Study Area (Fieldwork Survey)

Our fieldwork was conducted in Southern Tuscany (Provinces of Grosseto and Siena, Central Italy). We travelled once a week along the road linking the village of Prata (Province of Grosseto) to Grosseto and Siena for three years (2011–2013), for a total of about 101 km/week (Figure 1).

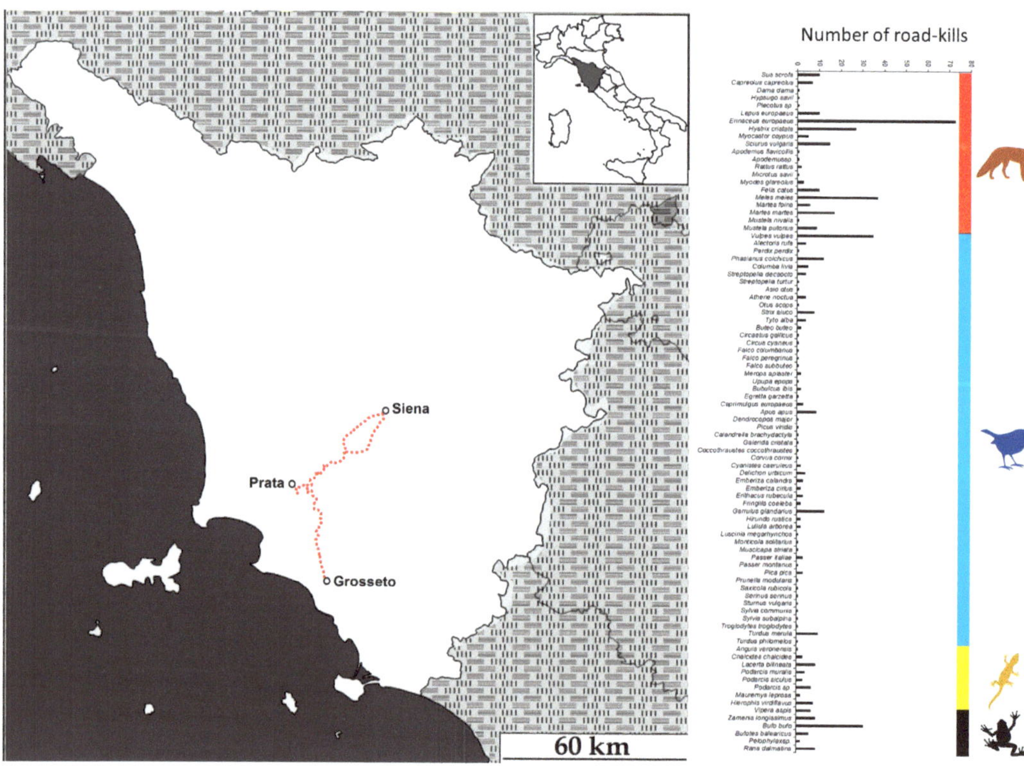

Figure 1. Study area and roadkilled species during the fieldwork survey. On the left is the location of the study area (Central Italy, Tuscany). Roadkill surveys started in the village of Prata (Province of Grosseto) in the direction of Grosseto and then Siena (101 km). On the right is the number of individuals of each roadkilled species.

The road skirts a high number of habitat types, including dense deciduous and mixed woodlands (about 27%), farmlands (about 43%), scrublands with hedgerows (16%), human settlements (9%) and fallows (5%). The whole road was paved, and an allowed speed limit was 70–90 km/h throughout most road sections, apart from 16 km (in total) of the beltway (speed limit: 110 km/h) in the surroundings of the main cities (Siena and Grosseto). The study area is characterised by the presence of at least 38 mammal species (represented among others by the wild boar *Sus scrofa*, the roe deer *Capreolus capreolus*, the red fox *Vulpes*

vulpes, the European badger *Meles meles* and the crested porcupine *Hystrix cristata*), about 100 bird species (also including the common pheasant *Phasianus colchicus*, the red-legged partridge *Alectoris rufa* and over 15 raptor species), 16 species of reptiles (e.g., the asp viper *Vipera aspis* and the Italian wall lizard *Podarcis siculus*) and eight species of Amphibians (e.g., the common toad *Bufo bufo* and the Italian crested newt *Triturus carnifex*) [79–81]. The local community of terrestrial vertebrates mainly consists of common species, but some threatened species are also present, such as the European turtle-dove *Streptopelia turtur* (globally listed as Vulnerable; [82]).

2.2. Data Collection and Analyses (Fieldwork Survey and Literature Review)

During our fieldwork survey, every time a roadkilled animal was detected, it was georeferenced and classified (when possible) at the species level. We also recorded whether the species was native or introduced (see Supplementary Materials S1). Using the list of roadkilled species resulting from our fieldwork survey, we calculated the percentage of introduced species on the total of roadkilled species in order to compare this percentage to the same value from all published and comparable lists from other regions of the world. Regarding introduced species, we also specified the percentage of recently introduced, paleo-introduced and domestic species (we considered all the species introduced before 1500 AD as paleo-introduced [83,84]).

We performed our first literature review by searching those lists of roadkilled terrestrial vertebrates on the Web of Science (WOS; [85]) by implementing the following search string: ("roadkill*" OR "road-kill*" OR ("collision" AND "road") AND ("vertebrate*" OR "wildlife" OR "animal*"). We retained articles focusing on the whole community of terrestrial vertebrates and including an explicit list of roadkilled species. Once we obtained the final list of retained articles, we extracted the study areas in order to perform a second literature review aimed at obtaining the percentage of introduced species on the total of species inhabiting those countries (always focusing on terrestrial vertebrates). We completed this second review also with information proceeding from institutional websites such as www.iucngisd.org, www.iucnredlist.org, www.griis.org (all accessed on 12 February 2023), and several national websites providing inventories of introduced and native species.

For each study area mentioned in those articles, we extracted the country, the biome (as defined by [86,87]), and three explanatory variables potentially affecting the number of introduced species in a given area, such as latitude, GDP, and the number of neighbouring countries, in addition to the aforementioned percentage of roadkilled introduced species. Considering that the analysis would probably have relatively little data, we modified some of the explanatory variables in order to have fewer levels within each of them. More concretely, we combined some biomes (i.e., "Temperate grasslands, savannas, and shrublands" and "Temperate broadleaf and mixed forests" in "Temperate biomes", "Tropical and subtropical grasslands, savannas, and shrublands" and "Tropical and subtropical moist broadleaf forests" in "Tropical biomes"), we implemented the latitude as a continuous variable without decimals, we transformed the GDP in an ordinal variable with four levels (i.e., less than USD one billion, between USD one and two billion, between USD two and three billion, and more than USD three billion), and we also transformed the number of neighbouring countries into an ordinal variable with four levels (i.e., without neighbouring countries, with one/two neighbouring countries, or with many neighbouring countries).

In order to test our first prediction (i.e., more introduced species in Mediterranean biomes), we used generalised linear mixed models (GLMMs; negative binomial error distribution and log link function) and evaluated their performance using Akaike Information Criterion (AIC). For each model, the response variable was the percentage of introduced species on the total of roadkilled species, considering the country of each retained study as a random factor. We compared all the possible univariate models, including the considered explanatory variables, with the full model and also the null model (i.e., only including the random factor). We selected the most supported model using AICc (i.e., AIC for small sample sizes) and calculated Akaike weights (wAICc) to estimate the relative support for every model (ranging from 0 to 1, with larger numbers indicating greater support; [88–90]). We also provided R^2 for each considered model. Additionally, we repeated exactly the same AIC model selection of the same GLMMs, but this time implementing as a response variable the percentage of introduced species on the total of existing species.

In order to test our second prediction (i.e., more introduced species in Italy than in other countries of Mediterranean biomes), we performed a generalised linear model (GLM) in which the response variable was again the percentage of introduced species on the total of roadkilled species, and the explanatory variable was the country of each retained study. We again used negative binomial error distribution and log link function. We evaluated model performance by comparing the AIC of the considered model with the AIC of the related null model.

3. Results

3.1. Fieldwork Survey

In Central Italy, we collected a total of 511 roadkill specimens belonging to 89 species (Figure 1; Supplementary Materials S1). Most roadkill belonged to Mammals (N = 273, for a total of 22 species), followed by Birds (N = 137, for a total of 53 species), Reptiles (N = 53, for a total of ten species) and Amphibians (N = 48, for a total of four species). Introduced species included two domestic (domestic cat *Felis catus* and domestic pigeon *Columba livia domestica*), four paleo-introduced (fallow deer *Dama dama*, crested porcupine *Hystrix cristata*, black rat *Rattus rattus*, and common pheasant *Phasianus colchicus*) and two recently introduced species (Iberian pond turtle *Mauremys leprosa* and coypu *Myocastor coypus*). Two further species (i.e., red-legged partridge *Alectoris rufa* and grey partridge *Perdix perdix*), although native to our study area, are present today as the result of releases for hunting purposes. Among the mammals, the most detected roadkilled species were the European hedgehog *Erinaceus europaeus* (73 roadkilled individuals), the European badger *Meles meles* (37 roadkilled individuals), the red fox *Vulpes vulpes* (35 roadkilled individuals) and the crested porcupine *Hystrix cristata* (27 roadkilled individuals). For birds, the most detected roadkill were the Eurasian jay *Garrulus glandarius* (13 roadkilled individuals), the common pheasant *Phasianus colchicus* (12 roadkilled individuals) and the European blackbird *Turdus merula* (ten roadkilled individuals). For reptiles, the most detected roadkill were the Western green lizard *Lacerta bilineata* (nine roadkilled individuals), the Aesculapian ratsnake *Zamenis longissimus* (nine roadkilled individuals) and the green whip snake *Hierophis viridiflavus* (eight roadkilled individuals). To conclude, the most detected amphibians were the common toad *Bufo bufo* (31 roadkilled individuals), the agile frog *Rana dalmatina* (nine roadkilled individuals) and the green toad *Bufotes balearicus* (six roadkilled individuals). Roadkill rates for each recorded species are provided in Supplementary Material S1.

3.2. Literature Review and Analysis

In our first literature review, we retrieved a total of 2314 works from WOS. Among those, only 41 (including our original data) fitted our requirements (i.e., including a total list of roadkilled terrestrial vertebrates, including the explicit numbers of dead individuals for each species) and were included in our analysis. Retained studies included up to 20.00% of introduced species (mean $\pm$ SD = 5.30 $\pm$ 5.69%). Most retained studies were conducted in tropical and subtropical areas (i.e., 20 studies from 11 countries), with a peak in American countries (i.e., ten studies from three countries; Figure 2). All retained studies included 5434 species (mean $\pm$ SD per study = 153.53 $\pm$ 432.36) for a total of 276,960 roadkilled individuals (mean $\pm$ SD per study = 7177.80 $\pm$ 25,451.81). Among those species, up to 17.85% were domestic (mean $\pm$ SD = 3.03 $\pm$ 4.45%), up to 4.62% were paleo-introduced (mean $\pm$ SD = 0.86 $\pm$ 1.56%) and up to 12.90% (mean $\pm$ SD = 1.24 $\pm$ 2.49%) were recent introductions. Roadkill rates for each retained study are provided in Supplementary Materials S2. As a result of our second literature review, we focused on the countries where the 40 retained studies were carried out, obtaining the percentage of introduced species on the total of existing species for all of them (Supplementary Material S2).

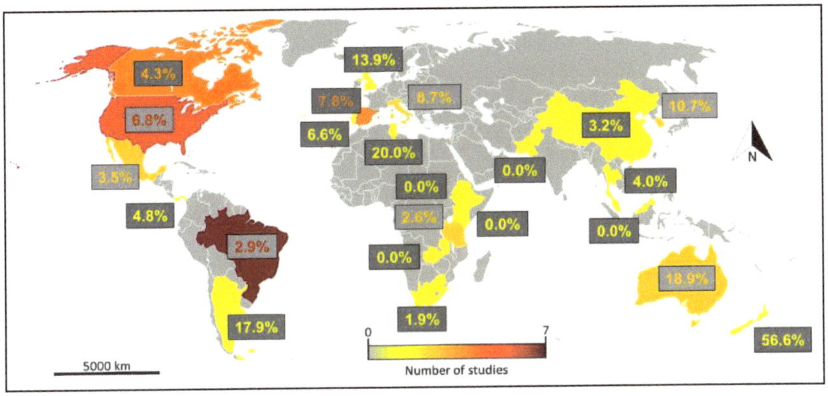

Figure 2. Global literature review of roadkill studies. Country location of retained studies (N = 41). We retained articles focusing on the whole community of terrestrial vertebrates and including an explicit list of roadkilled species. For each considered country, we provided the percentage of introduced species on the total of roadkilled species.

Concerning our first hypothesis, the biome was one of the most relevant explanatory variables in both model selections. In fact, the univariate model including this factor obtained plausible support in the roadkill model selection (according to [90]) and was the only supported model in the all-species model selection (Table 1). Overall, in both model selections, the percentage of introduced species in Mediterranean and Temperate biomes was higher than in Tropical and Desert biomes (Figure 3). On the other hand, other variables were also supported in the roadkill model selection, especially the latitude, which was the best-supported model (Table 1; with more introduced species at higher latitudes, roughly corresponding to Mediterranean and Temperate biomes; Supplementary Material S3). The number of neighbouring countries obtained plausible support in the roadkill model selection (according to [90]), being the second supported model (Table 1; with more introduced species in island countries than in nations having neighbouring countries).

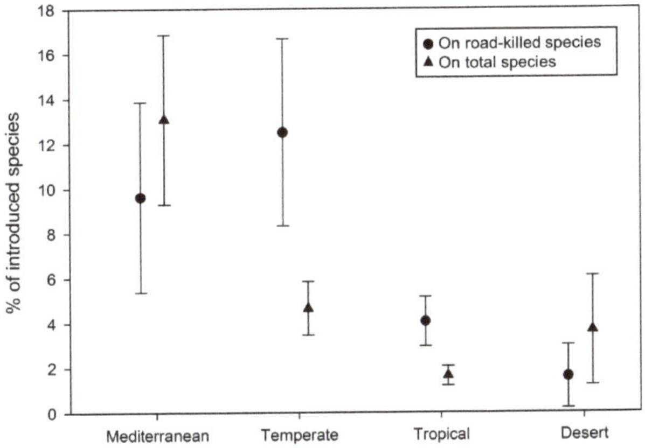

Figure 3. Percentage of introduced species in different biomes, including both the percentage of introduced species on the total of roadkilled species (circles) and on the total of existing species (triangles).

Table 1. Factors affecting the percentage of introduced species on the total of roadkilled species (upper table) and the percentage of introduced species on the total of existing species (lower table), model ranks by AICs. The upper table shows the model selection for the roadkill dataset (i.e., literature review on roadkilled terrestrial vertebrates plus our own field data), whereas the lower table shows the model selection for the all-species dataset (i.e., literature review on introduced terrestrial vertebrates in different countries). K is the number of model parameters. ΔAICc is the relative difference between a given AIC value compared to the smallest AICc value. AICc weights (wAICc) indicate the relative support for every model (the weights of all the models in the candidate set have the sum of 1). Evidence ratio (ER) is the ratio of wAIC, comparing the best-supported model with every competing one. Model rank is provided only for the best-supported models. R^2 is a measure of the global fit of each model. Into the model column, Nc are neighbouring countries.

			Roadkill Model Selection				
Model	**K**	**AICc**	**ΔAICc**	**wAICc**	**ER**	**Rank**	R^2
Latitude	3	244.6	0.0	0.78	1.0	1	0.15
Nc	4	247.9	3.3	0.15	5.0	2	0.21
Biome	5	250.5	5.9	0.04	19.0	3	0.16
Null	2	251.7	7.1	0.02	34.7	-	0.00
GDP	4	256.2	11.6	0.01	324.5	-	0.01
Full	10	258.3	13.6	0.00	918.0	-	0.32
			All-Species Model Selection				
Model		**AICc**	**ΔAICc**	**wAICc**	**ER**	**Rank**	R^2
Biome	5	187.9	0.0	0.99	1.0	1	0.19
Latitude	3	197.3	9.3	0.01	107.0	-	0.54
Full	10	199.2	11.2	0.00	276.7	-	0.57
Nc	4	205.5	17.6	0.00	6555.3	-	0.23
Null	2	210.2	22.3	0.00	70,829.4	-	0.00
GDP	4	214.1	26.1	0.00	475,931.8	-	0.02

Concerning the analysis focusing on Mediterranean countries (i.e., Italy, Spain, Portugal, and Tunisia), we retained only seven studies (three for Spain, two for Italy, one for Portugal, and one for Tunisia). No differences were observed in the percentage of introduced species on the total of roadkilled species among most of them, but Tunisia did show higher values (i.e., Figure 4). The country model was more supported than the null one (Supplementary Materials S4).

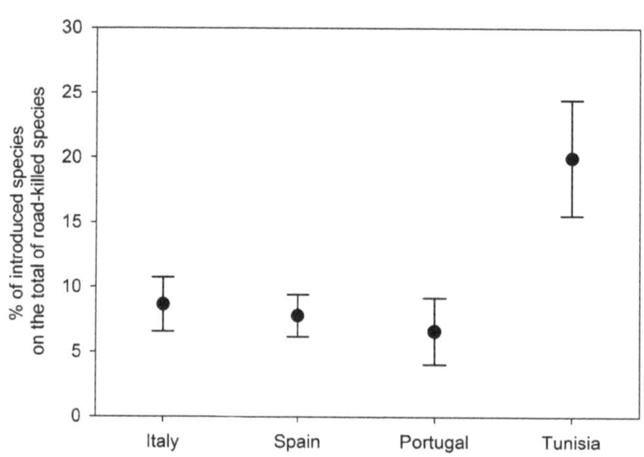

Figure 4. Percentage of introduced species on the total of roadkilled species in different countries, from the model considering all retained studies focusing on Mediterranean biomes.

4. Discussion

In this work, we compared our original data collected on roadkilled terrestrial vertebrates from Central Italy with 40 analogous lists proceeding from published studies from throughout the world. This is, to the best of our knowledge, the first attempt at geographical analysis implementing only roadkill data, and in order to investigate the reliability of this kind of data, we also present an analysis implementing more general data proceeding from biological invasion datasets. We found that the percentage of introduced roadkilled species was higher in both the Mediterranean and Temperate biomes compared to other biomes. This is only a first assessment, implementing relatively simple analyses, but in our opinion, this case study can represent an interesting example of the possible value of roadkill datasets for investigating questions related to research fields other than road ecology, such as biological invasions and biogeography in this case.

Roadkill surveys usually represent a reliable information source regarding the common species that can be found in a given area [91–93], although remarkable exceptions have been described for species actively avoiding roads [21,94,95]. Nevertheless, invasive vertebrates are often generalist species able to use roads and roadsides as suitable habitats and to take advantage of the related resources (i.e., roadkilled carcasses, garbage; [96,97]), consequently paying the price of high roadkill rates [15,98,99]. Therefore, it is relatively safe to assume that alien species would be well represented in the roadkill records of a given area [34,100] and, consequently, that such data would be suitable for geographical analyses focused on biological invasions. According to this rationale, our analyses showed that the percentage of introduced species (on both the total of roadkilled species and the total of existing species) is higher in Mediterranean and Temperate biomes than in Tropical and Desert biomes. These findings confirm our prediction that the overall amount of introduced species is higher in the Mediterranean biome, as previously suggested by the available scientific literature [101–103]. However, the Temperate biome also suffered high levels of species introduction, especially in Central Europe and North America, both of which regions characterised by suitable climates for most vertebrate species and consequently also suitable for their invasions [73,104,105]. These biomes are more anthropised than others [106–108], probably because they are widely distributed in countries with early human presence, mostly in the Mediterranean Basin and, more generally, in Europe. In addition, early anthropization can be a major driver of the presence of established populations of introduced species, especially the paleo-introduced ones, such as the fallow deer *Dama dama* and the crested porcupine *Hystrix cristata* in our fieldwork area (e.g., [109,110]). Both the Mediterranean and Temperate biomes also include some of the countries with suitable latitudes (and consequently climates) for allowing growth and reproduction of introduced terrestrial vertebrates [72,73], such as Italy and Spain, but also USA and UK, with all of them providing published studies to our literature review.

The Mediterranean biome, spanning from coastal and interior portions of the USA, Mexico, Chile, South Africa, Australia and the Mediterranean Basin, is also recognised as one of the most imperilled in terms of biodiversity loss due to its sensitivity to biological invasions, land use and climate change [101,111,112]. In this biome, at least in the most sensitive areas such as national parks, monitoring roadkill should be a priority action, in order to detect conservation issues related to road impacts (see, for example, [113], but also [114,115] for examples in other biomes). Additionally, we showed here that these surveys can also provide valuable information about biological invasions, not only about cost-effective early detection in a given location (see a recent example regarding the American mink *Neovison vison* in Central Italy: [116]), but also to conduct geographical analyses and to estimate the percentage of common alien species in a given area. Government authorities and institutions (e.g., protected areas and environmental agencies) should include roadkill surveys in their long-term monitoring schemes, but citizen science can also be involved in collecting a remarkable amount of data [35,117,118]. Importantly, the use of ad hoc projects on online platforms (e.g., iNaturalist) may allow the collection of data throughout the years and from different sites [119–121] and to analyse temporal patterns of invasions. Finally, we would also highlight that less than 2% of roadkill studies focusing on all vertebrates (i.e., 40 on 2314) reported a full list of affected species, making it more difficult than expected to use this literature resource, so we would strongly recommend future roadkill studies to publish their species list (and data), in order to increase their interest and the possibility to be used and cited by other researchers.

Our second prediction was not supported by the specific model selection because there were no differences in the percentage of introduced species on the total of roadkilled species among most considered Mediterranean countries (i.e., Italy, Spain, and Portugal), although Tunisia showed higher values (mainly due to domestic species, possibly free-ranging more than in other Mediterranean countries; [122,123]). Thus, although Italy hosts a high number of alien species among terrestrial vertebrates [66,67,124,125], it does not seem to include a significantly higher number of alien species with respect to other Mediterranean countries included in our analysis. However, these results are probably due to the extremely small sample size (i.e., few studies in few Mediterranean countries), which may prevent achieving exhaustive conclusions, but also because the analysed studies from Italy were carried out in rural areas characterised by a relatively low amount of alien species [79–81,115], and are perhaps not completely representative of the whole country.

In addition to the small sample size (for both our fieldwork data and the datasets from literature reviews) and the potentially biased locations of the considered study areas, this work surely entails further limitations. Several factors have been described to drive the number of introduced species in a given country, such as the trade importance, the amount of different climatic niches and the number of potentially competing native species (e.g., [126–128]), but most of them were of difficult implementation considering our relatively small sample size. However, although we acknowledge these study limitations, we believe they do not affect the main idea of this work, which is that roadkill datasets can be used beyond the study of road impacts, and in this case, for a first assessment of biological invasions among different countries.

5. Conclusions

In this study, we used our fieldwork data collected on roadkilled terrestrial vertebrates from Central Italy and 40 analogous lists proceeding from published studies from all over the world. Our analysis represents the first attempt at biological invasion risk assessment implementing only roadkill data, and, as a result, the percentage of introduced roadkilled species was higher in both the Mediterranean and Temperate biomes compared to other biomes. This is only a first assessment implementing relatively simple analyses, and we acknowledge several study limitations, but, in our opinion, they do not affect the main idea of this study, which is that roadkill datasets can be used for investigating questions related to research fields other than road ecology, such as in this case biological invasions.

Supplementary Materials: The following supporting information can be downloaded at: https://www.mdpi.com/article/10.3390/biology12060850/s1, Supplementary Material S1: List and amount of roadkilled species in our fieldwork survey; Supplementary Material S2: List of retained studies from the literature review, and related variables; Supplementary Material S3: Figure depicting the relation between the percentage of introduced species (both on the total of roadkilled species and on the total of existing species) and the latitude of the country hosting the study areas; Supplementary Material S4: Table showing the AIC model selection focusing on the countries of Mediterranean biome.

Author Contributions: Conceptualization, M.D. and E.M.; methodology, A.V. and E.M.; formal analysis, M.D.; data curation, A.V. and E.M.; writing—original draft preparation, A.V.; writing—review and editing, A.V., M.D. and E.M.; visualization, A.V., M.D. and E.M.; funding acquisition, E.M. All authors have read and agreed to the published version of the manuscript.

Funding: A.V. and E.M. were supported by MUR-FOE-Project Capitale Naturale-Task "Biodiversità". MD was supported by a JdC-Inc post-doctoral grant (IJC2019-039662-I) from the Spanish Ministry of Science and Innovation (MICINN). EM was also funded by the National Recovery and Resilience Plan (NRRP), Mission 4 Component 2 Investment 1.4—Call for tender No. 3138 of 16 December 2021, rectified by Decree n.3175 of 18 December 2021 of Italian Ministry of University and Research funded by the European Union—NextGenerationEU; Project code CN_00000033, Concession Decree No. 1034 of 17 June 2022 adopted by the Italian Ministry of University and Research, CUP B83C22002930006, Project title "National Biodiversity Future Center—NBFC".

Institutional Review Board Statement: Not applicable.

Informed Consent Statement: Not applicable.

Data Availability Statement: All data generated or analyzed during this study are included in this published article.

Acknowledgments: Isis Elisabeth Cohen Nicholson and Emily McCluskey kindly reviewed the English (first and second version, respectively). We sincerely appreciate all valuable comments and suggestions by the three Reviewers, which helped us to improve the quality of the manuscript.

Conflicts of Interest: The authors declare no conflict of interest. The funders had no role in the design of the study; in the collection, analyses, or interpretation of data; in the writing of the manuscript; or in the decision to publish the results.

References

1. Ibisch, P.L.; Hoffmann, M.T.; Kreft, S.; Pe'er, G.; Kati, V.; Biber-Freudenberger, L.; Dellasala, D.A.; Vale, M.M.; Hobson, P.R.; Selva, N. A global map of roadless areas and their conservation status. *Science* **2016**, *354*, 1423–1427. [CrossRef] [PubMed]
2. Ascensão, F.; Fahrig, L.; Clevenger, A.P.; Corlett, R.T.; Jaeger, J.A.; Laurance, W.F.; Pereira, H.M. Environmental challenges for the Belt and Road Initiative. *Nat. Sustain.* **2018**, *1*, 206–209. [CrossRef]
3. Meijer, J.R.; Huijbregts, M.A.; Schotten, K.C.; Schipper, A.M. Global patterns of current and future road infrastructure. *Environ. Res. Lett.* **2018**, *13*, 064006. [CrossRef]
4. Trombulak, S.C.; Frissell, C.A. Review of ecological effects of roads on terrestrial and aquatic communities. *Conserv. Biol.* **2000**, *14*, 18–30. [CrossRef]
5. Forman, R.T.; Sperling, D.; Bissonette, J.A.; Clevenger, A.P.; Cutshall, C.D.; Dale, V.H.; Fahrig, L.; France, R.; Goldman, C.R.; Heanue, K.; et al. *Road Ecology: Science and Solutions*; Island Press: Washington, DC, USA, 2003; p. 481.
6. Van der Ree, R.; Smith, D.J.; Grilo, C. *Handbook of Road Ecology*; John Wiley & Sons: Oxford, UK, 2015; p. 522.
7. Gaston, K.J.; Bennie, J.; Davies, T.W.; Hopkins, J. The ecological impacts of nighttime light pollution: A mechanistic appraisal. *Biol. Rev.* **2013**, *88*, 912–927. [CrossRef]
8. Laurance, W.F.; Clements, G.R.; Sloan, S.; O'Connell, C.S.; Mueller, N.D.; Goosem, M.; Venter, O.; Edwards, D.P.; Phalan, B.; Balmford, A.; et al. A global strategy for road building. *Nature* **2014**, *513*, 229–232. [CrossRef]
9. Leonard, R.J.; Hochuli, D.F. Exhausting all avenues: Why impacts of air pollution should be part of road ecology. *Front. Ecol. Environ.* **2017**, *15*, 443–449. [CrossRef]
10. Christen, D.C.; Matlack, G.R. The habitat and conduit functions of roads in the spread of three invasive plant species. *Biol. Invasions* **2009**, *11*, 453–465. [CrossRef]
11. Benedetti, Y.; Morelli, F. Spatial mismatch analysis among hotspots of alien plant species, road and railway networks in Germany and Austria. *PLoS ONE* **2017**, *12*, e0183691. [CrossRef] [PubMed]
12. McDougall, K.L.; Lembrechts, J.; Rew, L.J.; Haider, S.; Cavieres, L.A.; Kueffer, C.; Milbau, A.; Naylor, B.J.; Nuñez, M.A.; Pauchard, A.; et al. Running off the road: Roadside non-native plants invading mountain vegetation. *Biol. Invasions* **2018**, *20*, 3461–3473. [CrossRef]
13. Brown, G.P.; Phillips, B.L.; Webb, J.K.; Shine, R. Toad on the road: Use of roads as dispersal corridors by cane toads (*Bufo marinus*) at an invasion front in tropical Australia. *Biol. Conserv.* **2006**, *133*, 88–94. [CrossRef]
14. D'Amico, M.; Rouco, C.; Russell, J.C.; Román, J.; Revilla, E. Invaders on the road: Synanthopic bird foraging along highways. *Oecol. Aust.* **2013**, *17*, 86–95. [CrossRef]
15. Ascensão, F.; Latombe, G.; Anadón, J.D.; Abellán, P.; Cardador, L.; Carrete, M.; Tella, J.L.; Capinha, C. Drivers of compositional dissimilarity for native and alien birds: The relative roles of human activity and environmental suitability. *Biol. Invasions* **2020**, *22*, 1447–1460. [CrossRef]
16. D'Amico, M.; Ascensão, F.; Fabrizio, M.; Barrientos, R.; Gortázar, C. Twenty years of Road Ecology: A Topical Collection looking forward for new perspectives. *Eur. J. Wildl. Res.* **2018**, *64*, 26. [CrossRef]
17. Collinson, W.; Davies-Mostert, H.; Roxburgh, L.; van der Ree, R. Status of road ecology research in Africa: Do we understand the impacts of roads, and how to successfully mitigate them? *Front. Ecol. Evol.* **2019**, *7*, 479. [CrossRef]

18. Barrientos, R.; Ascensão, F.; D'Amico, M.; Grilo, C.; Pereira, H.M. The lost road: Do transportation networks imperil wildlife population persistence? *Perspect. Ecol. Conserv.* **2021**, *19*, 411–416. [CrossRef]
19. Grilo, C.; Borda-de-Água, L.; Beja, P.; Goolsby, E.; Soanes, K.; le Roux, A.; Koroleva, E.; Ferreira, F.Z.; Gagné, S.A.; Wang, Y.; et al. Conservation threats from roadkill in the global road network. *Glob. Ecol. Biogeogr.* **2021**, *30*, 2200–2210. [CrossRef]
20. Ascensão, F.; Desbiez, A.L. Assessing the impact of roadkill on the persistence of wildlife populations: A case study on the giant anteater. *Perspect. Ecol. Conserv.* **2022**, *20*, 272–278. [CrossRef]
21. Ascensão, F.; Kindel, A.; Teixeira, F.Z.; Barrientos, R.; D'Amico, M.; Borda-de-Água, L.; Pereira, H.M. Beware that the lack of wildlife mortality records can mask a serious impact of linear infrastructures. *Glob. Ecol. Conserv.* **2019**, *19*, e00661. [CrossRef]
22. Hels, T.; Buchwald, E. The effect of road kills on amphibian populations. *Biol. Conserv.* **2001**, *99*, 331–340. [CrossRef]
23. Spencer, R.J.; Van Dyke, J.U.; Thompson, M.B. Critically evaluating best management practices for preventing freshwater turtle extinctions. *Conserv. Biol.* **2017**, *31*, 1340–1349. [CrossRef] [PubMed]
24. Liu, J. Effects of global household proliferation on ecosystem services. In *Landscape Ecology for Sustainable Environment and Culture*; Fu, B., Jones, K., Eds.; Springer Science+Business Media: Dordrecht, The Netherlands, 2013; p. 364.
25. Gilbert, S.L.; Sivy, K.J.; Pozzanghera, C.B.; DuBour, A.; Overduijn, K.; Smith, M.M.; Zhou, J.; Little, J.M.; Prugh, L.R. Socioeconomic benefits of large carnivore recolonization through reduced wildlife-vehicle collisions. *Conserv. Lett.* **2017**, *10*, 431–439. [CrossRef]
26. Martin, A.E.; Graham, S.L.; Henry, M.; Pervin, E.; Fahrig, L. Flying insect abundance declines with increasing road traffic. *Insect Conserv. Divers.* **2018**, *11*, 608–613. [CrossRef]
27. Conover, M.R.; Pitt, W.C.; Kessler, K.K.; DuBow, T.J.; Sanborn, W.A. Review of human injuries, illnesses, and economic losses caused by wildlife in the United States. *Wildl. Soc. Bull.* **1995**, *23*, 407–414.
28. Ascensão, F.; Barrientos, R.; D'Amico, M. Wildlife collisions put a dent in road safety. *Science* **2021**, *374*, 1208. [CrossRef] [PubMed]
29. Llagostera, P.; Comas, C.; López, N. Modeling road traffic safety based on point patterns of wildlife-vehicle collisions. *Sci. Total Environ.* **2022**, *846*, 157237. [CrossRef] [PubMed]
30. Calenge, C.; Chadoeuf, J.; Giraud, C.; Huet, S.; Julliard, R.; Monestiez, P.; Piffady, J.; Pinaud, D.; Ruette, S. The spatial distribution of *Mustelidae* in France. *PLoS ONE* **2015**, *10*, e0121689. [CrossRef] [PubMed]
31. Dyczkowski, J. Golden nightjar in Western Sahara, Morocco, in May 2015. *Dutch Bird.* **2016**, *38*, 80–86.
32. Lapini, L.; Dreon, A.L.; Caldana, M.; Luca, M.; Villa, M. Distribuzione, espansione e problemi di conservazione di *Canis aureus* in Italia (*Carnivora*: *Canidae*). *Quad. Mus. Civ. Stor. Nat. Ferrara* **2018**, *6*, 89–96.
33. Russo, L.F.; Barrientos, R.; Fabrizio, M.; Di Febbraro, M.; Loy, A. Prioritizing road-kill mitigation areas: A spatially explicit national-scale model for an elusive carnivore. *Divers. Distrib.* **2020**, *26*, 1093–1103. [CrossRef]
34. Schwartz, A.L.; Shilling, F.M.; Perkins, S.E. The value of monitoring wildlife roadkill. *Eur. J. Wildl. Res.* **2020**, *66*, 18. [CrossRef]
35. Vercayie, D.; Herremans, M. Citizen science and smartphones take roadkill monitoring to the next level. *Nat. Conserv.* **2015**, *11*, 29–40. [CrossRef]
36. Waetjen, D.P.; Shilling, F.M. Large extent volunteer roadkill and wildlife observation systems as sources of reliable data. *Front. Ecol. Evol.* **2017**, *5*, 89. [CrossRef]
37. Englefield, B.; Starling, M.; Wilson, B.; Roder, C.; McGreevy, P. The Australian roadkill reporting project—Applying integrated professional research and citizen science to monitor and mitigate roadkill in Australia. *Animals* **2020**, *10*, 1112. [CrossRef]
38. Bik, H.M.; Goldstein, M.C. An introduction to social media for scientists. *PLoS Biol.* **2013**, *11*, e1001535. [CrossRef]
39. Van Strien, A.J.; Van Swaay, C.A.; Termaat, T. Opportunistic citizen science data of animal species produce reliable estimates of distribution trends if analysed with occupancy models. *J. Appl. Ecol.* **2013**, *50*, 1450–1458. [CrossRef]
40. Mori, E.; Di Bari, P.; Coraglia, M. Interference between roe deer and Northern chamois in the Italian Alps: Are Facebook groups effective data sources? *Ethol. Ecol. Evol.* **2018**, *30*, 277–284. [CrossRef]
41. Chyn, K.; Lin, T.E.; Chen, Y.K.; Chen, C.Y.; Fitzgerald, L.A. The magnitude of roadkill in Taiwan: Patterns and consequences revealed by citizen science. *Biol. Conserv.* **2019**, *237*, 317–326. [CrossRef]
42. Heigl, F.; Teufelbauer, N.; Resch, S.; Schweiger, S.; Stückler, S.; Dörler, D. A dataset of road-killed vertebrates collected via citizen science from 2014–2020. *Sci. Data* **2022**, *9*, 504. [CrossRef] [PubMed]
43. Panzeri, M.; Mori, E.; Mazza, G.; Menchetti, M. Records of introduced stripe-necked terrapins (*Mauremys* species) in Italy. *Acta Herpetol.* **2014**, *9*, 97–100.
44. Caley, P.; Ramsey, D.S.; Barry, S.C. Inferring the distribution and demography of an invasive species from sighting data: The red fox incursion into Tasmania. *PLoS ONE* **2015**, *10*, e0116631. [CrossRef] [PubMed]
45. Balčiauskas, L.; Stratford, J.; Balčiauskienė, L.; Kučas, A. Roadkills as a method to monitor raccoon dog populations. *Animals* **2021**, *11*, 3147. [CrossRef] [PubMed]
46. Mori, E.; Amerini, R.; Mazza, G.; Bertolino, S.; Battiston, R.; Sforzi, A.; Menchetti, M. Alien shades of grey: New occurrences and relevant spread of *Sciurus carolinensis* in Italy. *Eur. J. Ecol.* **2016**, *2*, 13–20. [CrossRef]
47. Kwak, M.L. The first record of the introduced flea *Spilopsyllus cuniculi* (Dale, 1878) (*Siphonaptera*: *Pulicidae*) from the invasive red fox in Australia, with a review of the fleas associated with the red fox in Australia. *Aust. Entomol.* **2017**, *44*, 289–292.
48. Ham, C.H.; Park, S.M.; Lee, J.E.; Park, J.; Lee, D.H.; Sung, H.C. First report of the Black-headed python (*Aspidites melanocephalus* Krefft, 1864) found in the wild in the Republic of Korea. *BioInvasions Rec.* **2022**, *11*, 571–577. [CrossRef]
49. Boscherini, A.; Mazza, G.; Menchetti, M.; Laurenzi, A.; Mori, E. Time is running out! Rapid range expansion of the invasive northern raccoon in central Italy. *Mammalia* **2020**, *84*, 98–101. [CrossRef]

50. Santos, S.M.; Lourenco, R.; Mira, A.; Beja, P. Relative effects of road risk, habitat suitability, and connectivity on wildlife roadkills: The case of tawny owls (*Strix aluco*). *PLoS ONE* **2013**, *8*, e79967. [CrossRef]
51. Haigh, A.; O'Riordan, R.M.; Butler, F. Hedgehog *Erinaceus europaeus* mortality on Irish roads. *Wildl. Biol.* **2014**, *20*, 155–160. [CrossRef]
52. Meyrom, K.; Yosef, R.; Charter, M. Are roadkills density-dependent? Case study of the barn owl (*Tyto alba*). *Diversity* **2023**, *15*, 412. [CrossRef]
53. Lapini, L. Il cane viverrino *Nyctereutes procyonoides ussuriensis* Matschie, 1908 in Italia: Segnalazioni 1980–2005 (*Mammalia: Carnivora: Canidae*). *Boll. Mus. Civ. St. Nat. Venezia* **2006**, *57*, 235–239.
54. Barrientos, R. Adult sex-ratio distortion in the native European polecat is related to the expansion of the invasive American mink. *Biol. Conserv.* **2015**, *186*, 28–34. [CrossRef]
55. Chen, T.Y.; Richard, R.; Lin, T.E.; Huang, S.P. Landscape forest impacts the potential activity time of an invasive lizard and its possibilities for range expansion in Taiwan under climate warming. *J. Therm. Biol.* **2021**, *98*, 102948. [CrossRef]
56. Kowalczyk, R.; Zalewski, A.; Jędrzejewska, B.; Ansorge, H.; Bunevich, A.N. Reproduction and mortality of invasive raccoon dogs (*Nyctereutes procyonoides*) in the Białowieża Primeval Forest (eastern Poland). *Ann. Zool. Fenn.* **2009**, *46*, 291–301. [CrossRef]
57. Boncompagni, L.; Molfini, M.; Ciampelli, P.; Fazzi, P.; Lucchesi, M.; Mori, E.; Petralia, L.; Mazza, G. No country for native crayfish: Importance of crustaceans in the diet of native and alien Northern raccoon. *Ethol. Ecol. Evol.* **2021**, *33*, 576–590. [CrossRef]
58. Cuttelod, A.; García, N.; Malak, D.A.; Temple, H.J.; Katariya, V. The Mediterranean: A biodiversity hotspot under threat. In *Wildlife in a Changing World—An Analysis of the 2008 IUCN Red List of Threatened Species;* Viè, J.C., Hilton-Taylor, C., Stuart, S.N., Eds.; IUCN Editions: Gland, Switzerland, 2009; pp. 89–102.
59. Clavero, M.; Hermoso, V.; Levin, N.; Kark, S. Biodiversity research: Geographical linkages between threats and imperilment in freshwater fish in the Mediterranean Basin. *Divers. Distrib.* **2010**, *16*, 744–754. [CrossRef]
60. Cao Pinna, L.; Axmanová, I.; Chytrý, M.; Malavasi, M.; Acosta, A.T.; Giulio, S.; Attorre, F.; Bergmeier, E.; Biurrun, I.; Campos, J.A.; et al. The biogeography of alien plant invasions in the Mediterranean Basin. *J. Veg. Sci.* **2021**, *32*, e12980. [CrossRef]
61. Laubier, L. Mediterranean Sea and humans: Improving a conflictual partnership. In *The Mediterranean Sea*; Saliot, A., Ed.; Springer: Berlin, Germany, 2005; pp. 3–27.
62. De Jong, Y.; Verbeek, M.; Michelsen, V.; de Place Bjørn, P.; Los, W.; Steeman, F.; Hagedorn, G.; Wetzel, F.T.; Glöcker, F.; Kroupa, A.; et al. Fauna Europaea—All European animal species on the web. *Biodivers. Data J.* **2014**, *2*, e4034. [CrossRef] [PubMed]
63. Martín-Forés, I. Exotic plant species in the Mediterranean biome: A reflection of cultural and historical relationships. In *Mediterranean Identities—Environment, Society, Culture;* Fuerst-Bjelis, B., Ed.; IntechOpen Limited: London, UK, 2017; pp. 179–201.
64. Maiorano, L.; Amori, G.; Capula, M.; Falcucci, A.; Masi, M.; Montemaggiori, A.; Pottier, J.; Psomas, A.; Rondinini, C.; Russo, D.; et al. Threats from climate change to terrestrial vertebrate hotspots in Europe. *PLoS ONE* **2013**, *8*, e74989. [CrossRef] [PubMed]
65. Gippoliti, S.; Groves, C.P. Overlooked mammal diversity and conservation priorities in Italy: Impacts of taxonomic neglect on a Biodiversity Hotspot in Europe. *Zootaxa* **2018**, *4434*, 511–528. [CrossRef]
66. Loy, A.; Aloise, G.; Ancillotto, L.; Angelici, F.M.; Bertolino, S.; Capizzi, D.; Castiglia, R.; Colangelo, P.; Contoli, L.; Cozzi, B.; et al. Mammals of Italy: An annotated checklist. *Hystrix* **2019**, *30*, 87–106.
67. Lorenzoni, M.; Borghesan, F.; Carosi, A.; Ciuffardi, L.; De Curtis, O.; Delmastro, G.B.; Di Tizio, L.; Franzoi, P.; Maio, G.; Mojetta, A.; et al. Check-list dell'ittiofauna delle acque dolci italiane. *Ital. J. Freshw. Ichthyol.* **2019**, *1*, 239–254.
68. Saliot, A. *The Mediterranean Sea;* Springer: Berlin, Germany, 2005; p. 408.
69. Blondel, J. The 'design' of Mediterranean landscapes: A millennial story of humans and ecological systems during the historic period. *Hum. Ecol.* **2006**, *34*, 713–729. [CrossRef]
70. Oro, D.; Pueyo, Y.; Bauzà, J.; Errea, M.P.; Arroyo, A.I. Long transient response of vegetation dynamics after four millennia of anthropogenic impacts in an island ecosystem. *Glob. Chang. Biol.* **2022**, *28*, 6318–6332. [CrossRef] [PubMed]
71. Dijkstra, J.A.; Westerman, E.L.; Harris, L.G. Elevated seasonal temperatures eliminate thermal barriers of reproduction of a dominant invasive species: A community state change for northern communities? *Divers. Distrib.* **2017**, *23*, 1182–1192. [CrossRef]
72. Hulme, P.E. Climate change and biological invasions: Evidence, expectations, and response options. *Biol. Rev.* **2017**, *92*, 1297–1313. [CrossRef]
73. Pyšek, P.; Hulme, P.E.; Simberloff, D.; Bacher, S.; Blackburn, T.M.; Carlton, J.T.; Dawson, W.; Essl, F.; Foxcroft, L.C.; Genovesi, P.; et al. Scientists' warning on invasive alien species. *Biol. Rev.* **2020**, *95*, 1511–1534. [CrossRef]
74. Westphal, M.I.; Browne, M.; MacKinnon, K.; Noble, I. The link between international trade and the global distribution of invasive alien species. *Biol. Invasions* **2008**, *10*, 391–398. [CrossRef]
75. Hulme, P.E. Biological invasions in Europe: Drivers, pressures, states, impacts and responses. In *Biodiversity under Threat;* Hester, R.E., Harrison, R.M., Eds.; The Royal Society of Chemistry: Cambridge, UK, 2007; pp. 56–80.
76. Gallardo, B. Europe's top 10 invasive species: Relative importance of climatic, habitat and socio-economic factors. *Ethol. Ecol. Evol.* **2014**, *26*, 130–151. [CrossRef]
77. Margolis, M.; Shogren, J.F.; Fischer, C. How trade politics affect invasive species control. *Ecol. Econ.* **2005**, *52*, 305–313. [CrossRef]
78. Hulme, P.E. Invasion pathways at a crossroad: Policy and research challenges for managing alien species introductions. *J. Appl. Ecol.* **2015**, *52*, 1418–1424. [CrossRef]
79. Cantini, M.; Menchetti, M.; Vannini, A.; Bruni, G.; Borri, B.; Mori, E. Checklist of Amphibians and Reptiles in a hilly area of Southern Tuscany (Central Italy): An update. *Herpetol. Notes* **2013**, *6*, 223–228.

80. Vannini, A.; Menchetti, M.; Mori, E. L'avifauna del SIC "Poggi di Prata" (Grosseto, Italia Centrale): Analisi faunistica, quantitativa e considerazioni sulla gestione ambientale del sito. *Alula* **2013**, *20*, 101–112.
81. Mori, E.; Menchetti, M.; Dondini, G.; Biosa, D.; Vergari, S. Theriofauna of Site of Community Importance Poggi di Prata (Grosseto, Central Italy): Terrestrial mammals and preliminary data on *Chiroptera*. *Check List* **2014**, *10*, 718–723. [CrossRef]
82. BirdLife International. Streptopelia turtur. In *The IUCN Red List of Threatened Species*; BirdLife International: Cambridge, UK, 2019. [CrossRef]
83. Gippoliti, S.; Amori, G. Ancient introductions of mammals in the Mediterranean Basin and their implications for conservation. *Mammal Rev.* **2006**, *36*, 37–48. [CrossRef]
84. Kemp, M.E.; Mychajliw, A.M.; Wadman, J.; Goldberg, A. 7000 years of turnover: Historical contingency and human niche construction shape the Caribbean's Anthropocene biota. *Proc. R. Soc. B Biol. Sci.* **2020**, *287*, 20200447. [CrossRef]
85. Halevi, G.; Moed, H.; Bar-Ilan, J. Suitability of Google Scholar as a source of scientific information and as a source of data for scientific evaluation—Review of the literature. *J. Informetr.* **2017**, *11*, 823–834. [CrossRef]
86. Olson, D.M.; Dinerstein, E. The Global 200: A representation approach to conserving the Earth's most biologically valuable ecoregions. *Conserv. Biol.* **1998**, *12*, 502–515. [CrossRef]
87. Olson, D.M.; Dinerstein, E.; Wikramanayake, E.D.; Burgess, N.D.; Powell, G.V.N.; Underwood, E.C.; D'Amico, J.A.; Itoua, I.; Strand, H.E.; Morrison, J.C.; et al. Terrestrial ecoregions of the world: A new map of life on Earth. *Bioscience* **2001**, *51*, 933–938. [CrossRef]
88. Burnham, K.P.; Anderson, D.R. *Model Selection and Multimodel Inference. A Practical Information-Theoretic Approach*; Springer: New York, NY, USA, 2002; p. 488.
89. Burnham, K.P.; Anderson, D.R. Multimodel inference: Understanding AIC and BIC in model selection. *Sociol. Methods Res.* **2004**, *33*, 261–304. [CrossRef]
90. Burnham, K.P.; Anderson, D.; Huyvaert, K. AIC model selection and multimodel inference in behavioral ecology: Some background, observations, and comparisons. *Behav. Ecol. Sociobiol.* **2011**, *65*, 23–35. [CrossRef]
91. D'Amico, M.; Román, J.; de los Reyes, L.; Revilla, E. Vertebrate road-kill patterns in Mediterranean habitats: Who, when and where. *Biol. Conserv.* **2015**, *191*, 234–242. [CrossRef]
92. Grilo, C.; Koroleva, E.; Andrášik, R.; Bíl, M.; González-Suárez, M. Roadkill risk and population vulnerability in European birds and mammals. *Front. Ecol. Environ.* **2020**, *18*, 323–328. [CrossRef]
93. Medrano-Vizcaíno, P.; Grilo, C.; Silva Pinto, F.A.; Carvalho, W.D.; Melinski, R.D.; Schultz, E.D.; González-Suárez, M. Roadkill patterns in Latin American birds and mammals. *Glob. Ecol. Biogeogr.* **2022**, *31*, 1756–1783. [CrossRef]
94. Eberhardt, E.; Mitchell, S.; Fahrig, L. Road kill hotspots do not effectively indicate mitigation locations when past road kill has depressed populations. *J. Wildl. Manag.* **2013**, *77*, 1353–1359. [CrossRef]
95. Quiles, P.; Ascensão, F.; D'Amico, M.; Revilla, E.; Barrientos, R. Are road-kills representative of wildlife community obtained from atlas data? *Hystrix* **2021**, *32*, 89–94.
96. Delgado, J.D.; Arévalo, J.R.; Fernández-Palacios, J.M. Road and topography effects on invasion: Edge effects in rat foraging patterns in two oceanic island forests (Tenerife, Canary Islands). *Ecography* **2001**, *24*, 539–546. [CrossRef]
97. Hollings, T.; Jones, M.; Mooney, N.; McCallum, H. Disease-induced decline of an apex predator drives invasive dominated states and threatens biodiversity. *Ecology* **2016**, *97*, 394–405. [CrossRef] [PubMed]
98. Schwartz, A.L.; Williams, H.F.; Chadwick, E.; Thomas, R.J.; Perkins, S.E. Roadkill scavenging behaviour in an urban environment. *J. Urban Ecol.* **2018**, *4*, juy006. [CrossRef]
99. Dhiab, O.; D'Amico, M.; Selmi, S. Experimental evidence of increased carcass removal along roads by facultative scavengers. *Environ. Monit. Assess.* **2023**, *195*, 216. [CrossRef] [PubMed]
100. Ascensão, F.; D'Amico, M.; Martins, R.C.; Rebelo, R.; Barbosa, A.M.; Bencatel, J.; Barrientos, R.; Abellán, P.; Tella, J.L.; Cardador, L.; et al. Distribution of alien tetrapods in the Iberian Peninsula. *NeoBiota* **2021**, *64*, 1–21. [CrossRef]
101. Underwood, E.C.; Viers, J.H.; Klausmeyer, K.R.; Cox, R.L.; Shaw, M.R. Threats and biodiversity in the Mediterranean biome. *Divers. Distrib.* **2009**, *15*, 188–197. [CrossRef]
102. Pyšek, P.; Richardson, D.M. Invasive species, environmental change and management, and health. *Annu. Rev. Environ. Resour.* **2010**, *35*, 25–55. [CrossRef]
103. Pyšek, P.; Jarošík, V.; Hulme, P.E.; Pergl, J.; Hejda, M.; Schaffner, U.; Vilà, M. A global assessment of invasive plant impacts on resident species, communities and ecosystems: The interaction of impact measures, invading species' traits and environment. *Glob. Change Biol.* **2012**, *18*, 1725–1737. [CrossRef]
104. Shi, J.; Luo, Y.Q.; Zhou, F.; He, P. The relationship between invasive alien species and main climatic zones. *Biodivers. Conserv.* **2010**, *19*, 2485–2500. [CrossRef]
105. Weidema, I.R. *Introduced Species in the Nordic Countries*; Nord 2000 Editions: Aarhus, Denmark, 2020; p. 242.
106. Zhou, Y.; Smith, S.J.; Zhao, K.; Imhoff, M.; Thomson, A.; Bond-Lamberty, B.; Asrar, G.R.; Zhang, X.; He, C.; Elvidge, C.D. A global map of urban extent from nightlights. *Environ. Res. Lett.* **2015**, *10*, 054011. [CrossRef]
107. Kuemmerle, T.; Levers, C.; Erb, K.; Estel, S.; Jepsen, M.R.; Müller, D.; Plutzar, C.; Stürck, J.; Verkerk, P.J.; Reenberg, A. Hotspots of land use change in Europe. *Environ. Res. Lett.* **2016**, *11*, 064020. [CrossRef]
108. Richardson, H.W.; Bae, C.H.C. *Urban Sprawl in Western Europe and the United States*; Routledge: London, UK, 2017; p. 344.

109. Masseti, M.; Albarella, U.; De Grossi Mazzorin, J. The crested porcupine, *Hystrix cristata* L., 1758, in Italy. *Anthropozoologia* **2010**, *45*, 27–42. [CrossRef]
110. Esattore, B.; Saggiomo, L.; Sensi, M.; Francia, V.; Cherin, M. Tell me what you eat and I'll tell you ... where you live: An updated review of the worldwide distribution and foraging ecology of the fallow deer (*Dama dama*). *Mamm. Biol.* **2022**, *102*, 321–338. [CrossRef]
111. Moreira, F.; Allsopp, N.; Esler, K.J.; Wardell-Johnson, G.; Ancillotto, L.; Arianoutsou, M.; Clary, J.; Brotons, L.; Clavero, M.; Dimitrakopoulos, P.G.; et al. Priority questions for biodiversity conservation in the Mediterranean biome: Heterogeneous perspectives across continents and stakeholders. *Conserv. Sci. Pract.* **2019**, *1*, e118. [CrossRef]
112. Newbold, T.; Oppenheimer, P.; Etard, A.; Williams, J.J. Tropical and Mediterranean biodiversity is disproportionately sensitive to land-use and climate change. *Nat. Ecol. Evol.* **2020**, *4*, 1630–1638. [CrossRef]
113. Garriga, N.; Santos, X.; Montori, A.; Richter-Boix, A.; Franch, M.; Llorente, G.A. Are protected areas truly protected? The impact of road traffic on vertebrate fauna. *Biodivers. Conserv.* **2012**, *21*, 2761–2774. [CrossRef]
114. Kang, W.; Minor, E.S.; Woo, D.; Lee, D.; Park, C.R. Forest mammal roadkills as related to habitat connectivity in protected areas. *Biodivers. Conserv.* **2016**, *25*, 2673–2686. [CrossRef]
115. Collinson, W.J.; Marneweck, C.; Davies-Mostert, H.T. Protecting the protected: Reducing wildlife roadkill in protected areas. *Anim. Conserv.* **2019**, *22*, 396–403. [CrossRef]
116. Mori, E.; Bini, A.; Viviano, A.; Bartolommei, P.; Mazza, G. Distribution of introduced American mink in the Northern Apennine area (Central Italy). *Mammalia* **2022**, *86*, 266–270. [CrossRef]
117. Périquet, S.; Roxburgh, L.; Le Roux, A.; Collinson, W.J. Testing the value of citizen science for roadkill studies: A case study from South Africa. *Front. Ecol. Evol.* **2018**, *6*, 15. [CrossRef]
118. Valerio, F.; Basile, M.; Balestrieri, R. The identification of wildlife-vehicle collision hotspots: Citizen science reveals spatial and temporal patterns. *Ecol. Process.* **2021**, *10*, 6. [CrossRef]
119. Aristeidou, M.; Herodotou, C.; Ballard, H.L.; Young, A.N.; Miller, A.E.; Higgins, L.; Johnson, R.F. Exploring the participation of young citizen scientists in scientific research: The case of iNaturalist. *PLoS ONE* **2021**, *16*, e0245682. [CrossRef]
120. Mesaglio, T.; Callaghan, C.T. An overview of the history, current contributions and future outlook of iNaturalist in Australia. *Wildl. Res.* **2021**, *48*, 289–303. [CrossRef]
121. Wolf, S.; Mahecha, M.D.; Sabatini, F.M.; Wirth, C.; Bruelheide, H.; Kattge, J.; Martínez, A.M.; Mora, K.; Kattenborn, T. Citizen science plant observations encode global trait patterns. *Nat. Ecol. Evol.* **2022**, *6*, 1850–1859. [CrossRef]
122. Otranto, D.; Dantas-Torres, F.; Mihalca, A.D.; Traub, R.J.; Lappin, M.; Baneth, G. Zoonotic parasites of sheltered and stray dogs in the era of the global economic and political crisis. *Trends Parasitol.* **2017**, *33*, 813–825. [CrossRef]
123. Dong, S. Overview: Pastoralism in the World. In *Building Resilience of Human-Natural Systems of Pastoralism in the Developing World*; Dong, S., Kassam, K.A., Tourrand, J., Boone, R., Eds.; Springer International Publishing: Cham, Switzerland, 2016; p. 298.
124. Giuntini, S.; Ancillotto, L.; Mori, E.; Viviano, A. The Alexandrine parakeet *Psittacula eupatria* as a naturalized breeding species: A proposal of integration to the national official checklist. *Avocetta* **2021**, *45*, 177–179.
125. Di Nicola, M.; Cavigioli, L.; Luiselli, L.; Andreone, F. *Anfibi & Rettili d'Italia*; Belvedere, Ed.; Historia Naturae: Latina, Italy, 2021; p. 576.
126. McKinney, M.L. Influence of settlement time, human population, park shape and age, visitation and roads on the number of alien plant species in protected areas in the USA. *Divers. Distrib.* **2002**, *8*, 311–318. [CrossRef]
127. Cardador, L.; Carrete, M.; Gallardo, B.; Tella, J.L. Combining trade data and niche modelling improves predictions of the origin and distribution of non-native European populations of a globally invasive species. *J. Biogeogr.* **2016**, *43*, 967–978. [CrossRef]
128. Mazzamuto, M.V.; Wauters, L.A.; Koprowski, J.L. Exotic pet trade as a cause of biological invasions: The case of tree squirrels of the genus *Callosciurus*. *Biology* **2021**, *10*, 1046. [CrossRef] [PubMed]

biology

MDPI

Article

Agricultural and Ecological Resources Safeguarded by the Prevention of Wild Pig Population Expansion

Colin Jareb [1,*], Kim M. Pepin [1], Ryan S. Miller [2], Sarah Sykora [1,3], Stephanie A. Shwiff [1] and Sophie C. McKee [1,3]

[1] National Wildlife Research Center, United States Department of Agriculture, Animal and Plant Health Inspection Service, 4101 Laporte Avenue, Fort Collins, CO 80521, USA; sarah.sykora@usda.gov (S.S.); sophie.mckee@colostate.edu (S.C.M.)

[2] Center for Epidemiology and Animal Health, United States Department of Agriculture, Veterinary Services, Fort Collins, CO 80526, USA; ryan.s.miller@usda.gov

[3] Department of Economics, Colorado State University, Fort Collins, CO 80523, USA

* Correspondence: colin.jareb@usda.gov

Simple Summary: The spread of invasive wild pigs across the United States has been successfully slowed due to the establishment of a national management program. In this paper, we study the effectiveness of the national program by modeling the spread of wild pigs in the absence of intervention. Further, we use the models to assess the value of resources safeguarded from the threat of wild pigs. Our findings indicate that, on average, invasive wild pigs were prevented from spreading to 724 counties, and USD 40.2 billion in resource value was safeguarded over the first eight years of the program. Continuation of the program will deliver additional benefits and further research is critical to understand its comprehensive economic effects.

Abstract: Wild pigs (*Sus scrofa*) are one of the most destructive invasive species in the US, known for causing extensive damage to agricultural commodities, natural resources, and property, and for transmitting diseases to livestock. Following the establishment of the National Feral Swine Damage Management Program (NFSDMP) in 2014, the expansion of wild pig populations has been successfully slowed. This paper combines two modeling approaches across eight separate models to characterize the expansion of wild pig populations in the absence of intervention by the NFSDMP and forecasts the value of a subset of resources safeguarded from the threat of wild pigs. The results indicate that if wild pigs had continued spreading at pre-program levels, they would have spread extensively across the US, with significant geographic variation across modeling scenarios. Further, by averting the threat of wild pigs, a substantial amount of crops, land, property, and livestock was safeguarded by the NFSDMP. Cumulatively, between 2014 and 2021, wild pig populations were prevented from spreading to an average of 724 counties and an average of USD 40.2 billion in field crops, pasture, grasses, and hay was safeguarded. The results demonstrate that intervention by the NFSDMP has delivered significant ecological and economic benefits that were not previously known.

Keywords: agricultural resources; economic impact; feral swine; invasive species; population dynamics; *Sus scrofa*; United States; wild pigs

1. Introduction

Initially introduced to the United States in the 1500s, wild pigs (*Sus scrofa*), also known as feral swine, wild swine, wild boar, or feral hogs [1], are a successful invasive species [2]. For this study, the collective term wild pig will be used [1] unless it is appropriate to use one of the common names discussed above. Native to Europe, Asia, and Northern Africa [3], wild pigs have been repeatedly introduced into the United States, with populations observed in at least 38 states [4]. Further, high reproductive rates and generalist habitat use have led to transient population dynamics [5] and highly variable geographic rates of population spread [6,7].

The literature on the economic damages caused by the spread of wild pigs is growing. In a survey of producers of six crops in 11 states, McKee et al. [8] estimate that USD 203 million of crop production was lost in 2022. Using a survey of producers of additional crops across 12 states in 2019, McKee et al. [9] estimate crop losses of USD 272 million. Boughton et al. [10] studied the effects of wild pig rooting and concluded that damage to pastureland and loss of forage generated regional losses greater than USD 2 million annually. Further, Didero et al. [11] provide examples of a wide range of methodologies used to produce damage estimates. Lastly, Vercauteren et al. [12] contribute comprehensive treatment on a variety of issues regarding the extent of knowledge about wild pig control and damage.

As a result of the outlined population dynamics and increasing damage, US Congress appropriated funds in 2014 to establish the National Feral Swine Damage Management Program (NFSDMP). The objectives of the program include operational population reduction, such as removal and eradication, cooperative research, and outreach and communication campaigns. The impacts from wild pig damage have primarily been felt in areas with large wild pig populations, such as the Southeastern United States and California. As a result of this geographic concentration, the NFSDMP's resources have been focused on mitigating damage in a few states and eliminating wild pigs in other states. In states with significant invasive wild pig populations, surveys have been conducted to estimate the extent of damage to corn, wheat, peanuts, rice, sorghum, soybeans, cotton, hay, almonds, pecans, melons, sugarcane, sweet potatoes, and berries in these areas [8,9].

Focusing on invasive wild pig population control, the NFSDMP has eliminated feral swine from 11 states [13], reduced spread rates [5], and prevented the spread of wild pigs to regions with no contemporary or previous wild pig populations [7]. By reducing the spread of wild pigs, the introduction of the NFSDMP has alleviated the threat of damage to resources including crops, land, property, and livestock.

However, little is known about the damage prevented by management in areas where pig populations would have been established and grown in the absence of intervention [7]. Estimating the extent of wild pig geographic spread prevented by NFSDMP activities allows for the evaluation of resources safeguarded by management. In this paper, we use two ecological modeling frameworks, a dynamic occupancy approach [14,15] and boosted regression trees [16] to forecast the spread of wild pigs between 2014 and 2021 in the absence of any intervention by the NFSDMP. Further, we utilize the spread forecasts to quantify the primary resources safeguarded from the threat of wild pigs during this period.

Analyzing the full extent of damages caused by the spread of wild pigs requires information on population densities and the degree to which wild pigs damage resources. However, the models forecasting the spread of wild pigs focus on predicting the share of counties occupied by wild pigs without providing predictions on density. Little is known about the relationship between the presence and density of wild pigs and the damage they cause [12], nor of the extent of damage caused by wild pigs to a variety of other crop resources and to land resources, such as wetlands [11]. For these reasons, it has been intractable to ascertain the full economic impact of wild pig damage avoided by introducing the NFSDMP. Therefore, our main objective was to quantify the economic value of resources safeguarded by a national-scale wild pig damage management program. We estimate the total resource acreage (including agricultural crops, pasture, and wetlands) and livestock inventories safeguarded from the presence of wild pigs by linking forecasted geographic spread of wild pigs in the absence of NFSDMP to the value of selected commodities in those areas where spread was prevented. We then estimate a dollar value for a subset of the resources for which annual revenue per acre is available.

Our results find that from 2014 to 2021, the NFSDMP has successfully prevented the spread of wild pigs to additional counties. Further, our analysis demonstrates that significant acreage and value in field crops, pasture, and grasses has been successfully safeguarded by the NFSDMP. These results indicate that the additional benefits generated by implementing the NFSDMP in areas where wild pigs have not been observed are significant and should be accounted for when evaluating the benefits of the program. Our framework can be used to evaluate the value of resources safeguarded by a national damage management program.

2. Materials and Methods

We modeled the economic consequences of preventing the spatial spread of wild pigs using an ensemble approach [17]. We forecast the spread of wild pigs using two approaches; boosted regression tree modeling with step-ahead forecasting from these models and a dynamic occupancy approach. For each approach, we further include a variety of covariates, simulating each model using four distinct parameterizations. This yields a total of eight forecasts that predict the spread of wild pigs in the absence of NFSDMP intervention. We quantify the resources safeguarded by containing the spread of wild pigs by using the total acres or head of a resource and the proportion of the county forecasted as occupied by wild pigs. Further, we combine this amount with the economic value of the respective resource to quantify the value of resources safeguarded. The two modeling frameworks were implemented at the county and watershed scales. All final model predictions are provided at the county scale annually and provide both the presence/absence and proportion of the county occupied by wild pigs. We follow previous work [7,18] by including a variety of covariates to predict the rate of spatial spread.

2.1. Data and Processing

We modeled the geographic expansion of wild pigs using annual county-level and watershed-level mapping of wild pig presence/absence [6,7,18]. These data were expressed as a county area occupied by wild pigs in km^2 and are a proportion of the county occupied by wild pigs. We subdivided the data into counties where there have been wild pigs at least once during 2009–2013 ('transient counties', $N = 926$) and those where there have always been wild pigs during 2009–2013 ('persistent counties', $N = 376$) because we expected spread rates and covariate effects to differ in these two situations. For estimating the pre-NFSDMP spread rates, first, we estimated annual spread rates for the 5 years immediately prior to the start of NFSDMP (2009–2013) by subtracting the km^2 area or proportion occupied in the current year from that in the previous year. Then, we took the mean of these annual spread rates per county to generate the average pre-program spread rates to estimate covariate effects.

Watershed level response data were obtained using the following procedure. Polygons representing the observed distribution of wild pigs were aggregated to 12-digit watershed boundary data using methods described in Miller et al. [5] resulting in the presence/absence of wild pigs by 12-digit watersheds. These 12-digit watersheds were assumed to represent discrete populations. The number of the invaded and total number of 12-digit watersheds within each coarse scale 8-digit watershed was generated and used as a response variable.

We use data from the National Feral Swine Mapping System to estimate the rate of spread of wild pigs in the contiguous United States. These data have been compiled annually since 2008. Each year, wildlife professionals in state wildlife resource agencies and the United States Department of Agriculture report polygons (polygons are drawn by technicians and vary in size based on extent of observed wild pigs) representing their opinion of the geographic extent of established wild pig populations to the National Feral Swine Mapping System. The definition of pig presence in this methodology includes populations present for two or more years with evidence of reproduction. The polygons are then processed to determine the proportion of each county that has wild pigs.

There are two major sources of error in the data that are difficult to account for because we did not have the requisite data: (1) error between the perception of where wild pigs are when experts draw the polygons and the true distribution of wild pigs; (2) differences in the reporting methodology over time and across states in terms of how detailed the polygons were and how much observational effort was allotted to their reporting. Because of these data limitations with interannual variation, we estimate the pre-program spread rate from 2009 to 2013 as an average rate across those 5 years. This was performed by first calculating the county-level annual rate for each year (2009–2013) as the difference in proportion occupied for each county between $year_t - year_{t-1}$, which ranged from -1 to 1. We also investigated models of the spread rates per km^2 because counties varied widely in their size. To estimate this second response variable, we multiplied the proportion differences by the county area to model the rate spread in km^2 and account for differences in county size. This response distribution ranged [−542.4, 3165.8], and as with the difference in proportion distribution, the change in km^2 response probability distribution was characterized by a high density near 0, a short tail on the negative-value side, and a long tail on the positive-value side. We investigate models of the change in proportions as well as models of the change in km^2 occupied.

We obtain crop, fruit, vegetable, grass, and pasture resource data at the county level from the National Agricultural Statistical Service (NASS) to calculate the quantity and value of resources safeguarded by containing and in some cases, reversing the spread of wild pigs. The data on acreage, yields, and prices were sourced from the 2017 Census of Agriculture. The agricultural census is conducted every five years and the analysis focuses on the period spanning 2014 to 2021; thus, we assume that crop data from 2017 are constant across all years of the analysis. Further, information on livestock counts was obtained from the 2017 Census of Agriculture; however, data covering the annualized values of livestock are not available from NASS, which prevented us from investigating the value of livestock that were unthreatened by the spread of wild pigs. Lastly, data gathered from the 2015 USGS National Land Cover Database were used to estimate the acreage of wetlands, which would have been threatened by the unmitigated spread of wild pigs. To calculate the resource acreage that would have been exposed to wild pigs, we interact the total acreage in each county with the forecasted proportion of the county occupied. Dollar values for crops and land were also obtained for the year 2017 and were used to quantify the value of resources safeguarded by the NFSDMP. After estimating the annualized value of resources that would have been threatened by the spread of wild pigs, the total value of resources safeguarded was computed by aggregating across the additional years in which a county was forecasted to be occupied.

2.2. Boosted Regression Tree Models

We estimated the covariate effects on response data using boosted regression trees with least-square boosting because the response variable did not follow a known parametric distribution and because we expected complex non-linear relationships between the response and covariates. Covariates for spread rates based on the difference in proportion of area occupied include the proportion occupied in the focal county in the previous year, the mean proportion of counties occupied in the adjacent counties (i.e., an occupancy pressure from adjacent counties), amount of wild pigs removed per km^2 in the focal county in the previous year, the mean count of removals per county in the adjacent counties in the previous year (log scale), the proportion of the focal county with active removal work in the previous year, the mean proportion of adjacent counties with active removal work in the previous year, the habitat suitability index (the habitat suitability index standardizes biotic and abiotic factors across counties and is obtained using the methods of Ref. [18] for use as a covariate to fit the boosted regression tree models) [18], median density estimates rescaled, the percentage of the focal county with hydrology, the percentage of the focal county with tree cover, the percentage of the focal county with crops, the percentage of the focal county with pasture, and the percentage of deciduous plant cover.

Covariates for spread rates based on the difference in the area (km^2) occupied in the focal county in the previous year, the mean proportion of counties occupied in the adjacent counties, the sum of removals in the focal county in the previous year, the mean count of removals per county in the adjacent counties in the previous year (log scale), the proportion of the focal county with active removal work in the previous year, the mean proportion of adjacent counties with active removal work in the previous year, the habitat suitability index [18] median density estimates rescaled, the percentage of the focal county with hydrology, the percentage of the focal county with tree cover, the percentage of the focal county with crops, the percentage of the focal county with pasture, and the percentage of deciduous plant cover. Several covariates are subject to the sources of error discussed in the previous section. Further, it is difficult to account for how wild pig management would have evolved in the absence of the NFSDMP in response to the continuously spreading wild pigs.

For each response, we fit models with and without the covariate describing status of the focal county in the previous year because this covariate was a very strong predictor. We took this approach to account for the extent of path dependency and uncertainty in forecasting the spread of wild pigs. When interpreting the results, it is important to emphasize that regardless of our modeling assumptions, the expansion of wild pig populations without intervention is considerable. We denote the first four models A through D for each possible response–covariate pair. Thus, we examined the following four models of wild pig spread: change in proportion occupied with proportion occupied in 2008 (model A), change in proportion occupied without proportion occupied in 2008 (model B), change in km^2 with km^2 occupied in 2008 (model C), and change in km^2 without km^2 occupied in 2008 (model D).

We use the fitted models to conduct step-ahead forecasting for years 2014–2021 (post-introduction of the NFSDMP), starting with the data in 2013. For each year, we use the Persistence and Transient models to predict the rate of change for each county given the current year's covariate values. We then add the change to the current value in each county (proportion occupied for A and B models, km^2 for C and D models). For A and B models, we bound the predictions between 0 and 1; for C and D models, we bound the predictions between 0 and the upper maximum of the county area in km^2. For each year, we recorded which counties became newly occupied and which counties became newly unoccupied. We used the updated proportion occupied values to update the mean proportion occupied in adjacent counties to be used in the next year's prediction.

2.3. *Dynamic Occupancy Models*

We modeled the occupancy dynamics using a two-step process. We modeled the occupancy rate within 8-digit watersheds using 12-digit watersheds as spatial replicates. We then used a stochastic framework to account for changes in the annual rate of introduction (invasion) and extinction due to stochastic processes. The dynamic occupancy models do not account for imperfect detection due to computational infeasibility. Covariates include a combination of annually time-varying and non-time-varying climate and land cover data obtained from Refs. [7,18]. In addition, the rate of invaded watersheds in the previous year and the distance to the nearest invaded watershed were also included. Based on previous work [6,7,18,19], we included the following non-time and time-varying covariates for predicting the rate of spatial spread.

Non-time-varying covariates include terrain ruggedness, predicted equilibrium density [18], and mast tree species richness. Time-varying covariates are shown in Table 1.

All models include the rate of invaded watersheds in the previous year and the distance to the nearest invaded watershed. Initial model screening was conducted using projection predictive variable selection that allows for large sets of candidate models to be evaluated quickly. We evaluated 108,537 possible candidate models using all possible combinations for eight sets of variables that did not have a large covariance. From that initial screening, a reduced set of 127 candidate models was evaluated using leave-one-out cross validation. Predictive capacity of the final model was evaluated using Bayesian version of R-squared and LOO-adjusted R-squared. Model convergence was evaluated by comparing the between- and within-chain predictions for model parameters as well as bulk and tail effective sample size. Evaluation of the top model indicated some potential regionally varying effects. To address this, we evaluated models with uncorrelated random intercepts and correlated intercepts with random slopes to account for regional variation in invasion rates. We considered four regions—west, plains, Great Lakes, and east. The final Bayesian model predicts the probability of a watershed being invaded annually given the covariates included in the model. Bayesian models and model evaluation were conducted using the rstanarm (rstanarm version 2.21.3 was used to estimate the model. The most recent update can be found at https://cran.r-project.org/web/packages/rstanarm/index.html, accessed on 14 June 2023) package within the R computing environment.

Table 1. Time-varying covariates used to predict spread in watershed-scale models. Land cover and indexed data are provided at an annual scale; thus, other covariates are also measured as an annual average.

Climate (Annual Average)	Land Cover (Annual Average)	3-Year Moving Average
Minimum Temperature Precipitation Snow Cover	Human Population Density Agriculture (proportion) Forest (proportion) Grassland (proportion) Wetland (proportion) Enhanced Vegetation Index Palmar Drought Severity Index	Enhanced Vegetation Index Palmar Drought Severity Index Minimum temp. Precipitation Snow cover Enhanced Vegetation Index

To predict the spread of wild pigs in each year from 2014 to 2021, we used step-ahead forecasting using a stochastic framework to account for changes in annual rate of invasion and extinction due to stochastic processes. To implement the simulation, we first predict the probability of invasion using the final Bayesian model for all uninvaded watersheds in year t using associated time-varying data for year t. This results in a probability of invasion for every uninvaded watershed. The number of invaded watersheds in year t was drawn from a beta distribution representing the range of observed invasion rates from 2004 to 2012. Watersheds are randomly assigned as invaded using the probability of invasion as a weight. The proportion of 12-digit watersheds within each 8-digit invaded watershed is updated and the distance to the nearest invaded watershed for all uninvaded watersheds is updated. In year t + 1, the probability of invasion and watersheds assigned as invaded was the same as in year t. Additionally, in year t + n, the number of watersheds with extinctions was drawn from a beta distribution representing the range of observed extinction rates from 2004 to 2012. This process of simulation was conducted until t + 8 (2021). The simulation was run 1000 times. Using these simulations, the median proportion of each 8-digit watershed invaded was calculated. These watershed-level proportions are used to estimate the proportion of each county invaded.

2.4. Resources Safeguarded

To evaluate the economic impacts of preventing the spread of wild pigs, we begin by accounting for all resources that would have been threatened under each modeling scenario. The resources that would have been threatened without intervention by the NFSDMP are also referred to as resources safeguarded due to intervention by the program. We interact the proportion of each county invaded by wild pigs with total resource acreage and livestock inventories. The acreage and value of resources were computed and visualized using the R computing environment (R version 4.3.1 and ggplot2 version 3.4.4 were used to generate maps and calculate acreage and values). A variety of resources are included under the following broad categories: field crops, fruits, vegetables, tree nuts, field and grass seeds, forage, and hay, land (including pasture and wetlands), and livestock. Details for each category can be found in Appendix A Table A1.

Studying the period running from 2014 to 2021, we allow for variation in the amount of time wild pigs invade a given county and changes over time in the proportion of a county occupied for each scenario. As a result of the caveats discussed above, there is a high degree of uncertainty within each scenario. Thus, aggregating across each scenario results in a high level of uncertainty when calculating the quantity of resources threatened by wild pigs in the absence of the NFSDMP.

2.5. Value of Resources Safeguarded

To quantify the value of resources that would have been threatened in the absence of intervention by the NFSDMP, we utilize information on the value of an acre for each resource for which data are available and scale by the additional proportion of a county that is forecasted to be occupied by wild pigs in each year.

Information on dollar values for crops is obtained using national-level prices. We then combine the price with information on crop yields at the state level, which produces a state-level dollar value per acre for each crop. Information on the dollar value for pasture comes from cash rents, measured in dollars per acre. This is available for 42 states and at the national level; the national-level value is used in states where cash rents are unavailable. For details on the dollar values used, see Appendix A Table A2.

For each model, we computed the total amount of resources for which the threat from wild pigs is averted. We refrained from estimating the value of damages to resources; though surveys have been conducted to estimate the rate of damages from wild pigs in 11 southern states [8,9], extrapolating these results to other regions is impracticable. Employing data on prices and yields, we then calculated the total value of resources safeguarded from wild pig presence due to the introduction of the NFSDMP. We calculated the value of resources safeguarded for each county and year and then aggregated these values across all years and counties for each model. The resulting value is presented for each broad category of resources and is available for each individual resource.

3. Results

3.1. Boosted Regression Tree Models

A summary of the data showed that there were 250 counties that became newly occupied between 2009 and 2013 (the five years immediately before the start of NFSDMP). This gives an average rate of invasion of 50 counties per year (250 counties/5 years). On average, the newly established populations in these counties occupied 447.5 km^2 (95% CI: [309.2, 585.7] km^2) equivalent to 0.17 proportion of the county (95% CI: [0.14, 0.21] proportion of the county). Also, 78 counties became newly unoccupied, giving a rate of 15.6 counties/year equivalent to 284.2 km^2/year (95% CI: [115.4, 453.0] km^2).

After the NFSDMP was introduced in 2014, 260 counties became newly occupied between 2014 and 2021, giving an average rate of 32.5 counties per year (260 counties/8 years). On average, the newly established populations in these counties occupied 226.3 km^2 (95% CI: [179.2, 273.4] km^2), equivalent to 0.20 proportion of the county (95% CI: [0.16, 0.24] proportion of the county). Also, 343 counties became newly unoccupied, giving a rate of 42.9/year equivalent to 166.0 km^2/year (95% CI: [120.4, 211.5] km^2). Thus, the rates of spread in terms of both the number of counties per year and the number of km^2 per county per year were higher before the program started (the NFSDMP worked to unify federal, state, and local wildlife managers as well as landowners and university collaborators to manage wild pigs. A variety of approaches including but not limited to population control as well as eradication were used in reducing the spread of wild pigs after 2014). The rate of elimination in terms of the number of counties per year where wild pigs were eliminated was almost three times higher after the program began, but the area in each county where wild pigs were eliminated tended to be lower after the program relative to before the program started (166 km^2/year vs. 284.2 km^2/year).

The forecasted annual dynamics of number of counties newly occupied and unoccupied for model A matched the pre-program rates most closely (~55 counties newly occupied/per year and 12 counties newly unoccupied/year) (Figure 1). Models C and D forecasted much higher rates for both becoming newly occupied or unoccupied than the rates before the program (Figure 1). For example, model D forecasted the most aggressive outcome with up to 895 newly occupied counties (Figure 2).

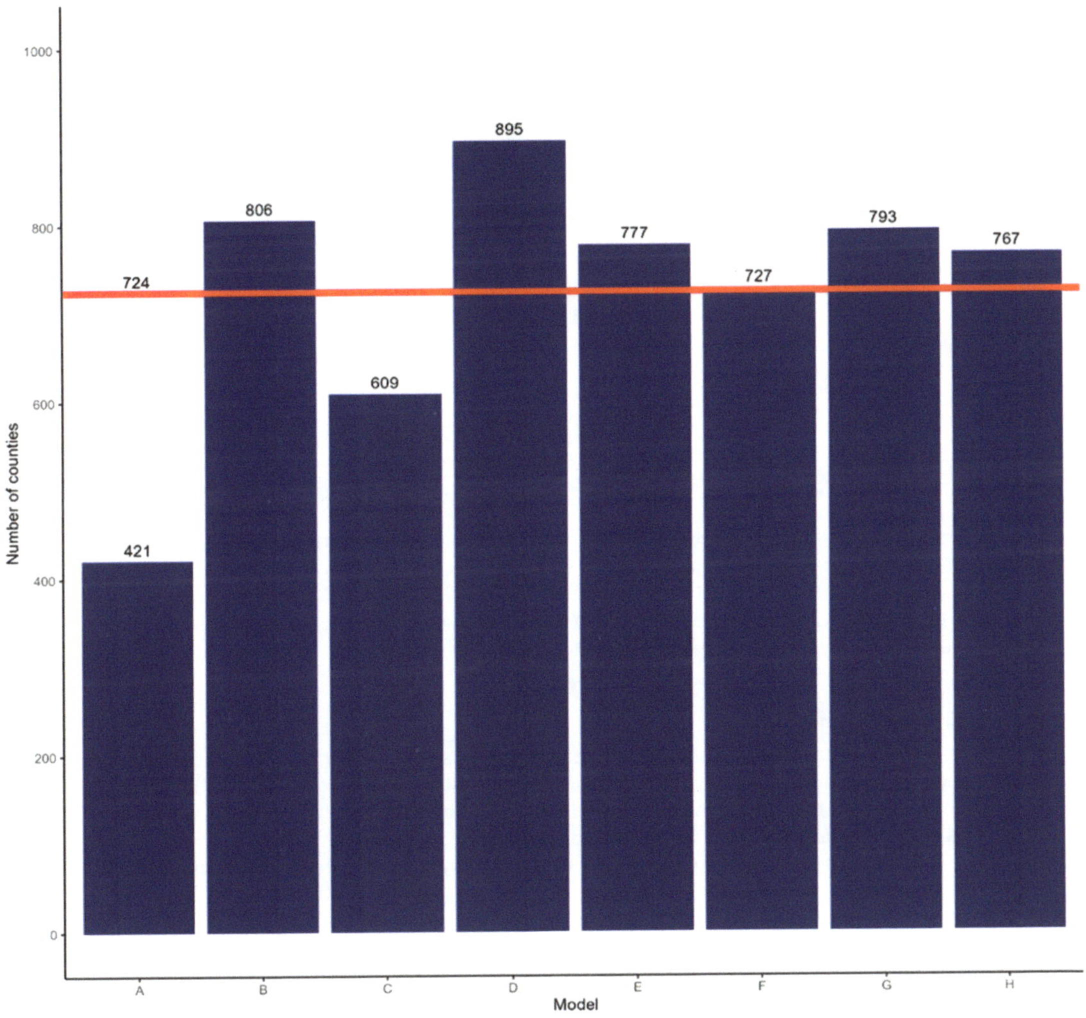

Figure 1. Number of counties forecasted to be newly occupied by wild pigs across all scenarios in the absence of the NFSDMP from 2014 to 2021. Across all frameworks, an average (red line) of 724 new counties are forecast to be occupied by wild pigs.

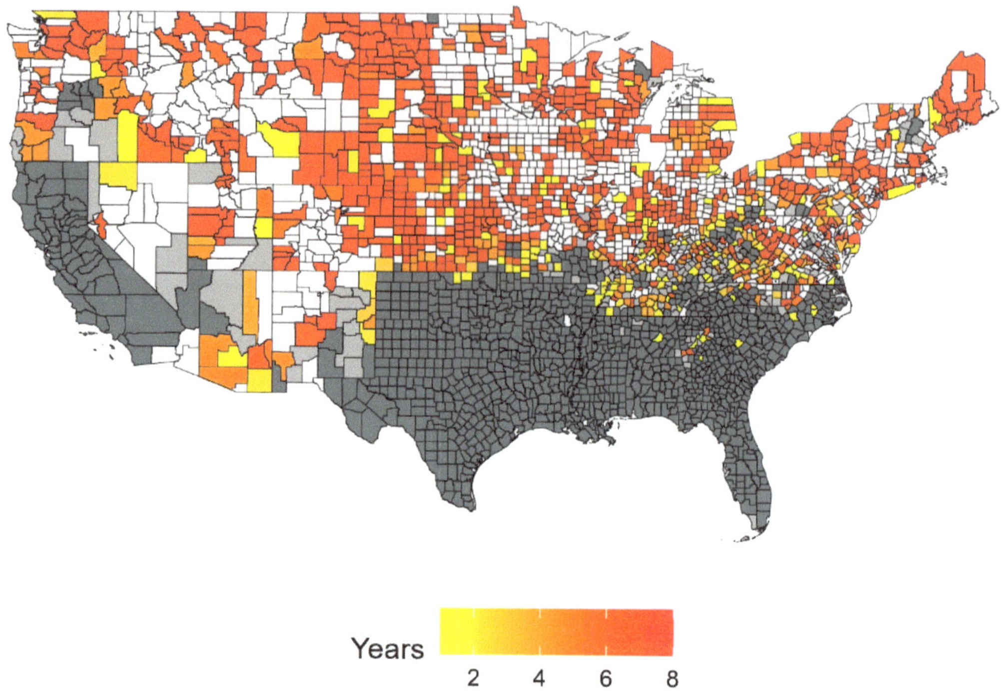

Figure 2. Counties where wild pigs are forecasted to spread under the assumptions of model D. This forecast is the most aggressive outcome. Dark gray counties were always occupied, light gray counties were partially occupied, and white counties were never occupied. Counties are forecasted to be occupied for as few as one additional year (yellow) up to eight additional years (red). Significant geographic variation is forecasted across other models.

3.2. Dynamic Occupancy Models

3.2.1. Data Fitting

The final Bayesian statistical model includes six predictors—terrain ruggedness, Lewis density (the Lewis density is the predicted density of wild pigs from Ref. [18]), annual temperature range, annual mean precipitation, rate of invaded watersheds in the previous year, and the distance to the nearest invaded watershed.

Evaluating models with only fixed effects, regional uncorrelated random intercepts, and regional random slopes with correlated intercepts found, all had generally good predictive performance. Further posterior predictive checks found no bias in the predictive capacity of the models. For detailed methods and results on the performance of the watershed scale models, see Appendix B.

3.2.2. Forecasting

The median forecasted spread of wild pigs varied among models ranging from 1999 to 2065 watersheds invaded by the year 2021 (lower and upper 95-percentiles are included in result files). The number of invaded watersheds annually varied due to both annual time-varying predictors of invasion as well as stochasticity in invasion as well as extinction. Additionally, there were differences in the geographic patterns of invasion. In general, forecasts resulted in generally small proportions of the invaded counties being occupied by wild pigs.

Across all models, an average of 724 additional counties would have been occupied by wild pigs in the absence of intervention. This reflects an overall high degree of transience in wild pig populations, as 1272 counties were observed to be occupied by wild pigs in 2013. In the most conservative forecast, 421 counties are forecast to be invaded by wild pigs, and in the most aggressive forecast, 895 counties are occupied by wild pigs. Details regarding the number of counties invaded for each year forecasted can be found in Appendix A Table A3.

3.3. Resources Safeguarded

There is wide variation across models in the quantity of resources safeguarded from wild pigs by the NFSDMP (see Figure 3). Taking an unweighted mean across each model, we find that an average of 60.12 million acres (243.3 thousand km^2) of field crops, 13.43 million acres (54.4 thousand km^2) of grasses and hay, 14.32 million acres (58 thousand km^2) of wetlands, 118.9 million acres (481.2 thousand km^2) of pasture, and 37.99 million heads of livestock would have been threatened by new wild pig populations without the NFSDMP. Further, the most significant crops that would have been threatened by wild pigs include 21.86 million acres (88.5 thousand km^2) of corn, 21.22 million acres (85.9 thousand km^2) of soybeans, and 11.84 million acres (47.9 thousand km^2) of wheat.

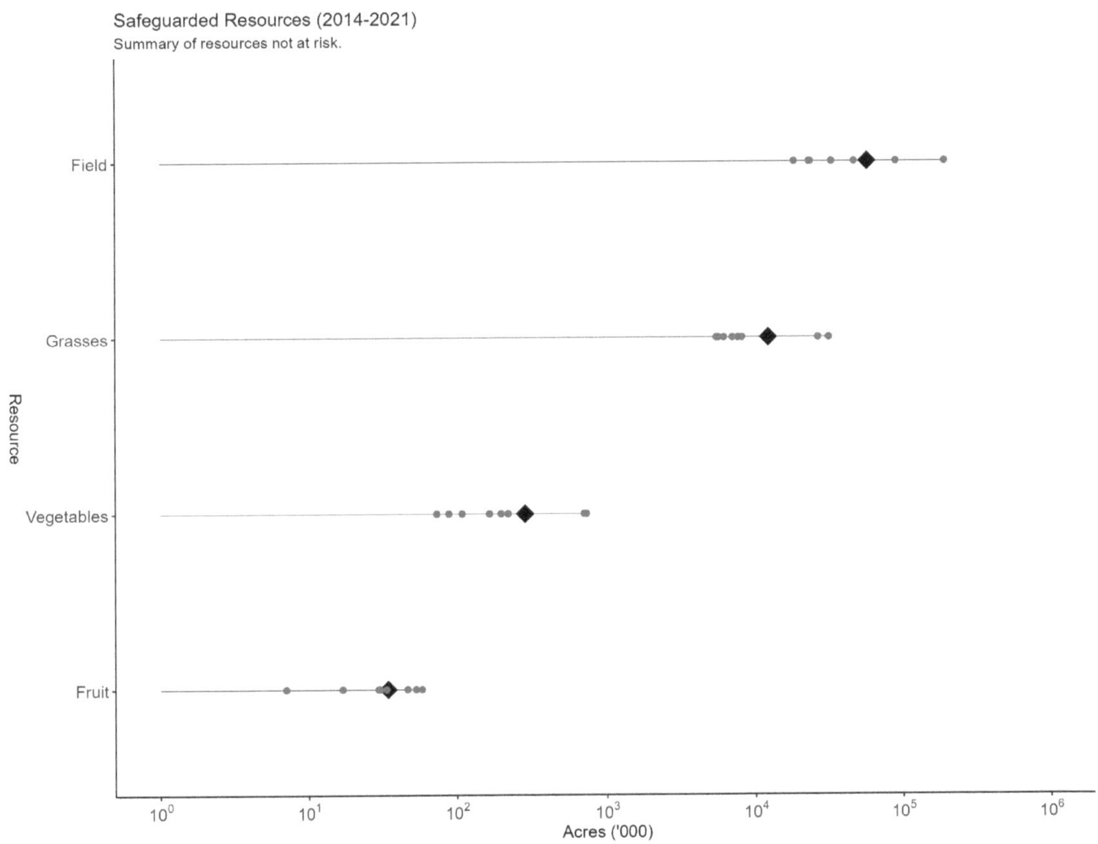

Figure 3. Summary of field crop acreage safeguarded from additional wild pig presence. Gray points represent each model forecast of resource acreage and black diamonds represent the mean acreage for each field crop. Acreage is scaled logarithmically. Acreage is cumulative over the period spanning 2014 to 2021.

Figure 4 plots all model outcomes for the quantity of field crops threatened by wild pigs without intervention. In the most conservative estimates, as few as 17.91 million acres [72.5 thousand km^2] of field crops and, in the most extreme estimate, as many as 198.3 million acres [802.5 thousand km^2] of field crops would have been threatened by the presence of wild pigs without the implementation of the NFSDMP. The high level of uncertainty in the estimates of field crops threatened by wild pigs extends to additional resources. Appendix A Figures A1–A4 plot all model outcomes for the quantity of other resources safeguarded. For example, the field and grass seed, forage, and hay acreage threatened by the potential spread of wild pigs ranges from 5.61 to 33.62 million acres [22.7 thousand km^2 to 136.1 thousand km^2]. Further, the threat from wild pigs to pasture range from 37.06 to 319.4 million acres [150 thousand km2 to 1.29 million km^2], wetlands range from 4.96 to 34.91 million acres [20.1 thousand km^2 to 141.3 thousand km^2], and livestock range from 12.05 to 117.4 million heads of livestock. This uncertainty is generated by sources ranging from the spread modeling assumptions, the speed at which wild pigs are forecast to spread, and the extent of and specific locations where wild pigs are forecast to invade.

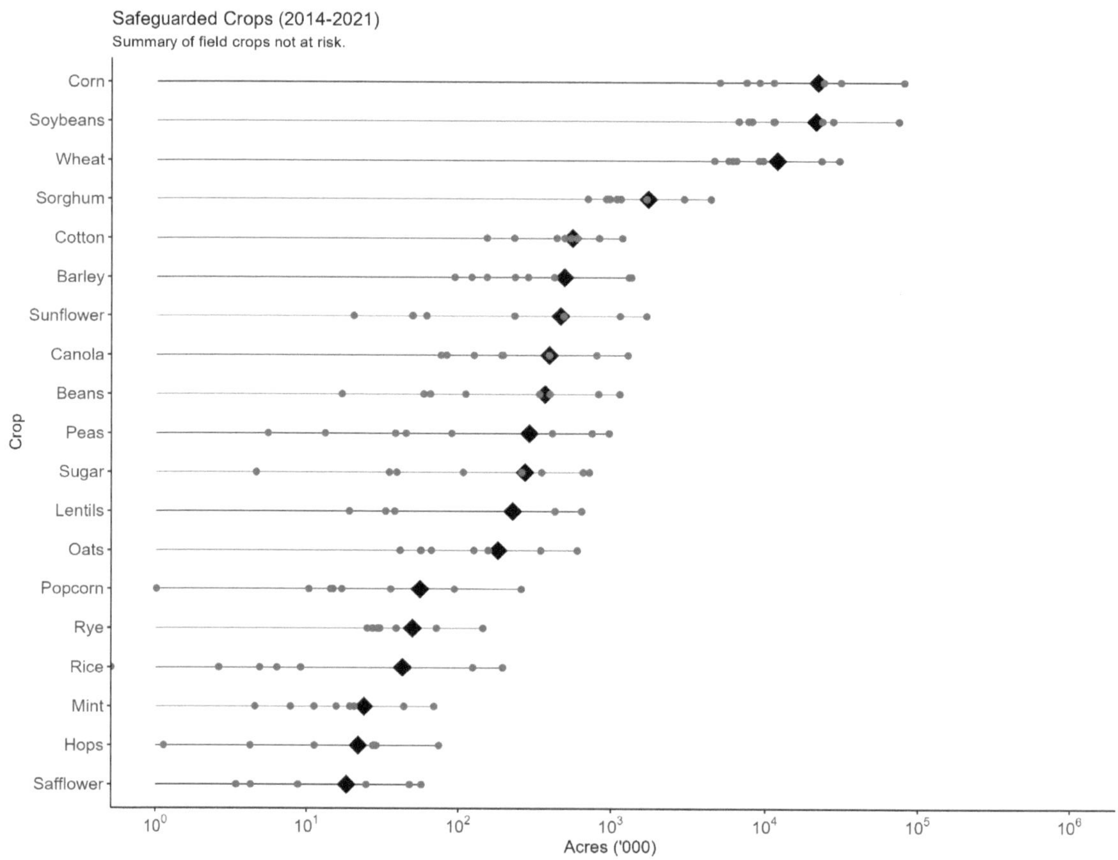

Figure 4. Summary of field crop acreage safeguarded from additional wild pig presence. Gray points represent each model forecast of crop acreage and black diamonds represent the mean acreage for each field crop. Acreage is scaled logarithmically. Acreage is cumulative over the period spanning 2014 to 2021.

3.4. Resources-Value of Resources Safeguarded

For a subset of resources, we use prices and rental rates to quantify the value of resources that were avoided by wild pigs through the implementation of the NFSDMP.

The assumption that crop acreage and value from 2017 can be extrapolated for the entire period spanning 2014 to 2021 continues to be utilized to quantify the value of resources safeguarded by preventing the spread of wild pigs. The results indicate that between USD 11 and USD 116 billion in value of field crops, grasses, and pasture have been safeguarded between 2014 and 2021 (Figure 5). Due to a lack of robust estimates for the economic value of wetlands and annualized values for livestock, the analysis of the value of resources focuses on crops, grasses, and pasture.

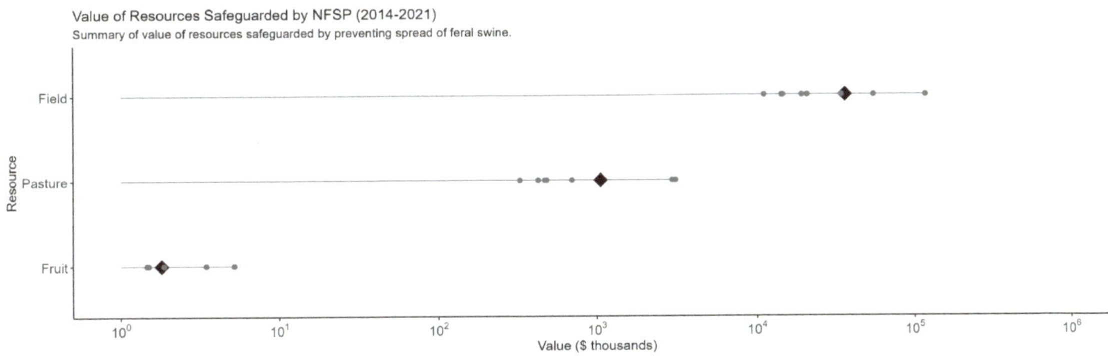

Figure 5. Summary of the value of resources safeguarded. The figure shows resource values for broad categories of resources. Gray points represent each model forecast of resource value safeguarded and black diamonds represent the mean value for each category of resource. Resource value is scaled logarithmically. Values are cumulative over the period spanning 2014 to 2021.

Figure 6 maps the spatial distribution of the value of crops, grasses, and pasture precluded from wild pigs across US counties for each model. Figure 7 maps the mean value of crops, grasses, and pasture resources for each county across all models of wild pigs spread, cumulated across the period from 2014 to 2021. The spatial distribution indicates that a key factor in determining the value of resources safeguarded is the specific counties where wild pigs are predicted to spread. For example, the forecasted spread in model D encompasses areas of states with significant agricultural production, such as Nebraska, Iowa, and Illinois, while the forecasted spread in model F indicates that wild pigs spread to counties in eastern states with lower levels of agricultural production.

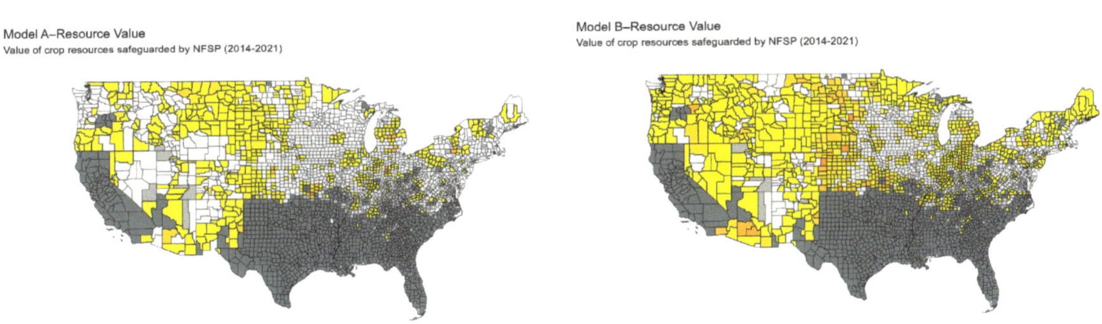

(a) (b)

Figure 6. *Cont.*

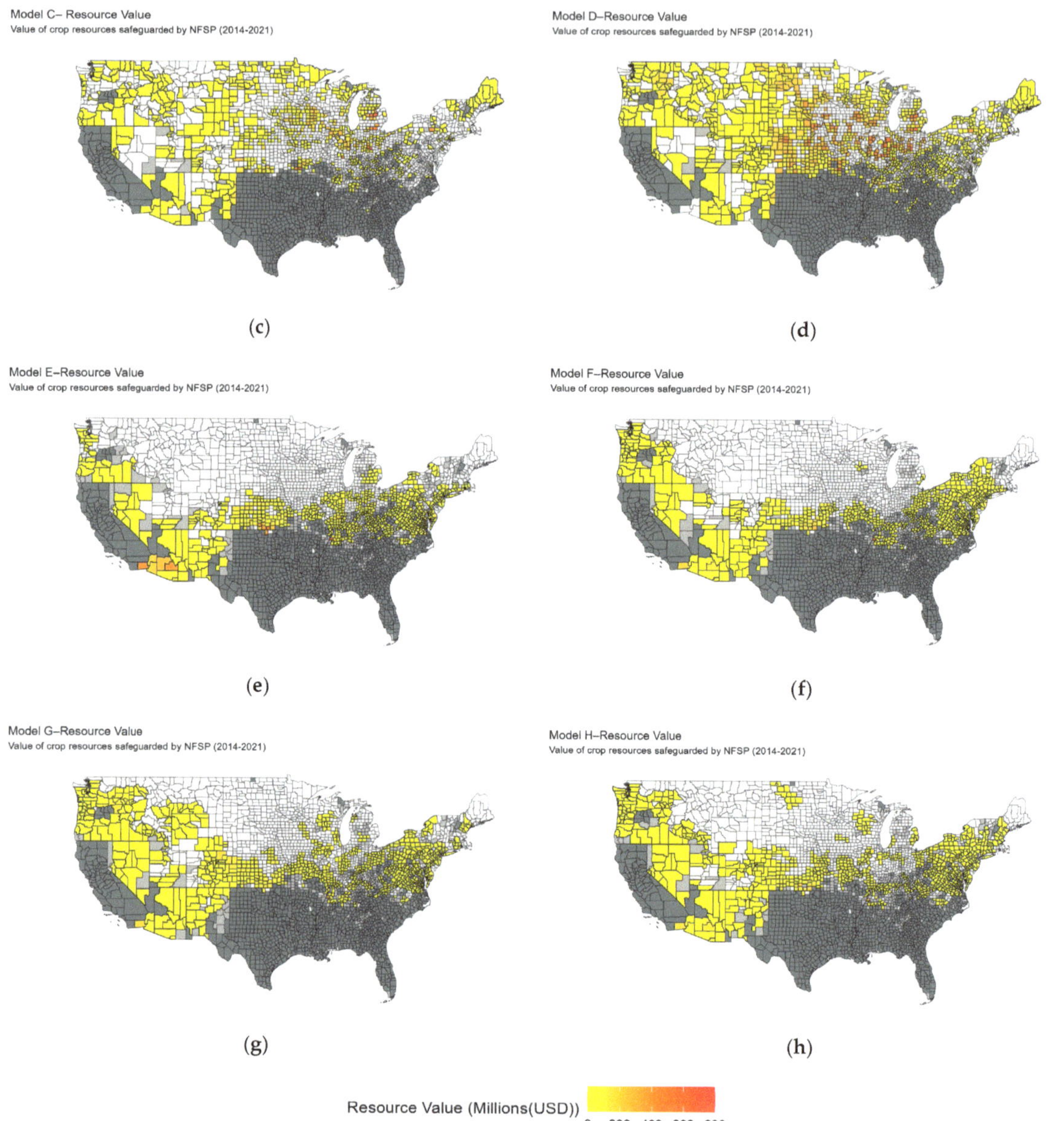

Figure 6. Spatial distribution of the value of resources threatened for each model. Each subfigure presents the value of resources safeguarded for one of eight model forecasts. Counties that are forecasted to be occupied by wild pigs are colored based on the amount of resource value safeguarded from lowest (yellow) to highest (red). Dark gray counties were always occupied, light gray counties were partially occupied, and white counties were never occupied and are not included in forecasting the value of resources safeguarded. Each model forecasts a large area that wild pigs would have occupied; however, in models C and D, wild pigs are forecasted to spread quickly to counties where many valuable resources are harvested, driving the large magnitudes that we estimate.

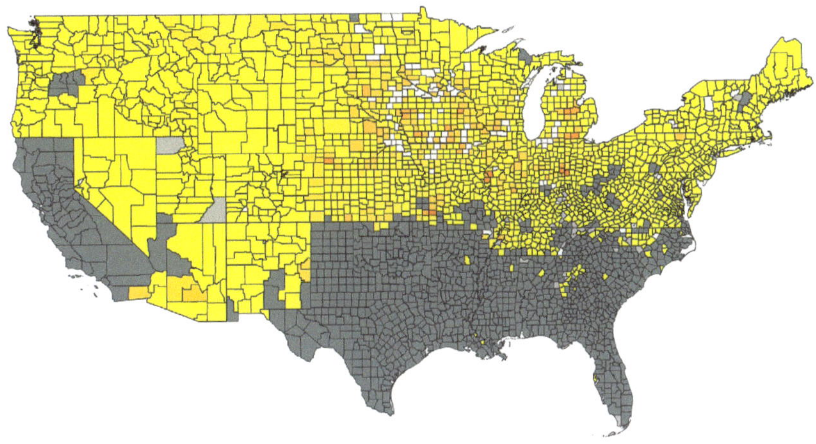

Figure 7. Average value of resources safeguarded across all models. In each model, counties that are never forecasted to be invaded by wild pigs are treated as zero. Counties that are forecasted to be occupied by wild pigs are colored based on the average amount of resource value safeguarded from lowest (yellow) to highest (red). Dark gray counties were always occupied, light gray counties were partially occupied, and white counties were never occupied and are not included in forecasting the value of resources safeguarded. The value of resources safeguarded is averaged across years for the period spanning 2014 to 2021.

4. Discussion

Areas in which invasive wild pig populations have been established or may emerge are currently subject to outreach and eradication efforts under the NFSDMP. Our results indicate that in addition to these areas, these efforts by the NFSDMP have effectively prevented the spread of wild pigs to counties where wild pig populations have not been observed. We demonstrate that between 421 and 895 counties would have experienced at least one additional year of wild pig presence. For each additional year in which wild pigs were prevented from occupying a county, we compute and aggregate the acreage of land resources and estimate the value of a subset of commodities safeguarded. This analysis shows that between USD 11 and USD 116 billion in value of field crops, pasture, grasses, and hay has been safeguarded because of intervention by the NFSDMP between 2014 and 2021. This value does not include the value of resources protected in counties that have experienced wild pig presence throughout the study period. These results indicate that the introduction of the NFSDMP delivered significant economic benefits to regions that have not experienced a wild pig invasion. When evaluating the efficacy of the NFSDMP and other similar programs, these benefits must be considered. Further, if control efforts are maintained over time, the benefits of safeguarding resources will continue to grow as wild pig invasions are prevented over a longer timeframe. Regardless, the large range of outcomes with respect to the area experiencing additional wild pig presence and the economic value of resources safeguarded can be attributed to high levels of uncertainty present in the underlying modeling framework and several limitations, which are outlined below. Despite this uncertainty, without relying on a single most accurate model, the models collectively suggest that the magnitude of our results is large and significant.

Uncertainty regarding the spread of wild pigs and the extent of the damage resulting from wild pig presence is an additional challenge. The environmental and climatic factors that drive wild pig population growth documented in McClure et al. [19] have been influential in predicting their northward expansion [6]. Once established, Miller et al. [5]

used watershed-scale modeling to uncover the determinants of wild pig population growth. Regardless, predicting the spread and growth of wild pig populations using ecological modeling utilizes pre-period data and a variety of assumptions. Due to errors in pre-period data, unpredictability in the ecological factors governing wild pig population dynamics, and assumptions inherent to a given modeling strategy, there is a high degree of uncertainty regarding the timing and location of wild pig population growth. Further, information on population densities in observed wild pig habitats is difficult to ascertain, which yields an additional source of uncertainty in spread forecasting. Lastly, the pre-period data used to inform the ecological models did not account for the international expansion of invasive wild pigs from Canada or Mexico. Notwithstanding, in all models we evaluated, the extent of the spread of wild pigs and the magnitude of the associated resources safeguarded are considerable.

There are significant challenges in ecological forecasting that result in potentially large uncertainties in predicted ecological states. To predict the spread of wild pigs in the contiguous US in the absence of the NFSDMP, we use a conservative approach with multiple modeling frameworks. This was performed to account for potential uncertainties and a diversity of different model assumptions. This has the benefit of providing multiple predictions that can be used in an ensemble framework for further assessment of potential economic impacts averted because of the NFSDMP.

For each boosted regression tree model, we acknowledge the following caveats. First, the predictions are deterministic. Thus, in the first year of forecasting, most counties become newly occupied and the number decreases thereafter. This is because the average spread rate from 2009 to 2013 is quite high. We tried adding noise by sampling the predictions from a normal distribution using the variance of the empirical distribution. This results in more spread overall). Second, management occurred before the program, but it is difficult to predict how that would continue as new counties become invaded. We tried adding a separate model in the forecasting as follows. We updated the amount of removal in the focal county using a separate model of the relationship between removal amount per km^2 or proportion of the county in the focal county in the current year and covariates (current km^2 occupied, removal amount per km^2 in the previous year, mean removal amount in adjacent counties in the previous year, and the habitat suitability and landcover variables as in the other models. We fit this model using the same methodology as the Persistence and Transient models, although we did not subset the data because there was no evidence that the relationship between take amount and covariates differed depending on historical wild pig presence. This model had an R^2 of 0.67 and 0.94 for in-sample predictions of models of number removed/km^2 and log numbers removed, respectively. At each time step, we used this model to update the amount of removal occurring in focal and adjacent counties in the previous year based on updated occupancy. This allowed us to maintain pre-program levels of removal as the spread to new areas progressed, but there is uncertainty in those estimates. These models produced very similar forecasts to their corresponding models without removal updates, except with slightly higher invasion rates overall). Lastly, there is a substantial amount of reporting variation both over time and space in the observed data. There are no data to explain it. Thus, our predictions are influenced by these data limitations.

The following caveats apply to the dynamic occupancy models. Firstly, while we accounted for time-varying factors influencing the probability of invasion and stochastic processes of invasion and extinction, there is uncertainty about which model might represent the invasion process best. We also did not explore non-linearities among covariates. Furthermore, watershed-level models did not explicitly account for control that likely would have continued to occur without a national program, although they did allow for extinction, which may have resulted from local or regional control activities. We assume that regional effects (random intercept and slopes) account for some of this variation. Additionally, associations with covariates are likely influenced greatly by the geographic location of the spread that occurred from 2004 to 2012. The majority of spread occurred in more arid regions and likely influenced covariate associations. This could be addressed by fitting models to different subsets of data. Finally, we assume perfect detection of watersheds with wild pigs, which is likely invalid.

Several caveats limit the scope of the economic analysis as it pertains to valuing the resources avoided by wild pigs due to the NFSDMP. First, comprehensive estimates of the annualized value of wetlands and livestock are not available to be used in the analysis. Specifically, previous evaluations of wetlands [20] have examined specific sites and estimated the value of specific features of wetlands; however, these estimates have not been applied to all wetland acres throughout the United States. Additionally, estimates of the stock value of livestock have been produced for cattle, domestic swine, sheep, and goats [21,22]; however, we are not aware of studies that convert these to annualized values that account for the production and depreciation of livestock. Lastly, Brown et al. [21] and Sauter-Louis et al. [23] demonstrate the extent of disease transmission risk from wild pigs to livestock, emphasize the degree of uncertainty regarding the economic consequences of disease risk, and highlight the self-sustaining cycles of disease transmission between wild pigs and domestic livestock. For example, the costs resulting from the death of livestock, culling exposed livestock, and potential trade restrictions due to a disease outbreak [24,25] have not been precisely estimated. For the above reasons, we focused primarily on field crops, fruits and vegetables, and pasture for the economic analysis.

Second, we completed our analysis under the assumption that resource acreage and prices would have been unaffected by the presence of wild pigs. In counties where the NFSDMP reduced or prevented wild pig populations from being established, it is possible that had wild pigs been present, producers would have responded by changing the mix of crop and livestock resources from what was observed in the 2017 Census of Agriculture. A reduction in planted crop acreage would have reduced the quantity of resources threatened by wild pigs; however, if this reduction would have generated an increase in price then the value of resources safeguarded would have increased. With limited information regarding changes in acreage and prices that would have resulted from a wild pig invasion, the effect of this assumption is ambiguous.

Lastly, the analysis assumed that the only way to control, mitigate, and eradicate wild pig presence and damage was through the NFSDMP. In the scenario where the NFSDMP was never introduced, it is likely that producers and landowners would have privately attempted to limit the spread of wild pigs. Further, we have ignored efforts that could have replicated the NFSDMP at the state and county levels. Though it is likely that effort would have been made to mitigate the impact of the spread of wild pigs in the absence of the NFSDMP, it is difficult to analyze the costs of implementing these measures at the sub-national level as well as their effectiveness in the absence of NFSDMP intervention. Acknowledging the above, interventions in the absence of the NFSDMP would have reduced the probability of a wild pig invasion relative to the model forecasts, and we are unable to account for this.

5. Conclusions

Our findings indicate that in the absence of intervention by the NFSDMP, invasive wild pig populations would have likely continued to spread widely across the United States. Various modeling scenarios demonstrate that there is large uncertainty regarding the exact locations where wild pigs would have spread. Regardless, across each scenario, we find that significant resources have been safeguarded due to intervention by the NFSDMP. As few as 421 and as many as 895 counties are invaded across the various model scenarios. Taking the average across each scenario, we show that an additional 60 million acres of field crops, 13 million acres of grasses and hay, 14 million acres of wetlands, 119 million acres of pasture, and 38 million heads of livestock would have been threatened by the spread of wild pigs in the absence of the NFSDMP. Lastly, we calculate that the value of field crops, grasses and hay, and pasture safeguarded from the spread of wild pigs by the NFSDMP is on average USD 40.2 billion over the period from 2014 to 2021.

These findings indicate that the economic benefits of introducing the NFSDMP spread far wider than previously described, although a significant portion of funding and resources provided by the NFSDMP has been focused on areas where wild pigs are already known to be present or have continued to impose damages to resources and property. A less commonly known outcome of the NFSDMP is the prevention of the spread of wild pigs and the associated threat to resources in regions where invasive feral swine populations may have emerged. The economic benefits that flow to these areas as a result should be considered as a component of the comprehensive benefits of introducing the NFSDMP. Future work that evaluates the effects of wild pig management should take into consideration the findings of this paper.

Author Contributions: Conceptualization, K.M.P., R.S.M., S.A.S. and S.C.M.; methodology, C.J., K.M.P., R.S.M. and S.C.M.; software, C.J., K.M.P. and R.S.M.; formal analysis, C.J., K.M.P. and R.S.M.; investigation, C.J., K.M.P., R.S.M. and S.C.M.; data curation, C.J., K.M.P. and R.S.M.; writing—original draft preparation, C.J.; writing—review and editing, C.J., K.M.P., R.S.M., S.S. and S.C.M.; visualization, S.S.; supervision, S.C.M.; funding acquisition, S.C.M. All authors have read and agreed to the published version of the manuscript.

Funding: This research was supported by the US Department of Agriculture, Animal and Plant Health Inspection Service, Wildlife Services, National Feral Swine Damage Management Program.

Institutional Review Board Statement: Not applicable.

Informed Consent Statement: Not applicable.

Data Availability Statement: The presence/absence of wild pig data presented in this study are available on request from the corresponding author due to privacy issues related to the National Feral Swine Mapping System. Crop acreage and values are openly available from https://www.nass.usda.gov/Quick_Stats/. The data was accessed on 29 November 2023.

Conflicts of Interest: The authors declare no conflict of interest.

Appendix A. Tables

Table A1. List of all resources analyzed, grouped by six broad categories.

Field Crops	Vegetables	Fruit and Tree Nuts	Field and Grass Seeds, Forage, and Hay	Land	Livestock
Corn	Melons	Almonds	Hay	Wetlands	Cattle
Wheat	Potatoes	Pecans	Fescue	Pastureland	Swine
Sorghum					
Peanuts					
Rice	Tomatoes	Grapes	Ryegrass		Sheep
Soybeans		Berries			Goat
Sugar					
Cotton					
Oats					
Barley					
Tobacco					
Beans					
Peas					
Flaxseed					
Lentils					
Hops					
Millet					
Mint					
Popcorn					
Rye					
Safflower					
Sunflower					

Table A2. National-level dollar values used to calculate county level value of resources safeguarded for a subset of resources.

Resource	Value (USD)
Corn	3.36/bushel
Wheat	4.44/bushel
Sorghum	3.05/bushel
Peanuts	0.21/lb
Rice	11.2/cwt
Soybeans	9.39/bushel
Sugar	31/ton
Cotton	0.72/lb
Cantaloup	19.20/cwt
Honeydew	28/cwt
Watermelon	14.70/cwt
Almonds	2.50/lb
Pecans	1.75/lb
Grapes	881/ton
Hay	137/ton

Table A3. Total counties forecasted to be occupied by wild pigs for each model and year.

Year	A	B	C	D	E	F	G	H
2013	1272	1272	1272	1272	1272	1272	1272	1272
2014	1473	1841	1633	1970	1485	1437	1474	1481
2015	1540	1890	1797	2090	1638	1594	1574	1623
2016	1558	1933	1754	2115	1708	1686	1747	1771
2017	1616	2055	1801	2154	1819	1785	1844	1842
2018	1651	2056	1861	2165	1871	1839	1883	1883
2019	1655	2059	1794	2131	1940	1884	1964	1956
2020	1676	2076	1858	2171	2011	1954	2013	2024
2021	1693	2078	1881	2167	2049	1999	2039	2065

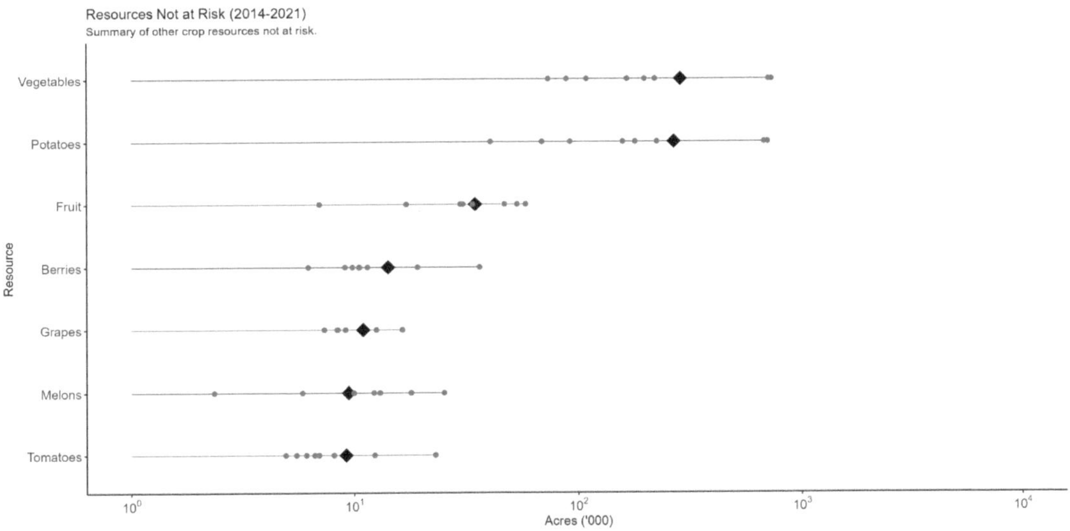

Figure A1. Summary of fruit and vegetable acreage safeguarded from additional years of wild pig presence. Gray points represent each model forecast of fruit and vegetable acreage and black diamonds represent the mean acreage for each fruit and vegetable resource. Acreage is scaled logarithmically. Acreage is cumulative over the period spanning 2014 to 2021.

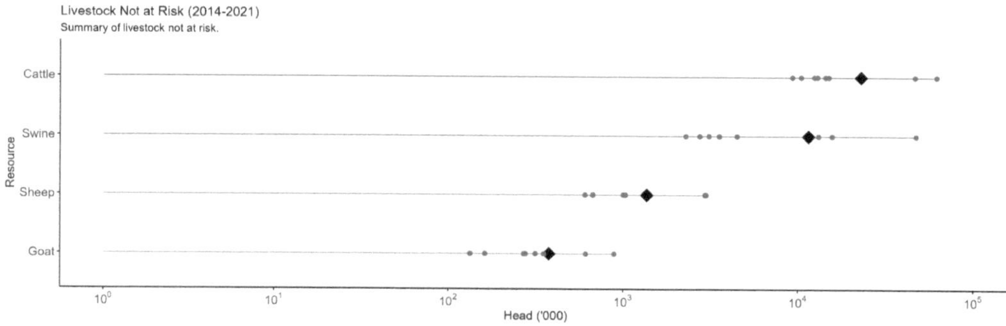

Figure A2. Summary of head of livestock safeguarded from additional wild pig presence. Gray points represent each model forecast of head of livestock and black diamonds represent the mean head of each type of livestock. Head of livestock is scaled logarithmically. Head of livestock is cumulative over the period spanning 2014 to 2021.

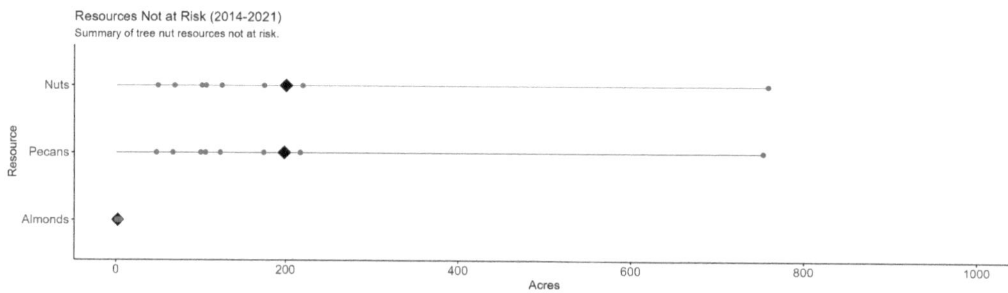

Figure A3. Summary of tree nuts safeguarded from additional wild pig presence. Gray points represent each model forecast of nut acreage and black diamonds represent the mean acreage for each type of tree nut. Acreage is scaled logarithmically. Acreage is cumulative over the period spanning 2014 to 2021.

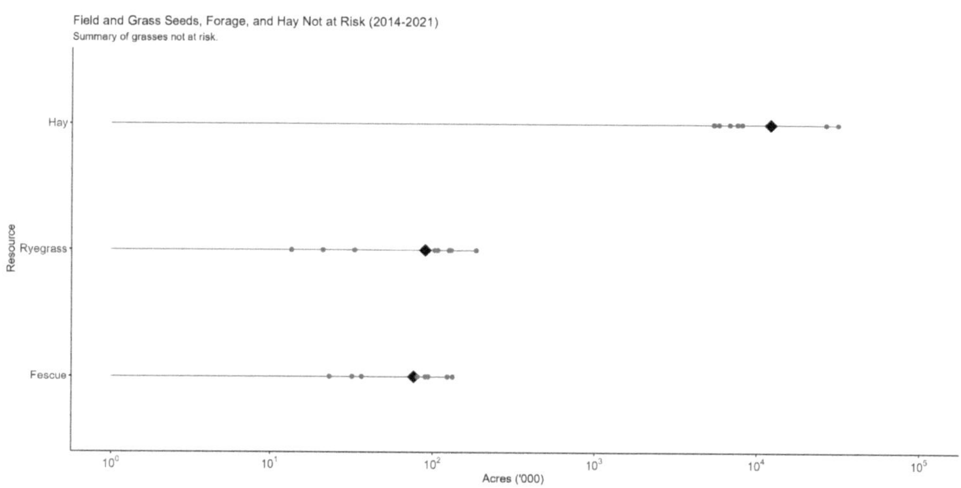

Figure A4. Summary of hay and grasses safeguarded from additional wild pig presence. Gray points represent each model forecast of hay and grass acreage and black diamonds represent the mean acreage for each type of hay or grass. Acreage is scaled logarithmically. Acreage is cumulative over the period spanning 2014 to 2021.

Appendix B

For the county-scale models, we estimated hyperparameters using Bayesian optimization with uniform prior distributions and mean squared error minimization as the criteria for parameter acceptance. The prior ranges were [50, 2000] for the number of learning cycles, [0.0001, 0.1] for the learning rate, [5 (Persistence models) or 10 (Transient models), 100] for the minimum leaf size, and [1, 50] for the maximum number of splits. We examined the fit of the models using R-squared (squared correlation of the observed and predicted response data). The fits of the models for each of the persistence (P) and transient (T) models with their respective most important predictors were as follows:

Model A: Change in proportion occupied with proportion occupied in 2008: P—0.83, T—1.0
P: Proportion occupied in $t-1$, Pasture cover, Proportion of adjacent counties worked in $t-1$
T: Proportion occupied in $t-1$
Model B: Change in proportion occupied without proportion occupied in 2008: P—0.55, T—0.74
P: Pasture cover, Proportion of adjacent counties worked in $t-1$
T: Pasture cover, Removal in adjacent counties in $t-1$
Model C: Change in km^2 with km^2 occupied in 2008: P—0.98, T—1.0
P: Pasture cover, Tree cover, Water cover, Deciduous, Removal in adjacent counties in $t-1$
T: Proportion occupied in $t-1$
Model D:Change in km^2 without km^2 occupied in 2008: P—0.88, T—1.0
P: Pasture cover, Tree cover, Water cover, Deciduous, Removal in adjacent counties in $t-1$
T: Pasture cover, Crop cover, Removal in adjacent counties in $t-1$

For the watershed-scale models, because there was uncertainty in the best model and visual evaluation indicated regional differences in predicted probability of invasion, we considered four models in our final simulations. These models included (a) fixed effects only, (b) uncorrelated random intercept, (c) random slope for ruggedness and Lewis density and a uncorrelated random intercept, (d) random slope for Lewis density with correlated intercept.

Table A4. Detailed model specifications and fit for models E, F, G, and H.

Model	ELPD Difference	Bayesian R^2 Conditional	Bayesian R^2 Marginal	LOO R^2 Conditional	LOOIC Weight	Performance Score
ruggedness | region + lewis.density | region + 1 | region	0.00	0.9632	0.9645	0.8775	3.2×10^{-1}	64.4%
lewis.density | region	−31.47	0.9632	0.9601	0.8765	3.6×10^{-7}	25.0%
lewis.density | region + 1 | region	−31.76	0.9632	0.9602	0.8764	3.7×10^{-2}	26.6%
ruggedness | region + 1 | region	−85.91	0.9633	0.9658	0.8755	8.0×10^{-6}	25.0%
ruggedness | region	−86.46	0.9633	0.9657	0.8756	2.0×10^{-5}	25.0%
1 | region	−179.18	0.9632	0.9607	0.8738	9.5×10^{-2}	29.1%
fixed only	−7071.29	0.9840	0.9840	0.8249	5.5×10^{-1}	50.0%

References

1. Keiter, D.A.; Mayer, J.J.; Beasley, J.C. What is in a "common" name? A call for consistent terminology for nonnative Sus scrofa. *Wildl. Soc. Bull.* **2016**, *40*, 384–387. [CrossRef]
2. VerCauteren, K.C.; Beasley, J.C.; Ditchkoff, S.S.; Mayer, J.J.; Roloff, G.J.; Strickland, B.K. (Eds.) *Invasive Wild Pigs in North America: Ecology, Impacts, and Management*, 1st ed.; CRC Press: Boca Raton, FL, USA, 2020.
3. Lowe, S.; Browne, M.; Boudjelas, S.; De Poorter, M. *100 of the World's Worst Invasive Alien Species: A Selection from the Global Invasive Species Database*; Invasive Species Specialist Group: Gland, Switzerland, 2020; pp. 1–12.
4. U.S Department of Agriculture. *Five Year Report FY 2014-2018, National Feral Swine Damage Management Program*; United States Department of Agriculture: Washington, DC, USA, 2018; pp. 1–22.
5. Miller, R.S.; Tabak, M.A.; Wolfson, D.W.; Burdett, C.L. Transient population dynamics drive the spread of invasive wild pigs and reveal impacts of management in North America. *Biol. Invasions* **2023**, *25*, 2461–2476. [CrossRef]
6. Snow, N.P.; Jarzyna, M.A.; VerCauteren, K.C. Interpreting and predicting the spread of invasive wild pigs. *J. Appl. Ecol.* **2017**, *54*, 2022–2032. [CrossRef]

7. Pepin, K.M.; Wolfson, D.W.; Miller, R.S.; Tabak, M.A.; Snow, N.P.; VerCauteren, K.C.; Davis, A.J. Accounting for heterogeneous invasion rates reveals management impacts on the spatial expansion of an invasive species. *Ecosphere* **2019**, *10*, e02657. [CrossRef]
8. McKee, S.; Mayer, J.; Shwiff, S. Comprehensive Economic Impacts of Wild Pigs on Producers of Six Crops in the South-Eastern US and California. *Agriculture* **2024**, *14*, 153. [CrossRef]
9. McKee, S.; Anderson, A.; Carlisle, K.; Shwiff, S.A. Economic estimates of invasive wild pig damage to crops in 12 US states. *Crop Prot.* **2020**, *132*, 105105. [CrossRef]
10. Bankovich, B.; Boughton, E.; Boughton, R.; Avery, M.L.; Wisely, S.M. Plant community shifts caused by feral swine rooting devalue Florida rangeland. *Agric. Ecosys. Environ.* **2016**, *220*, 45–54. [CrossRef]
11. Didero, N.M.; Ernst, K.H.; McKee, S.C.; Shwiff, S.A. A call and suggested criteria for standardizing economic estimates of wild pig damage. *Crop Prot.* **2023**, *165*, 106149. [CrossRef]
12. VerCauteren, K.C.; Pepin, K.M.; Cook, S.M.; McKee, S.; Pagels, A.; Kohen, K.J.; Messer, I.A.; Glow, M.P.; Snow, N.P. What is known, unknown, and needed to be known about damage caused by wild pigs. *Biol. Invasions* **2024**, *26*, 1313–1325. [CrossRef]
13. Glow, M.P.; Snow, N.P.; VerCauteren, K.C. *Wildlife Damage Management Technical Series: Feral Swine*; U.S Department of Agriculture: Washington, DC, USA, 2020; pp. 1–20.
14. MacKenzie, D.I.; Nichols, J.D.; Hines, J.E.; Knutson, M.G.; Franklin, A.B. Estimating site occupancy, colonization, and local extinction when a species is detected imperfectly. *Ecology* **2003**, *84*, 2200–2207. [CrossRef]
15. MacKenzie, D.I.; Nichols, J.D.; Royle, J.A.; Pollock, K.H.; Bailey, L.L.; Hines, J.E. *Occupancy Estimation and Modeling: Inferring Patterns and Dynamics of Species Occurrence*; Academic Press: Amsterdam, The Netherlands, 2006.
16. Elith, J.; Leathwick, J.R.; Hastie, T. A working guide to boosted regression trees. *J. Anim. Ecol.* **2008**, *77*, 802–813. [CrossRef] [PubMed]
17. Araújo, M.B.; New, M. Ensemble forecasting of species distributions. *Trends Ecol. Evol.* **2007**, *22*, 42–47. [CrossRef] [PubMed]
18. Lewis, J.S.; Farnsworth, M.L.; Burdett, C.L.; Theobald, D.M.; Gray, M.; Miller, R.S. Biotic and abiotic factors predicting the global distribution and population density of an invasive large mammal. *Sci. Rep.* **2017**, *7*, 44152. [CrossRef] [PubMed]
19. McClure, M.L.; Burdett, C.L.; Farnsworth, M.L.; Lutman, M.W.; Theobald, D.M.; Riggs, P.D.; Miller, R.S. Modeling and mapping the probability of occurrence of invasive wild pigs across the contiguous United States. *PLoS ONE* **2015**, *10*, e0133771. [CrossRef] [PubMed]
20. Engemen, R.M.; Stevens, A.; Allen, J.; Dunlap, J.; Daniel, M.; Teague, D.; Constantin, B. Feral swine management for conservation of an imperiled wetland habitat: Florida's vanishing seepage slopes. *Biol. Conserv.* **2007**, *134*, 440–446. [CrossRef]
21. Brown, V.R.; Marlow, M.C.; Gidlewski, T.; Bowen, R.; Bosco-Lauth, A. Perspectives on the past, present, and future of feral swine disease surveillance in the United States. *J. Anim. Sci.* **2020**, *98*, skaa256. [CrossRef] [PubMed]
22. McKee, S.; Mooney, D.; Miller, R. Type and frequency of wild pig-domestic livestock contacts and operator concern for disease spread. *Q Open* **2023**, *3*, qoad027. [CrossRef]
23. Sauter-Louis, C.; Conraths, F.J.; Probst, C.; Blohm, U.; Schulz, K.; Sehl, J.; Fischer, M.; Forth, J.H.; Zani, L.; Depner, K.; et al. African Swine Fever in Wild Boar in Europe—A Review. *Viruses* **2021**, *13*, 1717. [CrossRef] [PubMed]
24. Brown, V.R.; Miller, R.S.; Pepin, K.M.; Carlisle, K.M.; Cook, M.A.; Vanicek, C.F.; Holmstrom, L.K.; Rochette, L.T.; Smyser, T.J. African swine fever at the wildlife-livestock interface: Challenges for management and outbreak response within invasive wild pigs in the United States. *Front. Vet. Sci.* **2024**, *11*, 1348123. [CrossRef]
25. Brown, V.R.; Miller, R.S.; McKee, S.C.; Ernst, K.H.; Didero, N.M.; Maison, R.M.; Grady, M.J.; Shwiff, S.A. Risks of introduction and economic consequences associated with African swine fever, classical swine fever and foot-and-mouth disease: A review of the literature. *Transbound. Emerg. Dis.* **2021**, *68*, 1910–1965. [CrossRef] [PubMed]

biology MDPI

Article

Returning to the Wilderness: Potential Habitat Suitability of Non-Native Pet Birds in South Africa

Tinyiko C. Shivambu [1,*], Ndivhuwo Shivambu [1], Takalani Nelufule [1], Moleseng C. Moshobane [2], Nimmi Seoraj-Pillai [1] and Tshifhiwa C. Nangammbi [1]

1 Faculty of Science, Department of Nature Conservation, Tshwane University of Technology, Private Bag X680, Pretoria 0001, South Africa; shivambun@tut.ac.za (N.S.); nelufulet@tut.ac.za (T.N.); seorajpillayn@tut.ac.za (N.S.-P.); nangammbitc@tut.ac.za (T.C.N.)

2 South African National Biodiversity Institute, Pretoria National Botanical Garden, 2 Cussonia Avenue, Brummeria, Silverton 0184, South Africa; m.moshobane@sanbi.org.za

* Correspondence: shivambutc@tut.ac.za or shivambucavin@gmail.com; Tel.: +27-12-382-5337

Simple Summary: The global trade of pet birds has grown significantly, leading to more non-native birds being introduced into the wild. Our study examined the potential habitats and environmental factors that might allow pet birds lost and sighted in South Africa to survive outside captivity. Using geographic location data of pet birds, we predicted which areas might be suitable for various species such as the African grey parrot, Budgerigar, Cockatiel, Green-cheeked conure, Monk parakeet, and Rose-ringed parakeet. We found that a significant portion of South Africa could support these species, with some areas being highly suitable. The study highlighted how pet birds could escape or be released into the wild due to human activities, and that urban areas are particularly at risk. This research emphasizes the need for careful monitoring to prevent these birds from becoming invasive, which could impact local ecosystems and biodiversity.

Abstract: The global trade of non-native pet birds has increased in recent decades, and this has accelerated the introduction of invasive birds in the wild. This study employed ensemble species distribution modelling (eSDM) to assess potential habitat suitability and environmental predictor variables influencing the potential distribution of non-native pet bird species reported lost and sighted in South Africa. We used data and information on lost and found pet birds from previous studies to establish and describe scenarios of how pet birds may transition from captivity to the wild. Our study revealed that models fitted and performed well in predicting the suitability for African grey (*Psittacus erithacus*), Budgerigar (*Melopsittacus undulatus*), Cockatiel (*Nymphicus hollandicus*), Green-cheeked conure (*Pyrrhura molinae*), Monk parakeet (*Myiopsitta monachus*), and Rose-ringed parakeet (Psittacula krameri), with the mean weighted AUC and TSS values greater than 0.765. The predicted habitat suitability differed among species, with the suitability threshold indicating that between 61% and 87% of areas were predicted as suitable. Species with greater suitability included the African grey, Cockatiel, and Rose-ringed parakeet, which demonstrated significant overlap between their habitat suitability and reported lost cases. Human footprint, bioclimatic variables, and vegetation indices largely influenced predictive habitat suitability. The pathway scenario showed the key mechanisms driving the transition of pet birds from captivity to the wild, including the role of pet owners, animal rescues, adoption practices, and environmental suitability. Our study found that urban landscapes, which are heavily populated, are at high risk of potential invasion by pet birds. Thus, implementing a thorough surveillance survey is crucial for monitoring and evaluating the establishment potential of pet species not yet reported in the wild.

Keywords: biological invasions; ensemble models; establishment scenarios; invasion risk assessment; companion animals; surveillance monitoring

1. Introduction

Globalization and increased human movement have accelerated the introduction of non-native species into novel and critically threatened ecosystems [1–3]. This has presented considerable challenges to local biota and ecological balance [4,5]. In today's era of globalization, the commercialization of goods, including wildlife products such as animal products and live pets, contributes to billions of dollars in turnover profit [6–9]. This has led to a global increase in the trade of live animals across various taxa, including birds, mammals, reptiles, amphibians, insects, fishes, gastropods, and spiders [10–13]. This surge in trade exacerbates illicit activities involving endangered and unregulated non-CITES pet species and contributes to an increase in the introduction of invasive species [8,14–16]. In Europe alone, over 20 million pet reptiles were imported between 2004 and 2011 [17], and in Canada, more than 20 million wildlife were imported between 2007 and 2017 [18]. This complex chain of trading live animals is a global phenomenon linked to pet escapes and releases, leading to biological invasions worldwide [15,19,20].

Biological invasion has emerged as one of the widely discussed concepts, referring to the introduction of non-native species through various pathways into new regions, with the potential to cause environmental and socioeconomic impacts [21–23]. There have been several cases of non-native pet taxa reported invasive in non-native ranges, more especially those with the probability of escaping captivity [15,24,25]. For example, birds are among groups that have the potential to escape captivity and are more likely to establish invasive populations if released into the wild [25–27]. This is likely because the majority of bird species, particularly vagrant invasive, are not sensitive to changing environments and have displayed the phenomenon of plasticity in feeding behavior, enabling them to enhance their adaptability [28–30]. Examples of such species include global invasive pet birds such as Common myna (*Acridotheres tristis*), Common starling (*Sturnus vulgaris*), Monk parakeet (*Myiopsitta monachus*), Rock dove (*Columba livia*), and the Rose-ringed parakeet (*Psittacula krameri*) [28,30,31]. These bird species are among the most kept pets in the world, and many of them are designated as of least concern or non-CITES, linking them to relaxed regulation, although their impacts are of concern to native biodiversity [9,10,12,32]. For example, the establishment of a feral population of Monk parakeets has been associated with infrastructure damage, while Rose-ringed parakeets have been associated with displacement of native cavity-nesters from nests [33–35]. In this regard, the results on homing mechanisms and movement routes of different species of animals in an unknown habitat are significant [36,37].

Given that most pet birds either escape or are released from captivity [27], there is a lack of information regarding their distribution patterns, including the size of suitable home ranges and the environmental factors contributing to their distribution and potential invasion. The solution to this question lies in utilizing various ecological models, such as species distribution modelling (SDM) [38,39]. SDM has emerged as a pivotal tool in ecological research, encompassing risk assessment [40], climate change studies [41], and early detection of alien invasive species and pathways management [42]. It provides a quantitative framework to predict and analyze the spatial distribution of native and non-native species, relying on environmental variables [43,44]. SDM involves using statistical and computational techniques to analyze environmental variables and their relationship to the presence or absence of a particular species [45,46]. Assessing the environmental conditions where the species occurs, SDM aims to generate models that can predict suitable habitats for the species across a larger geographic scale [46,47]. The importance of utilizing SDM for invasive species lies in its ability to provide valuable insights and predictions about non-native organisms' potential spread and impact in new environments [48–50]. SDM has been applied to causative vagrant invasive bird species across the nations, e.g., the Rose-ringed parakeet in Marseille, France [51], House crow in South Africa (*Corvus splendens*) [52], and Asian pied starling (*Sturnus contra*) globally [53]. The SDM modelling approach has become valuable for understanding invasive species' ecological niche, conservation planning, and assessing the potential impacts of environmental changes on biodiversity [40,50].

South Africa is home to 971 recorded bird species, including 98 endemics, representing 9% of the global avifauna [54]. Additionally, there are 14 known alien invasive bird species in the region, although further quantification is needed (https://invasives.org.za/national-list-of-invasive-bird-species/, accessed on 10 May 2024). These bird species are kept as pets and traded for financial gain [12]. Some of these species, initially involved in the pet trade or held in captivity for companionship, end up in the wild, forming a self-sustaining population that poses a threat to native biodiversity, e.g., the Rose-ringed parakeet, Rock dove, and Common myna [55–57]. This includes the Monk parakeet, which is recognized as one of the most problematic invasive species established in over 16 countries [31]. There have been incidents of lost and sighted pet birds, including invasive species, reported in South Africa [58]. However, there is insufficient information regarding their potential spread and the factors that might facilitate their establishment in both invaded and previously uninvaded areas. This study employed the ensemble Species Distribution Modelling (eSDM), aiming to explore the potential distribution patterns, habitat suitability, and factors likely to facilitate the establishment of non-native lost and sighted pet birds. Furthermore, we also described pathways scenarios of lost and found pet birds, associating them with species that are likely to establish populations in the wild. Employing the eSDM technique on non-native pet birds reported lost and sighted will help uncover the ecological factors shaping the potential distribution patterns of these species.

2. Materials and Methods

2.1. Geographical Study Areas

The geographical focus area of this study is South Africa, a country located in the southernmost part of Africa, encompassing the entire southern tip of the continent (Figure 1). It shares its borders with Namibia to the northwest, Botswana and Zimbabwe to the north, Mozambique, and two landlocked countries (Lesotho, and Eswatini, formerly known as Swaziland). It also shares borders with the Indian and Atlantic Oceans to the east and west, respectively ([59], Figure 1). The country features varied topography, including coastal plains, mountains, and plateaus. The Drakensberg Mountain range runs along the eastern border, with mountain peaks reaching over 3400 m above sea level and a variety of species adapted to this landscape ([59], Figure 1). The country has varied climatic conditions, ranging from Mediterranean, which occurs mostly in Western Cape Province, with mild, wet winters and warm, dry summers and heavy rainfall in winter. This is followed by the Subtropical climate, which mostly occurs in eastern parts of the country, including coastal regions of KwaZulu-Natal Province, with hot and humid summers, mild winters, and rainfall throughout the year, with a peak in summer. The northwestern parts, including the Kalahari Desert and parts of Northern Cape Province, experience a Desert and Semi-desert climate with low rainfall in summer and variation in temperatures between day (hot) and night (cold) [59]. The interior plateau climate (highveld) occurs in the central parts of the country, particularly in Johannesburg and Pretoria, with rainfall concentrated in summer and moderate temperatures, with colder winters and warmer summers. Finally, the Savannah climate is concentrated in the country's northeastern regions, including parts of Limpopo and Mpumalanga provinces, with distinct wet and hot seasons in summer and a dry season in winter [59].

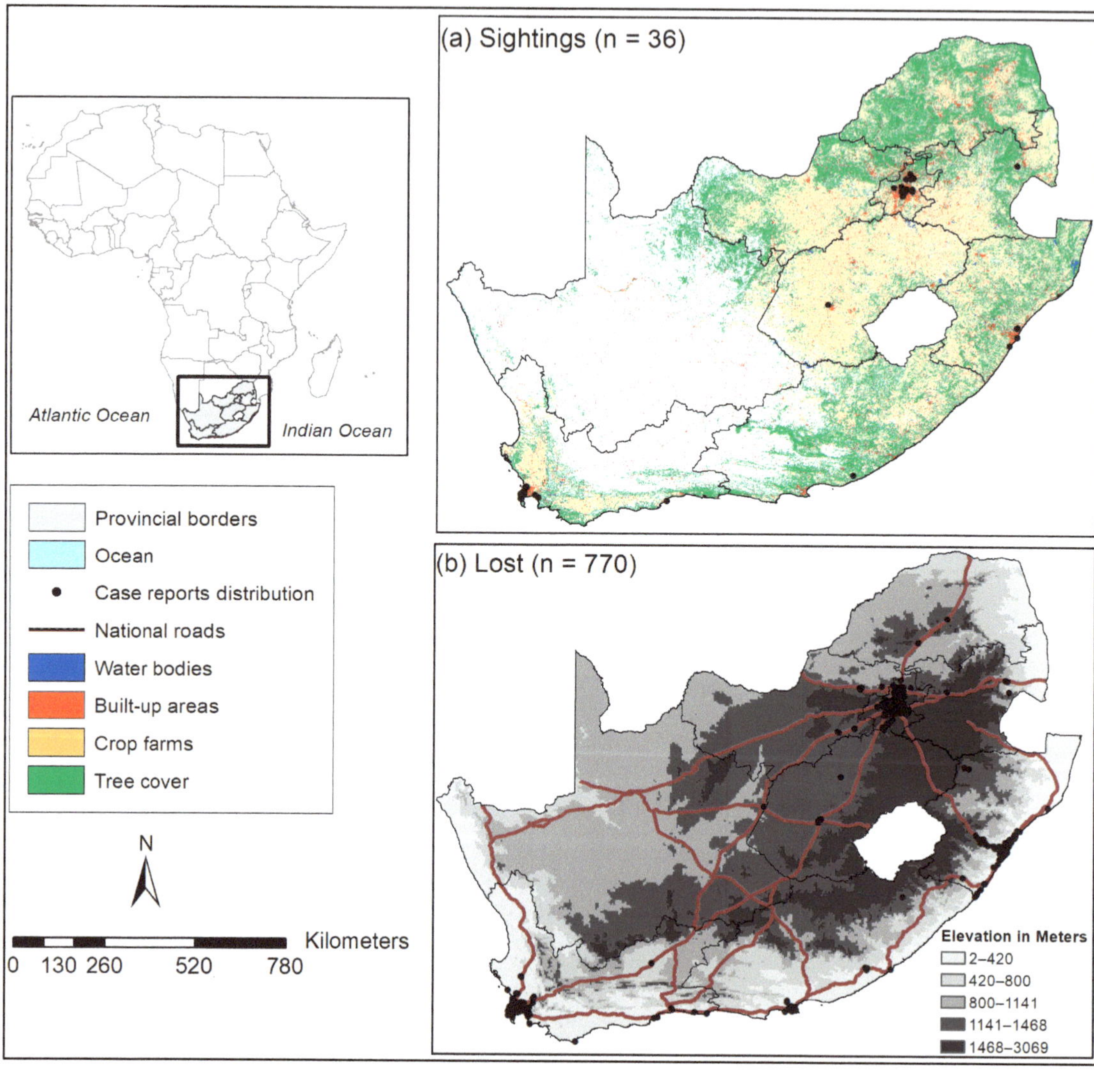

Figure 1. A map illustrating the geographic study area showing the land-use cover types and elevation in regions where six non-native pet birds have either been (**a**) observed outside captivity or (**b**) reported as lost in South Africa [58]. It includes national roads (N1–N4, N7–N10, N12, N14) and built-up areas as indicators of human footprint activities.

2.2. Species Selection Criteria

For this study, we used a dataset of non-native pet bird species reported lost, sighted, and found across South Africa between September 2011 and December 2023, covering nine provinces ([58], Figure 1). These are case reports from pet traders or owners (enthusiasts or hobbyists) who reported their pet birds lost (missing) or sighted to the dedicated lost-and-found advertisement platforms in South Africa (Figure 2), e.g., ParrotAlert (https://www.parrotalert.com/lost-and-found.php, accessed on 13 October–31 December 2023). We selected lost and sighted pet bird species from a study by Shivambu et al. [58] for SDM based on the following criteria: (1) species with a history of invasion elsewhere and introduction pathway, (2) sighted or distribution records in both or either introduced or native ranges [60], (3) potential environmental and socio-economic impacts [31,52,61], (4) a high number of lost and sighted case reports, and (5) availability in the South African pet trade [12], following selection criteria by Chucholl [62]. The availability of pet birds was evaluated based on Chucholl's [62] selection criteria: (1) "very common"—refers to species consistently accessible across various trade sources, spanning over four provinces or involving more than three online platforms, and generally available in substantial quantities; (2) "common"—pertains to species regularly found in at least one trade source, across multiple provinces or online platforms, and in significant quantities; (3) "rare"—species occasionally available in either one source of trade, a limited number of provinces, or online platforms, and typically in low quantities; and (4) "very rare"—refers to species that are available for a brief duration in either a single trade source (online or at a pet store), in fewer than four provinces, or across less than three online platforms, typically in limited quantities. As a result, six reported lost and sighted non-native pet bird species were selected for this study: African grey, Budgerigar, Cockatiel, Green-cheeked conure, Monk parakeet, and Rose-ringed parakeet (Figure 2). Despite being reported as lost and sighted, species such as Common starling, Common myna, and Rock dove were excluded from this study because they have fewer reports and were already considered in the previous study conducted in South Africa [52].

Figure 2. Photographs of pet bird species selected for SDM and reported lost and seen outside captivity in the field and residential areas for (**a**) African grey parrot (*Psittacus erithacus*), (**b**) Budgerigar (*Melopsittacus undulatus*), (**c**) Cockatiel (*Nymphicus hollandicus*), (**d**) Green-cheeked conure (*Pyrrhura molinae*), (**e**) Monk parakeet (*Myiopsitta monachus*), (**f**) Rose-ringed parakeet (*Psittacula krameri*), (**g**) flock of Rose-ringed parakeet found in a nest built in dead wood, and (**h**) those found dead from bee bites [top to bottom: Harlequin macaw (cross-breed), Blue-and-yellow macaw (*Ara ararauna*), Blue-winged macaw (*Primolius maracana*), Green-cheeked conure (*Pyrrhura molinae*), and Lord Derby's parakeet (*Psittacula derbiana*)]. Photographs sourced from lost and sighted pet bird advertisements [58].

2.3. Species Occurrence Records Download and Cleaning

To develop the SDM, it is necessary to utilize occurrence records from either native or introduced ranges, as indicated by Steiner et al. [63]. In this study, we downloaded current global spatial occurrence records for six selected pet bird species reported lost and sighted in South Africa from the Global Biodiversity Information Facility (GBIF: https://www.gbif.org/, accessed on 17 January 2024, [64–69]), using the 'rgbif' package in R statistical environment [70,71] and manual download. Here, we refer to "spatial occurrence records" as georeferenced data points that indicate the locations where individuals of a particular species have been observed. The GBIF encompasses the most extensive occurrence dataset compiled from verified data from accredited sources worldwide, including international herbariums, museums (including virtual), institutions, and published literature [72–74]. We used GBIF records from museums, zoological collections, ornithological databases, and published scientific literature for this study, given that species identification and locations were identified and confirmed. The present occurrence records ranged between 6383 and 492,196 and were downloaded between 2010 and 2024. The quality assessment and cleaning of distribution records were conducted using 'Biogeo' [75] and 'CoordinateCleaner' [76] packages in R and using ArcGIS [77]. This process involved removing records with duplicated or overlapping occurrences and those with missing latitude or longitude coordinates within a 10 arc-minute grid cell. The outcome ensured that only one occurrence point was retained per 10 arc-minute pixels [78]. Additionally, occurrences extending beyond contiguous political boundaries were excluded using the ArcGIS, 'Biogeo', and 'CoordinateCleaner' packages in R, following the approach outlined by Robertson et al. [75]. This procedure prevents autocorrelation and spatial bias of occurrence records, substantially improving model precision and prediction reliability [46,79,80]. As a result, a subset of between 924 and 8185 occurrence records was used for modelling. Although records for species such as Budgerigar and Cockatiel were low, their inclusion in this study was important due to their successful establishment from lost captives in other studies [81].

2.4. Climatic Data and Environmental Variable Selections

We employed the 'sdm' package [46] in the R statistical environment to produce ensemble ecological niche models for six selected pet bird species reported lost and sighted in South Africa. We downloaded a set of 19 bioclimatic variables (https://www.worldclim.org/, accessed on 13 January 2024, [82,83]) with a spatial resolution of 10 arc-min in GeoTIFF format. In addition, we downloaded water vapor pressure (kPa) from WorldClim (https://www.worldclim.org/, accessed on 14 January 2024) along with elevation variables at a 10 arc-min spatial resolution in GeoTIFF format, derived from the Shuttle Radar Topography Mission [84]. The Normalized Difference Vegetation Index (NDVI) for the year 2024, serving as a measure of biome greenness, was obtained from Copernicus Global Land Service version V2.0.1 (tile NDVI300_202401010000; [85]). The initial resolution at 300 m was transformed to a spatial resolution of 10 arc-min to align with the resolution employed for other variables in this study. We employed these environmental variables as predictors to assess the suitability of each pet bird species. These variables were selected due to their potential influence on direct physiological and ecological processes that could impact the survival and distribution of the bird species [86–89]. In addition, ambient humidity linked to water vapor pressure is crucial for nest conditions, as nest humidity is influenced by how water vapor is retained within the nest [90–92]. We tested correlations among 19 bioclimatic variables using the Variance Inflation Factor function in R (VIF; [93]) and Pearson (r) correlation coefficients to identify multicollinearity. This is because multicollinearity has an impact on model performance and prediction, as well as effects on SDM, such as model uncertainty and overfitting [46,94]. A threshold value of $\rho = |0.7|$ was considered as a cut-off point to choose variables that were not highly collinear, i.e., variables with $\rho < 0.7$ [94]. As a result, eight non-collinear bioclimatic variables were included when building the model for each species (Table 1).

In addition to environmental predictor variables, we included the global map of the human influence/impact (human footprint) on the earth's climate, ecosystems, wildlife, and human health and welfare on a global scale [95,96]. Human footprint variables encompass a wide range of factors, including population density, land use and land cover, infrastructure presence (e.g., roads and buildings), access to natural areas, electric power infrastructure, light pollution, and agricultural activities. Each of these factors contributes to quantifying the extent and intensity of human impact on the environment [96]. Incorporating the human footprint in this study was deemed essential, as most non-native invasive pet bird species have demonstrated a tendency to be widely distributed across the globe in connection with human activities [28,30,56,97]. The human footprint index for the year 2020 (tile HF3-2) (https://storage.googleapis.com/hii-export/2020-01-01/hii_2020-01-01.tif, accessed on 10 January 2024; [96]) in GeoTIFF format was downloaded and integrated into the model at a 10 arc-minute spatial resolution.

Table 1. A set of non-collinear environmental variables used as predictor variables for modelling six non-native pet bird species reported lost and sighted outside captivity.

Variables Code	Variable Description	Sources
BIO2	Mean Diurnal Range (Mean of monthly (max temp − min temp))	(WorldClim, [82,83])
BIO3	Isothermality (BIO2/BIO7) (×100)	(WorldClim, [82,83])
BIO5	Max Temperature of Warmest Month	(WorldClim, [82,83])
BIO8	Mean Temperature of Wettest Quarter	(WorldClim, [82,83])
Bio9	Mean Temperature of Driest Quarter	(WorldClim, [82,83])
Bio15	Precipitation Seasonality (Coefficient of Variation)	(WorldClim, [82,83])
Bio18	Precipitation of Warmest Quarter	(WorldClim, [82,83])
Bio19	Precipitation of Coldest Quarter	(WorldClim, [82,83])
WVP	Water Vapor Pressure (kPa)	(WorldClim, [83])
Elv	Elevation	(WorldClim, [83,84])
NDVI	Normalized Difference Vegetation Index	[85]
HF3	Human footprint	[95,96]

2.5. Model Fitting and Performance

The potential distribution models for six lost and sighted pet bird species were assessed using the R statistical software [71]. SDMs were produced using three algorithms, correlating occurrence records of six pet bird species with environmental predictor variables using the 'sdm' package [46]. Several studies have emphasized the importance of including both presence occurrence and absence records for species in fitting distribution models [46,47,98]. Thus, in this study, 1000 pseudo-absence records were randomly drawn from a defined background of present records at average runs of 100 bootstrap replications [47,99,100]. The algorithms employed in this study included a combination of both popular machine-learning and regression-based methods capable of handling varying levels of complexity in determining the relationship between the occurrence records and the employed environmental variables. The algorithms used included the Generalized Linear Model, Random Forest, and Support Vector Machine (Table 2). Calibration for each of the three methods was conducted using the default settings embedded in the packages from R libraries associated with the 'sdm' package. Models were calibrated using 70% occurrence records split for training and 30% records for testing, while predictive performance was evaluated at 100 runs of bootstrap replication to give the model adequate time to converge.

Table 2. A set of three algorithms employed as a statistical method for modelling six selected non-native pet bird species that went missing or were observed outside captivity.

Method Full Names	Code	Dependent Packages	References
Generalized Linear Model	GLM	STATS	[71,101]
Random Forest	RF	randomForest	[102,103]
Support Vector Machine	SVM	e1071	[104]

Two widely used independent and dependent-threshold statistics, including the area under the receiver operating characteristic curve (AUC) and the true skill statistic (TSS), were employed to assess the predictive performance of individual pet bird species and construct ensemble SDMs [105,106]. The AUC and TSS were specifically used to assess the level of agreement between the presence and pseudo-absence records in predicting species suitability. The AUC threshold values range from 0 to 1, with values below 0.7 considered poor, those between 0.7 and 0.9 considered good, and values greater than 0.9 deemed excellent [107,108]. The TSS threshold values are computed using a confusion matrix and range from −1 (indicating worse performance than random) to +1 (indicating excellent performance) [105]. To predict habitat suitability for six pet bird species, an ensemble model comprising three combined algorithm methods was used to improve prediction accuracy, as demonstrated in previous studies [109,110]. To construct a single consensus ensemble model for each species, we employed a weighted method to average the output of all three methods for individual species. We utilized the Maximum Training Specificity plus Sensitivity threshold to establish the presence–absence threshold [111,112]. This threshold minimizes the average error rate for both positive and negative observations and demonstrates superior performance compared to alternative thresholds in delivering precise presence predictions [112,113]. As a result, the omission rate, sensitivity, specificity, proportion of model correctness, and model calibration were assessed to determine the habitat suitability model performance's satisfactory accuracy, reliability, and performance levels. The habitat suitability threshold was calibrated, indicating the percentage of suitable areas for each modelled species. The percentage represented as the suitability threshold indicates the predicted extent of suitable habitat for each of the six species in South Africa.

2.6. Establishment Pathway Scenarios

Based on observations and documented occurrences [58], we developed a pathways scenario describing the mechanisms through which pet birds transition from captivity to instances of loss. The inclusion of these pathway scenarios was necessary for this study as they illustrate a range of mechanisms linked to the found-lost-establishment of pet birds in the country, aiding in the selection of suitable species for modelling. The pathway scenario highlights the key factors driving the transition of pet birds from captivity to the wild, including the roles of pet owners, animal rescues, adoption practices, and environmental suitability. This association is fundamental for understanding the processes that influence the establishment of these species and ensures that our models accurately reflect real-world dynamics. To construct the stepwise pathway scenarios, we used yEd Graph Editor software version 3.23.2 [114].

3. Results

3.1. Model Performance

The models fitted in this study performed well in predicting the habitat suitability for six pet bird species reported lost and sighted, with a weighted mean AUC value of 0.938 (±0.033; mean ± SD) and weighted mean TSS value of 0.765 (±0.093; mean ± SD) (Table 3, Supplementary Materials Table S1). Of the six pet bird species, the model for independent threshold, AUC, performed well for all pet birds, with the model for Budgerigar being the least performing (Table 3). The model for dependent threshold (TSS) performed well for Green-cheeked conure with a mean TSS of 0.891 (±0.019; mean ± SD). For the remaining five pet bird species, the mean TSS value ranged from 0.64 to 0.788 (Table 3, Supplementary Materials Table S1). The three algorithm methods used in fitting the model performed well for the six bird species selected (Supplementary Materials Table S1). The assessment of model performance indicates satisfactory levels of accuracy, reliability, and performance, as reflected by the omission rate (0.104 ± 0.074, mean ± SD), sensitivity, specificity, proportion of model correctness (0.892 ± 0.072), and model calibration (0.805 ± 0.065) (Supplementary Materials Table S1). For all six pet bird species, GLM, RF, and

SVM methods performed significantly in fitting model prediction for both independent and dependent thresholds (Supplementary Materials Table S1).

Of the six selected pet bird species, only the Rose-ringed parakeet has known invasive populations outside captivity, and it has been reported as lost and as an escapee (Supplementary Materials Table S2). The occurrence records for Rose-ringed parakeets are largely concentrated in Gauteng, KwaZulu-Natal, and Western Cape provinces (Supplementary Materials Figure S1). The remaining five pet bird species have varied lost case reports, ranging from 20 to 314 reports (Table 3, Supplementary Materials Figures S2–S6).

Table 3. A summary of the ensemble model performance for three algorithm methods generated using presence records of six non-native pet bird species reported as lost or sighted in the urban and natural ecosystems of South Africa. The number of lost case reports represents instances of bird species reported outside of captivity in South Africa. These reports were extracted from Shivambu et al. [58].

Species	Lost Case Reports	Variable		Variable of Importance (%)											
		AUC	TSS	BIO2	BIO3	BIO5	BIO8	BIO9	BIO15	BIO18	BIO19	Humidity	Elv	NDVI	HF3
African grey	314	0.968	0.826	8.30	9.94	5.67	6.22	7.85	10.52	13.81	4.63	4.73	5.34	8.3	14.69
Budgerigar	26	0.889	0.64	7.52	10.15	3.30	2.21	13.33	15.42	10.85	6.62	5.40	8.33	5.3	11.57
Cockatiel	220	0.912	0.68	5.77	9.32	3.01	2.32	14.82	8.18	10.22	9.21	7.30	7.94	4.2	17.71
Green-cheeked conure	41	0.977	0.891	9.01	6.41	3.73	3.41	6.87	7.93	16.71	12.62	5.04	4.15	10.33	11.79
Monk parakeet	20	0.943	0.788	9.41	7.42	3.64	2.75	8.63	10.11	8.54	11.71	13.02	4.41	8.11	12.25
Rose-ringed parakeet	149	0.938	0.762	5.91	7.13	3.34	1.64	12.87	11.63	9.13	14.05	7.15	3.26	9.4	14.49

3.2. Habitat Suitability and Importance Variables

The prediction threshold showed that 0.61 (61%) of South Africa has suitable habitat for the African grey (Supplementary Materials Table S1). The predicted suitability for this species was largely in eastern areas of South Africa, including Gauteng, Mpumalanga, KwaZulu-Natal, and Eastern Cape provinces (Figures 3a and S2). This includes the coastal regions of the Western Cape Province and central and western parts of Limpopo extending towards the eastern regions of Nort West Province (Figures 3a and S2). The eastern regions in the Free State Province towards the Drakensberg Mountains had suitable areas (Figures 3a and S2). The reported lost cases of African grey overlapped with predicted suitability in the Gauteng, KwaZulu-Natal, Eastern Cape, and Western Cape provinces (Figure S2). The predictor environmental variables that contributed the most to the African grey model predictions included Human footprint, Bio 18, Bio 15, Bio 3, Bio 2, and NDVI (Table 3).

The prediction suitability threshold for Budgerigar showed that 0.858 (86%) of areas in South Africa are likely suitable for this species (Supplementary Materials Table S1). The model predictions indicate that nearly all regions of the country are predicted to be suitable for the species, including areas where the species was largely reported lost, particularly in Gauteng and Western Cape provinces (Figures 3b and S3). The predictor variables best describing the Budgerigar suitability included Bio 15, Bio 9, Human footprint, Bio 18, Bio 3, and elevation (Table 3).

The predicted threshold suitability showed that 0.723 (72%) of areas are likely suitable for the Cockatiel in South Africa (Supplementary Materials Table S1). All nine provinces in the country were predicted to be suitable, with less predicted suitability in Northern Cape arid regions extending towards the border with Eastern Cape and Western Cape provinces (Figures 3c and S4). The reported lost case report for Cockatiel overlapped with the predicted suitability in Gauteng and Western Cape provinces (Figure S4). The predicted suitability for this species was best described by human footprint, Bio 9, Bio 18, Bio 3, Bio 19, and elevation (Table 3).

Our study showed that 0.673 (67%) of habitat is likely suitable for Green-cheeked conure (Supplementary Materials Table S1). The model predictions showed that only areas in the eastern regions of the country are suitable for this species, particularly areas in Limpopo, Gauteng, Mpumalanga, KwaZulu-Natal, and eastern coastal regions of Eastern Cape Province (Figure 3c). The predicted suitable areas overlapped with documented case reports in Gauteng and KwaZulu-Natal provinces (Figure S5). Bio 18, Bio 19, human footprint, NDVI, and Bio 2 were the most contributing environmental predictor variables in predicting the suitability for Green-cheeked conure (Table 3).

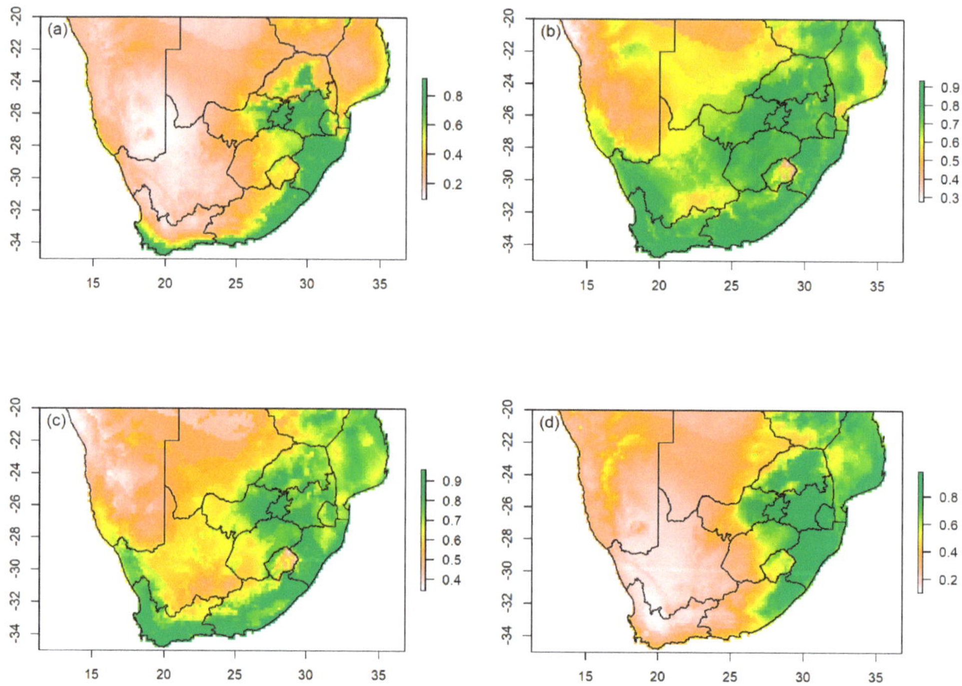

Figure 3. The species distribution modelling showing predicted potential suitability for (**a**) African grey parrot (*Psittacus erithacus*), (**b**) Budgerigar (*Melopsittacus undulatus*), (**c**) Cockatiel (*Nymphicus hollandicus*), and (**d**) Green-cheeked conure (*Pyrrhura molinae*) in South Africa. The suitability measure is represented by a threshold with a color ramp on the right of the map; the greener the threshold, the greater the suitability. The level of overlap between suitability and lost case report for each species is shown in Supplementary Materials Figures S2–S5.

The threshold suitability showed that 0.866 (87%) of areas are likely to be suitable for the Monk parakeet in South Africa (Supplementary Materials Table S1). Almost all provinces in the country were predicted to be suitable, with seven provinces having greater suitability (Figure 4a). Provinces such as the Northern Cape and Limpopo had reduced suitability in western and northern regions, respectively (Figure 4a). There was an overlap between reported lost case reports and the predicted suitable areas, particularly in Gauteng and Western Cape provinces (Figure S6). The environmental predictor variables that best describe the predicted suitability for the Monk parakeet included humidity, human footprint, Bio 19, Bio 15, Bio 2, and Bio 9 (Table 3).

The predicted suitability threshold for Rose-ringed parakeets showed that 0.717 (72%) of predicted areas are likely suitable for this species in South Africa (Supplementary Materials Table S1). The predicted suitability for this species is largely in the eastern coastal regions of South Africa, particularly in the Eastern Cape and Western Cape provinces (Figure 4b). The suitability for this species extends to inland regions of the Mpumalanga, Free State, Gauteng, North West, and Limpopo provinces (Figure 4b). Our study showed an overlap between predicted suitable areas and lost case distribution, particularly in the concentrated regions of Gauteng, KwaZulu-Natal, and Western Cape provinces (Figure S6). The predictor variables, such as human footprint, Bio 19, Bio 9, Bio 15, NDVI, and Bio 18, contributed the most in predicting suitability for the Rose-ringed parakeet in South Africa (Table 3).

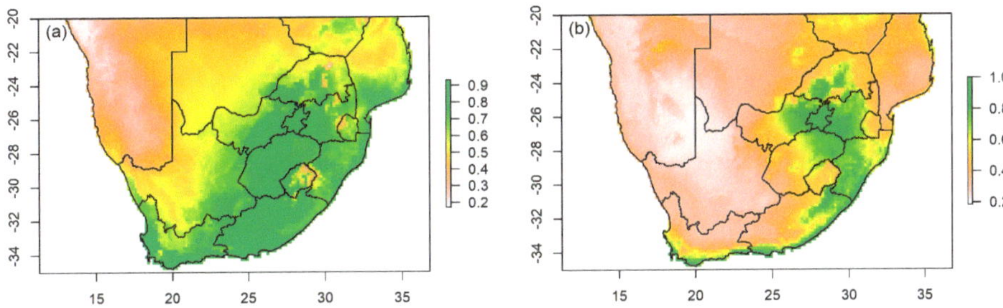

Figure 4. The species distribution modelling showing predicted potential suitability for (**a**) Monk parakeet (*Myiopsitta monachus*) and (**b**) Rose-ringed parakeet (*Psittacula krameri*) in South Africa. The suitability measure is represented by a threshold with a color ramp on the right of the map; the greener the threshold, the greater the suitability. The level of overlap between suitability and lost case report for each species is shown in Supplementary Materials Figures S1 and S6.

3.3. Pathway Scenario of Lost Pet Bird Leading to Potential Establishment

This study describes two-way pathways (found and lost) transition mechanisms leading to the potential establishment of some pet bird species (Figure 5). The stepwise mechanisms are described below:

Found pet birds: Pet birds, whether deceased (dead) or alive, can be found in various locations, including private captivity, pet stores, animal rescues, and zoos (Figure 2a–h). Often, pet birds are found alive either by their pet owners or individuals not affiliated with the ownership (hereafter referred to as common person) within the fields or residential areas (i.e., sports fields and parks).

Response of pet owners: If a pet bird is found alive by its owner, it is typically kept in secure confinement to prevent further escape. However, if the confinement is relaxed, the bird may escape again. In some cases, owners may release unwanted pet birds into the wild, contributing to the pool of stray or feral individuals, where some of them may be found and taken to animal rescues.

Response of common person: Lost birds found by individuals other than the owners in the wild or within residential areas (i.e., including sports fields and parks) may be taken to animal rescues, especially if they are identified as micro-chipped or tagged pet birds.

Role of animal rescues: Animal rescues face challenges in housing a large number of pet birds, requiring manpower, medical attention, and specific dietary needs. While some birds may be successfully reunited with their owners, others may remain unclaimed and may be donated to alleviate this burden.

Adoption and release: Unclaimed pet birds in rescue facilities may be adopted or donated by breeders, pet traders, hobbyists, or common persons. However, this process may inadvertently lead to accidental escapes and intentional releases, particularly by inexperienced individuals. This often occurs when the adopted bird is no longer wanted, the adopter relocates, or the bird is intentionally released (i.e., to form part of native wildlife).

Lost species: Lost species may either die due to harsh environmental conditions or survive in the wild, often within residential areas or urban green spaces (Figure 2h).

Survival: Those lost species that survive typically do so in environments where favorable abiotic (e.g., suitable climate, water availability, and appropriate ambient temperature for breeding) and biotic (e.g., food sources, nesting sites, and potential breeding partners) factors are present (Figure 5).

Adaptation: In the absence of external pressures such as predators, competitors, and diseases, most lost species are likely to adapt due to favorable environmental conditions at the point of loss or release.

Establishment of propagule pressure: Species already breeding in the wild, such as Rose-ringed parakeets and lovebirds, are likely to find breeding partners, integrate with the feral populations, and expand their populations in the wild, leading to high propagule pressure (Figure 2g).

Breeding potential: Pet birds capable of breeding and successfully producing propagule may either become invasive or fail to do so due to unsuccessful invasion debt.

Management interventions: Pet birds with successful establishment are likely to spread and occupy uninvaded territories, leading to management intervention to minimize their spread and prevent future invasions.

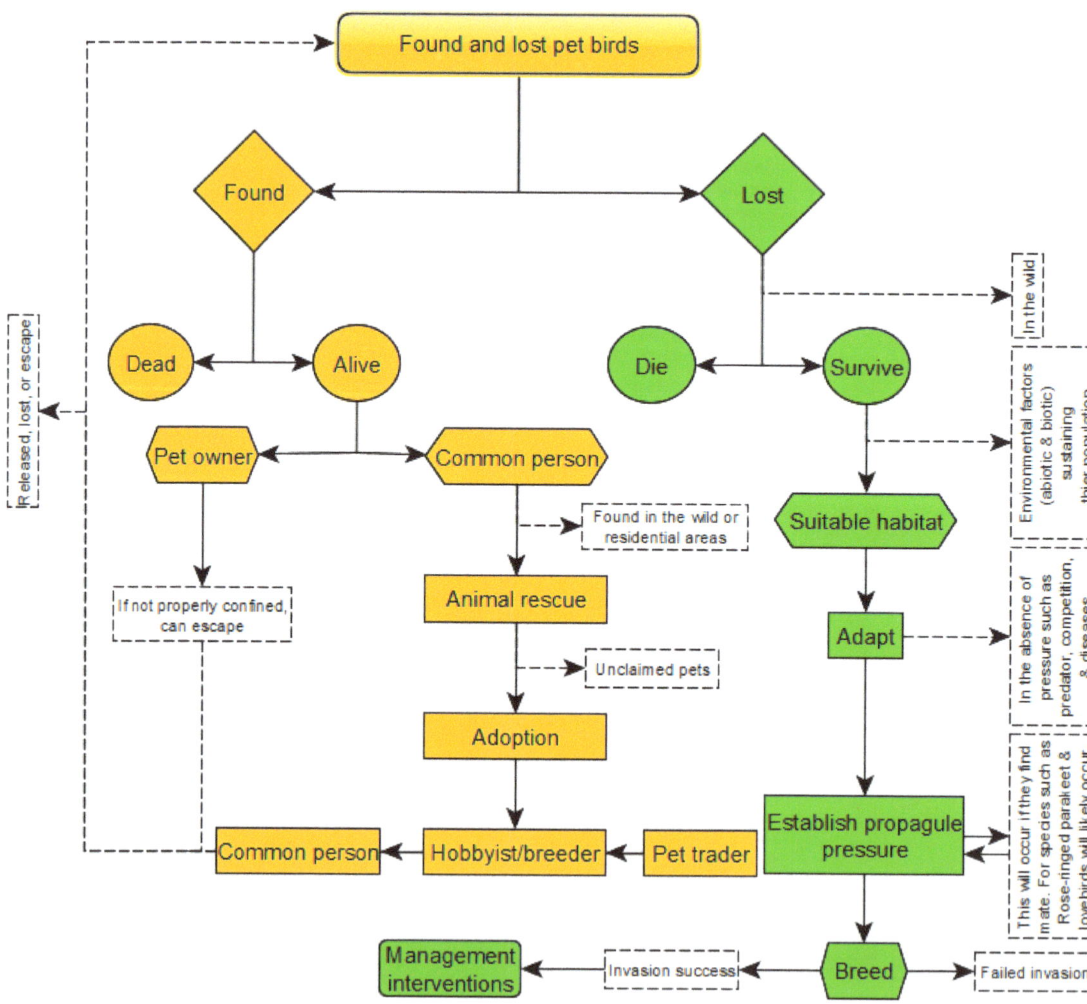

Figure 5. A stepwise pathways scenario describing the mechanisms by which pet birds transition from captivity to instances of found or lost, based on observations and documented occurrences in Shivambu et al. [58] and Figure 1. The yEd-constructed figure also illustrates the sequential process by which lost pet birds transition into potential invasive populations.

4. Discussion

4.1. Model Performance and Predictive Power

The model prediction in this study provides an important insight into the distribution, habitat suitability, and potential establishment of non-native pet birds reported lost and sighted in South Africa. Through the application of ensemble SDM, the model accurately predicted the habitat suitability for six non-native pet bird species, namely the African grey, Budgerigar, Cockatiel, Green-cheeked conure, Monk parakeet, and Rose-ringed parakeet. Our results demonstrated robust model performance, with satisfactory levels of accuracy, reliability, and performance across different algorithm methods. The model for each species exhibited high predictive power, as indicated by the AUC and TSS values. Evidently, the application of eSDM combined with varied environmental variables and algorithm methods has been reported to enhance model performance and predictive reliability in many studies [109,115–117]. The application of SDM has become a common predictive tool for several taxa, including alien invasive pet birds [51] and those that have gone wild from the cages and become invasive [118]. Thus, the utilization of these predictive techniques in this study proved to be appropriate for non-native pet birds reported lost and sighted, incorporating a variety of applied methodologies, robust techniques, and predictor variables in alignment with the recommendations of Araújo and New [109] is necessary. As a result, our study highlighted the efficacy of eSDM in capturing the complex relationship between the environmental variables used and the distribution patterns of six selected pet bird species.

4.2. Environmental Variables Shaping Species Suitability

The predicted habitat suitability maps revealed distinct distribution patterns for each species, influenced by a combination of environmental factors such as bioclimatic variables, human footprint, vegetation indices, and elevation. Our study revealed that the human footprint contributed the most in predicting suitability for the African grey, Cockatiel, Monk parakeet, and Rose-ringed parakeet. These results are not surprising given that these pet bird species are commensal and depend on human environments [28,56,58,81,119]. Therefore, their overrepresentation in urban areas may be attributed to their tendency to be released there or their inability to find suitable habitats in non-native ranges. Most pet species assessed in this study are non-migratory, exhibit localized tendencies, and rely on aided movement. For example, the invasive Monk parakeet in Barcelona and Spain has a mean home range of 12.4 $\pm$ 1.22 ha and long-distance dispersions of 10 km [120,121]. Similarly, the observed travel distance of Rose-ringed parakeets in Brussels was up to 9 km between roosting and nesting sites [122]. In addition, Budgerigar, a newly emerging pet bird species, is locally establishing its population in Japan's urban landscapes [81]. In Texas, USA, the Monk parakeet, an invasive localized to the area, shows a preference for using structures such as electricity utility for breeding [123]. A study by Banha et al. [97] applied a range of environmental variables, including the human footprint, which contributed highly to the distribution of invasive pet commensal Red-eared sliders (*Trachemys scripta elegans*) in the Iberian Peninsula. This suggests that the human footprint is highly probable to be the primary factor in predicting the distribution of various localized non-native pet species associated with human presence, thus serving as a valuable component for incorporation into habitat suitability models for alien invasive species [97,124].

Other predictor variables found to contribute to the habitat suitability model of the six bird species were Bio 9, Bio 18, and Bio 15. This is likely because these variables have a potential influence on direct physiological and ecological processes that could impact the survival and distribution of these bird species [86–89]. For example, Bio 9 can directly affect the availability of resources such as water and food, influencing habitat suitability for these bird species. Bio 18 can indicate the availability of water sources crucial for these bird species during periods of high temperature and potential drought stress. Lastly, Bio 15 influences the predictability and consistency of precipitation patterns, as it affects resource availability and habitat suitability for birds' survival and breeding [125–127]. In addition to this, a study conducted in Puerto Rico used bioclimatic variables to predict the suitability of habitat for 10 lost and sighted pet birds, including the invasive Monk parakeet [118]. The study identified possible range expansion for these birds in both human-dominated areas and natural environments [118]. These concurrent trends and the application of similar models suggest the importance of applying such variables in studies such as this to facilitate a better understanding of potential range expansions.

Our study found that vegetation indices contributed largely to the habitat suitability models for the Green-cheeked conure, Monk parakeet, and Rose-ringed parakeet. This could be explained by the tendency of these species to inhabit areas characterized by specific vegetation types offering a variety of food options (fruits, seeds, and foliage) important for their daily dietary needs. Additionally, these vegetation types create microhabitats suitable for nesting and roosting activities, resulting in increased species richness [125,128–130]. A case in point is the invasive Rose-ringed parakeets in Gauteng and KwaZulu-Natal regions, where they have invaded habitats characterized by tall, dense trees like Natal fig (*Ficus natalensis*), Natal mahogany (*Trichilia emetica*), Rose gum (*Eucalyptus grandis*), and White milkwood (*Sideroxylon inerme*), providing them with food sources, secondary nest cavities, and roosting sites [56,57]. It is, therefore, imperative to note that vegetation indices are valuable predictors for the distribution of these lost pet birds in South Africa as they provide an insight into habitat suitability, food availability, nesting sites, microclimate conditions, and water availability, all of which are important factors influencing the distribution of these bird species.

Our study also showed that elevation may contribute to the potential distribution of lost pet birds in South Africa. Although elevation did not have much influence on the distribution of modelled species, it is known to have an influence on the environmental factors that affect habitat suitability for bird distributions [131,132]. It influences factors such as climate variation, habitat diversity, availability of nesting and roosting sites, dispersal patterns, and protection from predators [132,133]. Several studies indicated that most invasive taxa predominantly occur in subalpine and lower altitudes, e.g., the Feral cat (*Felis catus*), Common myna, Common Starling, Monk parakeet, Rock dove, and Rose-ringed parakeet [28,30,31,52,132,133]. In addition, the population of Rose-ringed parakeets in KwaZulu-Natal Province roost in elevated areas such as Umhlanga Rocks and Cowey's Park, while during the day, they breed and forage in lower elevated areas [56]. This behavior is probably influenced by the presence of raptor species such as the Black sparrowhawk (*Accipiter melanoleucus*) and Yellow-billed kite (*Milvus aegyptius*), prompting parakeets to select roost sites in elevated locations to avoid potential predation [56]. In this study, the distribution of several lost pet bird species was least influenced by elevation, although it contributed more than some of the bioclimatic variables. This highlights the importance of including elevation as an environmental variable in the habitat suitability model to describe its potential influence on the distribution patterns of modelled non-native pet bird species.

On the other hand, our study included humidity in the model, but its contribution was low across three species and moderate across three species, namely the Rose-ringed parakeet, Cockatiel, and Monk parakeet. The inclusion of this variable was influenced by the fact that humidity plays a significant role in bird breeding and nest conditions due to its physiological influences on various aspects of their reproductive biology and habitat suitability [91,92,134]. Humidity at optimal levels is crucial for the development of bird eggs, improving incubation conditions, facilitating feather maintenance, and supporting the bird's microclimate [135]. Thus, the potential distribution of these non-native pet bird species is likely to be influenced by the availability of optimal humidity needed during the breeding season, particularly for those species already establishing and invasive elsewhere and in South Africa, e.g., Budgerigars, Rose-ringed parakeets, and Monk parakeets [35,56,81,130]. Our study emphasizes the continued use of humidity in habitat suitability modelling, given its critical physiological role in the reproductive success and survival of bird species.

4.3. Potential Invasion Hotspots and Risk Areas

The output maps generated by the habitat suitability model identified potential hotspots of invasions and high-risk areas for the establishment of non-native pet bird species in South Africa. Through the integration of eSDM and analysis of lost and sighted case reports [58], our study highlighted regions where environmental conditions are favorable for the establishment of non-native populations of pet birds. We found that all six modelled pet bird species have exhibited predicted habitat suitability in the country, with threshold suitability ranging from 61% to 87%. This suggests that large regions projected to be suitable habitats for any of these lost and sighted pet bird species in the country present a potential establishment threat.

Species with extensive suitability coverage greater than 70% included the Monk parakeet, Budgerigar, Cockatiel, and Rose-ringed parakeet. On the other hand, the Green-cheeked conure and African grey had suitability thresholds below 70%. Species of great concern are those exhibiting greater overlap between their habitat suitability maps and a large number of reported cases, such as the African grey, Cockatiel, and Rose-ringed parakeet. The predicted suitability for these species highlights regions with a high likelihood of establishment. The suitable areas identified as invasion hotspots are characterized by a combination of factors such as favorable climatic conditions, the presence of suitable habitats, and proximity to urban centers where pet bird trade and ownership are prevalent [12,58]. The previous study conducted in South Africa on pet bird trade indicated 147 species, with Budgerigar, African grey, Cockatiel, Monk parakeet, and Rose-ringed parakeet among the top 20 most traded species [12]. Of these, the latter has already been declared an invasive species in the country [136], displacing native birds from their nests, and its feral population is established in urban cities where the majority of pet shops are located [56,57]. Therefore, the predicted habitat suitability for this species is likely to facilitate its expansion beyond known established ranges, given its recent lost and sighted case reports [58]. In addition, this species is likely to thrive given that predators tend to ignore it in areas where it is established. For instance, our observations indicated that predators such as the Black sparrowhawk and Yellow-billed kite primarily prey on Rock doves rather than Rose-ringed parakeets in Durban, KwaZulu-Natal Province [56]. Conversely, the Monk parakeet, although not currently invasive, is likely to thrive in the wild, considering it is reported lost in the wild and has large habitat suitability in the country despite its invasion status elsewhere. This raises concerns, given that the species is known for its ability to outcompete native species for food and nesting sites, cause damage to infrastructure, predate crops, and create noise disturbance in urban areas [119,123,130,137].

Despite a large number of lost reports and favorable habitat suitability for the Cockatiel and African grey, there have been no invasion reports from locations inside and outside South Africa. It is worth noting that the Cockatiel is known to be a reservoir of zoonotic pathogens, some of which can spill over to other birds and pet owners [138]. If the majority of lost Cockatiels and African greys are not found by their owners and instead remain in the wild, there is a likelihood that they may find breeding mates, adapt, and potentially establish populations in areas where suitable habitats are available. A recent study by Nishida and Kitamura [81] indicated that the lost pet Budgerigar has transitioned into an invasive species and is now establishing itself in urban areas across Japan. Interestingly, this same species, which is among the commonly lost pet species, was also predicted to have a suitable habitat in South Africa, with an 85% suitability threshold. Although the Budgerigar is among the most kept birds worldwide, there have been many escape cases, resulting in a few self-sustaining populations in the wild [139–141]. Although this warrants further ground truthing to assess whether the species is established in the wild, it is important to acknowledge the potential for this species to establish a self-sustaining population in the wild, given that the predicted habitat suitability and environmental variables are associated with its preferences. Given that modelling showed suitability for all six pet bird species, it is important to note that lost pet birds, such as the African grey and Cockatiel, are less likely to survive in the wild. This is primarily because they originate from captive breeding, with most being descendants of captive individuals from several generations, which may affect their ability to adapt and survive in natural environments.

4.4. Lost Birds Pathways Scenarios to Species Establishment

Our study showed a comprehensive scenario analysis which reveals the pathways by which lost and sighted pet birds are likely to transition from captivity to potential establishment in the wild. This information is crucial for predicting and managing potential invasive populations. We described the key mechanisms driving this transition, including pet owners' role, animal rescues, adoption practices, and environmental suitability. Our pathway scenario involves initial encounters with pet owners or rescue facilities, followed by potential release into the environment, adaptation to new conditions, and the eventual establishment of self-sustaining populations. This tendency is not new for pet birds; several studies have touched on some of these mechanisms, although their approach was based on specific factors of pet trade pathways that influence the release events of amphibians, birds, and reptiles [81,142,143]. Conversely, our study exclusively used proposed establishment pathway scenarios based on the habitat suitability model output for adaptation and actual reported events of pet birds reported lost and sighted in South Africa [58], with the aforementioned key mechanisms guiding the establishment process. While quantification of adaptation remains pending for other lost and sighted pet birds, species such as the Rose-ringed parakeet, which is already invasive in the country, and those established elsewhere like the Budgerigar, Cockatiel, and Monk parakeet in various regions [56,81,118,119], are considered worrisome. The adaptation and establishment scenarios of lost pet birds may be possible given that the habitat suitability model suggested that several environmental predictor variables influenced most of the modelled species' distribution. This may lead to the emergence of problematic pet bird species with an invasion debt [144], resulting in the establishment of new populations, new invasions, and more areas invaded with greater impacts.

Consequently, lost pet bird species exhibiting predicted habitat suitability, fulfilling establishment pathway scenarios, and demonstrating the ability to breed in the wild require adaptive management interventions. Management interventions included developing targeted strategies to prevent further spread and minimize ecological impacts on native biodiversity through invasive species detection, assessment, and eradication planning [22,145,146]. Although much effort is placed into invasive plant management [147], escapees or lost pet animal taxa need to be incorporated into the management intervention and included in the NEM:BA alien and invasive species regulations list [148], or at least be included in national watchlist for non-native species before they become invasives (see [146]). This may assist in identifying four invasion indicators—introduction debt, establishment debt, spread debt, and impact debt [149]—which can be used to forecast and identify the most likely lost and sighted problematic non-native pet birds. Through tracing the pathway processes of establishment scenarios from initial loss to potential establishment, our study offers insights into the dynamics of invasion processes and highlights the importance of addressing multiple entry points to prevent further spread.

5. Conclusions

Our study demonstrated that all six selected pet bird species reported lost and sighted outside captivity have potentially suitable habitats in South Africa. The ensemble model performed well, predicting a large area of habitat suitability (61% to 87% threshold) overlapping with lost and sighted cases. Species showing the highest suitability overlap with lost and sighted reports may pose an invasion threat. Of most concern is the invasive Rose-ringed parakeet, which showed significant overlap between its habitat suitability and the lost and sighted reports, with its populations dominating in three provinces and persisting in the wild [58]. The predictor variables Bio 18, Bio 19, Bio 9, and Bio 3 contributed significantly to our model predictions, with the human footprint being the largest factor due to the majority of these species being commensal and urban invaders with humans' assistance [56,118]. Many of the lost pet birds with larger habitat suitability, such as Budgerigars and Monk parakeets, have been reported to establish self-sustaining populations in the wild elsewhere [81,118,141]. As lost or sighted pet birds primarily originate from captive breeding, with most being descendants of captive individuals from several generations, this lineage may affect their ability to survive in the wild. Therefore, we recommend thorough surveillance to monitor and evaluate the establishment potential of the five remaining non-native pet birds (excluding the Rose-ringed parakeet) reported as lost and sighted but lacking documented invasion history in the country. Our study describes potential pathways through which lost and sighted pet birds transition from captivity to the wild, demonstrating various establishment scenarios which can help facilitate alien invasive management interventions. Therefore, these species should be prioritized for inclusion in the watchlist, to be potentially incorporated into the existing NEM:BA A&IS regulations list. This approach will likely promote managed and responsible pet ownership practices in the country.

Supplementary Materials: The following supporting information can be downloaded at: https://www.mdpi.com/article/10.3390/biology13070483/s1, Figure S1. Ensemble species distribution modelling showing the potential distribution of the Rose-ringed parakeet (*Psittacula krameri*) in South Africa. Black dots resemble the distribution localities of Rose-ringed parakeets reported lost in the country; Figure S2. Ensemble species distribution modelling showing the potential distribution of the African grey parrot (*Psittacus erithacus*) in South Africa. Black dots resemble the distribution localities of African grey reported lost in the country; Figure S3. Ensemble species distribution modelling showing the potential distribution of the Budgerigar (*Melopsittacus undulatus*) in South Africa. Black dots resemble the distribution localities of Budgerigar reported lost in the country; Figure S4. Ensemble species distribution modelling showing the potential distribution of the Cockatiel (*Nymphicus hollandicus*) within South Africa. Black dots resemble the distribution localities of Cockatiels reported lost in the country; Figure S5. Ensemble species distribution modelling showing the potential distribution of the Green-cheeked conure (*Pyrrhura molinae*) in South Africa. Black dots resemble the distribution localities of Green-cheeked conure reported lost in the country; Figure S6. Ensemble species distribution modelling showing the potential distribution of the Monk parakeet (*Myiopsitta monachus*) in South Africa. Black dots resemble the distribution localities of Monk parakeets reported lost in the country; Table S1. A model performance for three ensembled algorithm methods generated using presence records of eight non-native pet bird species reported as having escaped from captivity in South Africa; Table S2: Geographic distribution of lost and found cases for six selected pet bird species in South Africa.

Author Contributions: T.C.S. and T.C.N. conceptualized the study. T.C.S. and N.S. collected datasets and spatially analyzed the data, while T.C.S. drafted the original manuscript. N.S., T.N., M.C.M., N.S.-P. and T.C.N. edited the manuscript prior to submission. All authors have read and agreed to the published version of the manuscript.

Funding: This research was funded by the Department of Nature Conservation, Faculty of Sciences, and the Directorate of Research and Innovation at Tshwane University of Technology (ZA).

Data Availability Statement: All data used for this study are included in the Supplementary Materials.

Acknowledgments: We are most grateful to the Department of Nature Conservation and the Directorate of Research and Innovation at Tshwane University of Technology (ZA) for supporting this study. Their financial assistance and logistical support have been invaluable in successfully completing our research. We are most grateful to the reviewers for their constructive feedback, which greatly contributed to the improvement of our manuscript.

Conflicts of Interest: The authors declare that they have no conflicts of interest.

References

1. Meyerson, L.A.; Mooney, H.A. Invasive alien species in an era of globalization. *Front. Ecol. Environ.* **2007**, *5*, 199–208. [CrossRef]
2. Ladin, Z.S.; Eggen, D.A.; Trammell, T.L.; D'Amico, V. Human-mediated dispersal drives the spread of the spotted lanternfly (*Lycorma delicatula*). *Sci. Rep.* **2023**, *13*, 1098. [CrossRef] [PubMed]

3. Searcy, C.A.; Howell, H.J.; David, A.S.; Rumelt, R.B.; Clements, S.L. Patterns of Non-Native Species Introduction, Spread, and Ecological Impact in South Florida, the World's Most Invaded Continental Ecoregion. *Annu. Rev. Ecol. Evol. Syst.* **2023**, *54*, 195–218. [CrossRef]
4. Williams, S.L.; Grosholz, E.D. The invasive species challenge in estuarine and coastal environments: Marrying management and science. *Estuaries Coasts* **2008**, *31*, 3–20. [CrossRef]
5. Pyšek, P.; Jarošík, V.; Hulme, P.E.; Pergl, J.; Hejda, M.; Schaffner, U.; Vilà, M. A global assessment of invasive plant impacts on resident species, communities, and ecosystems: The interaction of impact measures, invading species' traits and environment. *Glob. Chang. Biol.* **2012**, *18*, 1725–1737. [CrossRef]
6. Chardonnet, P.; Clers, B.D.; Fischer, J.; Gerhold, R.; Jori, F.; Lamarque, F. The value of wildlife. *Sci. Tech. J. Int. Off. Epiz.* **2002**, *21*, 15–52. [CrossRef] [PubMed]
7. Merem, E.C.; Twumasi, Y.; Wesley, J.; Isokpehi, P.; Fageir, S.; Crisler, M.; Romorno, C.; Hines, A.; Ochai, G.S.; Leggett, S.; et al. Assessing the menace of illegal wildlife trade in the sub-Saharan African region. *Adv. Life Sci.* **2018**, *8*, 1–25. [CrossRef]
8. Collard, R.C. *Animal Traffic: Lively Capital in the Global Exotic Pet Trade*; Duke University Press: Durham, NC, USA, 2020.
9. Andersson, A.A.; Tilley, H.B.; Lau, W.; Dudgeon, D.; Bonebrake, T.C.; Dingle, C. CITES and beyond: Illuminating 20 years of global, legal wildlife trade. *Glob. Ecol. Conserv.* **2021**, *26*, e01455. [CrossRef]
10. Bush, E.R.; Baker, S.E.; Macdonald, D.W. Global trade in exotic pets 2006–2012. *Conserv. Biol.* **2014**, *28*, 663–676. [CrossRef] [PubMed]
11. Nelufule, T.; Robertson, M.P.; Wilson, J.R.; Faulkner, K.T.; Sole, C.; Kumschick, S. The threats posed by the pet trade in alien terrestrial invertebrates in South Africa. *J. Nat. Conserv.* **2020**, *55*, 125831. [CrossRef]
12. Shivambu, T.C.; Shivambu, N.; Downs, C.T. An assessment of avian species sold in the South African pet trade. *Afr. J. Ecol.* **2022**, *60*, 980–995. [CrossRef]
13. Sinclair, J.S.; Stringham, O.C.; Udell, B.; Mandrak, N.E.; Leung, B.; Romagosa, C.M.; Lockwood, J.L. The international vertebrate pet trade network and insights from US imports of exotic pets. *BioScience* **2021**, *71*, 977–990. [CrossRef] [PubMed]
14. Caldas, A.T.M.; Dias, M.A.; Peres, M. Invertebrate (Araenae: Mygamolomorphae) illegal trade: An ignored side of wildlife trafficking. *Am. J. Zool.* **2018**, *7*, 20–23. [CrossRef]
15. Lockwood, J.L.; Welbourne, D.J.; Romagosa, C.M.; Cassey, P.; Mandrak, N.E.; Strecker, A.; Leung, B.; Stringham, O.C.; Udell, B.; Episcopio-Sturgeon, D.J.; et al. When pets become pests: The role of the exotic pet trade in producing invasive vertebrate animals. *Front. Ecol. Environ.* **2019**, *17*, 323–330. [CrossRef]
16. Mozer, A.; Prost, S. An introduction to illegal wildlife trade and its effects on biodiversity and society. *Forensic Sc. Int. Anim. Environ.* **2023**, *3*, 100064. [CrossRef]
17. Auliya, M.; Altherr, S.; Ariano-Sanchez, D.; Baard, E.H.; Brown, C.; Brown, R.M.; Cantu, J.C.; Gentil, E.G.; Gildenhuys, P.; Henningheim, E.; et al. Trade in live reptiles, its impact on wild populations, and the role of the European market. *Biol. Conserv.* **2016**, *204*, 103–119. [CrossRef]
18. Toland, E.; Bando, M.; Hamers, M.; Cadenas, V.; Laidlaw, R.; Martínez-Silvestre, A.; van der Wielen, P. Turning negatives into positives for pet trading and keeping: A review of positive lists. *Animals* **2020**, *10*, 2371. [CrossRef] [PubMed]
19. Hulme, P.E. Unwelcome exchange: International trade as a direct and indirect driver of biological invasions worldwide. *One Earth* **2021**, *4*, 666–679. [CrossRef]
20. Toomes, A.; Moncayo, S.; Stringham, O.C.; Lassaline, C.; Wood, L.; Millington, M.; Drake, C.; Jense, C.; Allen, A.; Hill, K.G.; et al. A snapshot of online wildlife trade: Australian e-commerce trade of native and non-native pets. *Biol. Conserv.* **2023**, *282*, 110040. [CrossRef]
21. Simberloff, D.; Martin, J.L.; Genovesi, P.; Maris, V.; Wardle, D.A.; Aronson, J.; Courchamp, F.; Galil, B.; García-Berthou, E.; Pascal, M.; et al. Impacts of biological invasions: What's what and the way forward. *Trends Ecol. Evol.* **2013**, *28*, 58–66. [CrossRef] [PubMed]
22. Faulkner, K.T.; Robertson, M.P.; Rouget, M.; Wilson, J.R. Understanding and managing the introduction pathways of alien taxa: South Africa as a case study. *Biol. Invasions* **2016**, *18*, 73–87. [CrossRef]
23. Sagoff, M. What is invasion biology? *Ecol. Econ.* **2018**, *154*, 22–30. [CrossRef]
24. Krishnakumar, K.; Raghavan, R.; Prasad, G.; Bijukumar, A.; Sekharan, M.; Pereira, B.; Ali, A. When pets become pests–exotic aquarium fishes and biological invasions in Kerala, India. *Curr. Sci.* **2009**, *97*, 474–476.
25. Stanley, M.C.; McNaughton, E.J.; Fewster, R.M.; Galbraith, J.A. Cumulative propagule pressure exerted by escaped pet parrots. *J. Appl. Ecol.* **2023**, *60*, 384–392. [CrossRef]
26. Shieh, B.; Lin, Y.; Lee, T.; Chang, C.; Cheng, K. Pet trade as sources of introduced bird species in Taiwan. *Taiwania* **2006**, *51*, 81–86.
27. Vall-Llosera, M.; Cassey, P. Leaky doors: Private captivity as a prominent source of bird introductions in Australia. *PLoS ONE* **2017**, *12*, e0172851. [CrossRef]
28. Strubbe, D.; Matthysen, E. Establishment success of invasive ring-necked and monk parakeets in Europe. *J. Biogeogr.* **2009**, *36*, 2264–2278. [CrossRef]
29. Su, S.; Cassey, P.; Blackburn, T.M. The wildlife pet trade as a driver of introduction and establishment in alien birds in Taiwan. *Biol. Invasions* **2016**, *18*, 215–229. [CrossRef]

30. Postigo, J.L.; Carrillo-Ortiz, J.; Domènech, J.; Tomàs, X.; Arroyo, L.; Senar, J.C. Dietary plasticity in an invasive species and implications for management: The case of the monk parakeet in a Mediterranean city. *Anim. Biodivers. Conserv.* **2021**, *44*, 185–194. [CrossRef]
31. Downs, C.T.; Hart, L.A. *Invasive Birds: Global Trends and Impacts*; CABI International: Wallingford, UK, 2020; 400p. [CrossRef]
32. Martin-Albarracin, V.L.; Amico, G.C.; Simberloff, D.; Nuñez, M.A. Impact of non-native birds on native ecosystems: A global analysis. *PLoS ONE* **2015**, *10*, e0143070. [CrossRef]
33. Avery, M.L.; Greiner, E.C.; Lindsay, J.R.; Newman, J.R.; Pruett-Jones, S. Monk parakeet management at electric utility facilities in South Florida. *Proc. Vertebr. Pest Conf.* **2002**, *20*, 140–145. [CrossRef]
34. Hernández-Brito, D.; Carrete, M.; Ibáñez, C.; Juste, J.; Tella, J.L. Nest-site competition and killing by invasive parakeets cause the decline of a threatened bat population. *R. Soc. Open Sci.* **2018**, *5*, 172477. [CrossRef] [PubMed]
35. Battisti, C.; Fraticelli, F. When 'bullies' come out of cities: Mobbing among introduced Rose-Ringed Parakeets *Psittacula krameri* and native species of conservation concern in extra-urban habitats with suggestions for a framework of PSIR indicators. *Biodiversity* **2023**, *24*, 96–101. [CrossRef]
36. Bovet, J. Homing in red squirrels (*Tamiasciurus hudsonicus*): The importance of going straight. *Ethology* **1995**, *101*, 1–9. [CrossRef]
37. Andreychev, A.V.; Kiyaykina, O.S. Homing of the forest dormouse (*Dryomys nitedula*, Rodentia, Gliridae). *Biol. Bull. Russ. Acad. Sci.* **2020**, *47*, 1227–1234. [CrossRef]
38. McCartney, K.R.; Kumar, S.; Sing, S.E.; Ward, S.M. Using invaded-range species distribution modeling to estimate the potential distribution of Linaria species and their hybrids in the US northern Rockies. *Invas. Plant Sci. Manag.* **2019**, *12*, 97–111. [CrossRef]
39. Gong, S.; Gao, Y.; Duan, H.; Ge, Y.; Wei, Y. Incorporating physiological data into species distribution models to predict the potential distribution range of the red-eared slider in China. *Ecol. Indic.* **2023**, *154*, 110749. [CrossRef]
40. El-Barougy, R.F.; Dakhil, M.A.; Halmy, M.W.; Gray, S.M.; Abdelaal, M.; Khedr, A.H.A.; Bersier, L.F. Invasion risk assessment using trait-environment and species distribution modelling techniques in an arid protected area: Towards conservation prioritization. *Ecol. Indic.* **2021**, *129*, 107951. [CrossRef]
41. Adhikari, P.; Lee, Y.H.; Poudel, A.; Hong, S.H.; Park, Y.S. Global spatial distribution of *Chromolaena odorata* habitat under climate change: Random forest modeling of one of the 100 worst invasive alien species. *Sci. Rep.* **2023**, *13*, 9745. [CrossRef] [PubMed]
42. Jarnevich, C.; Engelstad, P.; LaRoe, J.; Hays, B.; Hogan, T.; Jirak, J.; Pearse, I.; Prevéy, J.; Sieracki, J.; Simpson, A.; et al. Invaders at the doorstep: Using species distribution modeling to enhance invasive plant watch lists. *Ecol. Inform.* **2023**, *75*, 101997. [CrossRef]
43. Naimi, B.; Capinha, C.; Ribeiro, J.; Rahbek, C.; Strubbe, D.; Reino, L.; Araújo, M.B. Potential for invasion of traded birds under climate and land-cover change. *Glob. Chang. Biol.* **2022**, *28*, 5654–5666. [CrossRef] [PubMed]
44. O'Mahony, J.; Vanmechelen, A.; Holloway, P. Quantifying the distribution and potential biotic interactions between deer and flora using species distribution modelling. *Ann. GIS* **2023**, *29*, 553–568. [CrossRef]
45. Miller, J. Species distribution modeling. *Geogr. Comp.* **2010**, *4*, 490–509. [CrossRef]
46. Naimi, B.; Araújo, M.B. sdm: A reproducible and extensible R platform for species distribution modelling. *Ecography* **2016**, *39*, 368–375. [CrossRef]
47. Guisan, A.; Thuiller, W. Predicting species distribution: Offering more than simple habitat models. *Ecol. Lett.* **2005**, *8*, 993–1009. [CrossRef] [PubMed]
48. Srivastava, V.; Lafond, V.; Griess, V.C. Species distribution models (SDM): Applications, benefits and challenges in invasive species management. *CABI Rev.* **2019**, 1–13. [CrossRef]
49. Vilizzi, L.; Copp, G.H.; Hill, J.E.; Adamovich, B.; Aislabie, L.; Akin, D.; Al-Faisal, A.J.; Almeida, D.; Azmai, M.A.; Bakiu, R.; et al. A global-scale screening of non-native aquatic organisms to identify potentially invasive species under current and future climate conditions. *Sci. Total Environ.* **2021**, *788*, 147868. [CrossRef] [PubMed]
50. Britton, J.R.; Lynch, A.J.; Bardal, H.; Bradbeer, S.J.; Coetzee, J.A.; Coughlan, N.E.; Dalu, T.; Tricarico, E.; Gallardo, B.; Lintermans, M.; et al. Preventing and controlling nonnative species invasions to bend the curve of global freshwater biodiversity loss. *Environ. Rev.* **2023**, *31*, 310–326. [CrossRef]
51. Le Louarn, M.; Clergeau, P.; Strubbe, D.; Deschamps-Cottin, M. Dynamic species distribution models reveal spatiotemporal habitat shifts in native range-expanding versus non-native invasive birds in an urban area. *J. Avian Biol.* **2018**, *49*, jav-01527. [CrossRef]
52. Shivambu, T.C.; Shivambu, N.; Downs, C.T. Impact assessment of seven alien invasive bird species already introduced to South Africa. *Biol. Invasions* **2020**, *22*, 1829–1847. [CrossRef]
53. Farashi, A.; Alizadeh-Noughani, M. Global distribution modelling, assessment of the spatial invasion risk of Asian pied starling (*Sturnus contra* Linnaeus, 1758) under climate change. *Eur. J. Wildl. Res.* **2023**, *69*, 83. [CrossRef]
54. Chittenden, H.; Davies, G.; Weiersbye, I. *Roberts Bird Guide: Illustrating nearly 1,000 Species in Southern Africa*; John Voelcker Bird Book Fund: Cape Town, South Africa, 2016.
55. Measey, J.; Hui, C.; Somers, M.J. Terrestrial vertebrate invasions in South Africa. In *Biological Invasions in South Africa*; Van Wilgen, B., Measey, J., Richardson, D., Wilson, J., Zengeya, T., Eds.; Springer Nature: Cham, Switzerland, 2020; pp. 115–151. [CrossRef]
56. Shivambu, T.C.; Shivambu, N.; Downs, C.T. Population estimates of non-native rose-ringed parakeets *Psittacula krameri* (Scopoli, 1769) in the Durban Metropole, KwaZulu-Natal Province, South Africa. *Urban Ecosyst.* **2021**, *24*, 649–659. [CrossRef]
57. Shivambu, T.C.; Shivambu, N.; Downs, C.T. Breeding status of invasive Rose-ringed Parakeets *Psittacula krameri* in Durban, South Africa. *Ostrich* **2022**, *93*, 257–270. [CrossRef]

58. Shivambu, T.C.; Shivambu, N.; Nelufule, T.; Moshobane, M.C.; Seoraj-Pillai, N.; Nangammbi, T.C. Evaluating the status of lost, found and sighted non-native pet bird species in South Africa. *Diversity* **2024**, *16*, 283. [CrossRef]
59. Finch, J.M.; Meadows, M.E. South African biomes and their changes over time. In *The Geography of South Africa*; Knight, J., Rogerson, C., Eds.; Springer Nature: Cham, Switzerland, 2019; pp. 57–69. [CrossRef]
60. IUCN. International Union for Conservation of Nature. 2024. Available online: https://www.iucnredlist.org/ (accessed on 5 December 2023).
61. GISD. The Global Invasive Species Database. 2024. Available online: https://www.iucngisd.org/gisd/ (accessed on 12 January 2024).
62. Chucholl, C. Invaders for sale: Trade and determinants of introduction of ornamental freshwater crayfish. *Biol. Invasions* **2013**, *15*, 125–141. [CrossRef]
63. Steiner, F.M.; Schlick-Steiner, B.C.; Vanderwal, J.; Reuther, K.D.; Christian, E.; Stauffer, C.; Crozier, R.H. Combined modelling of distribution and niche in invasion biology: A case study of two invasive *Tetramorium* ant species. *Divers. Distrib.* **2008**, *14*, 538–545. [CrossRef]
64. GBIF.org. GBIF Occurrence of African Grey *Psittacus erithacus* (Linnaeus, 1758). 2024. [CrossRef]
65. GBIF.org. GBIF Occurrence of Budgerigars *Melopsittacus undulatus* (Shaw, 1805). 2024. [CrossRef]
66. GBIF.org. GBIF Occurrence of Cockatiel *Nymphicus hollandicus* (Kerr, 1792). 2024. [CrossRef]
67. GBIF.org. GBIF Occurrence of Green Cheeked Conure *Pyrrhura molinae* (Massena & Souance, 1854). 2024. [CrossRef]
68. GBIF.org. GBIF Occurrence of Monk parakeet *Myiopsitta monachus* (Boddaert, 1783). 2024. [CrossRef]
69. GBIF.org. GBIF Occurrence of Rose-ringed parakeet *Psittacula krameria* (Scopoli, 1769). 2024. [CrossRef]
70. Chamberlain, S.; Boettiger, C.; Ram, K.R.; Barve, V. rgbif: Interface to the Global Biodiversity Information Facility API Methods. R Package Version 0.3.0. 2014. Available online: http://CRAN.R-project.org/package=rgbif (accessed on 12 January 2024).
71. R Core Team. *R: A Language and Environment for Statistical Computing*; R Foundation for Statistical Computing: Vienna, Austria, 2023. Available online: http://www.R-project.org/ (accessed on 24 November 2023).
72. Li, G.; Xu, G.; Guo, K.; Du, S. Mapping the global potential geographical distribution of black locust (*Robinia pseudoacacia* L.) using herbarium data and a maximum entropy model. *Forests* **2014**, *5*, 2773–2792. [CrossRef]
73. James, S.A.; Soltis, P.S.; Belbin, L.; Chapman, A.D.; Nelson, G.; Paul, D.L.; Collins, M. Herbarium data: Global biodiversity and societal botanical needs for novel research. *Appl. Plant Sci.* **2018**, *6*, e1024. [CrossRef] [PubMed]
74. Huettmann, F. The forgotten data: A rather short but deep story of museums and libraries in HKH and similar information sources in support of the Global Biodiversity Information System (GBIF.org) and model-predictions for improved conservation management. In *Hindu Kush-Himalaya Watersheds Downhill: Landscape Ecology and Conservation Perspectives*; Regmi, G., Huettmann, F., Eds.; Springer Nature: Cham, Switzerland, 2020; pp. 497–520. [CrossRef]
75. Robertson, M.P.; Visser, V.; Hui, C. Biogeo: An R package for assessing and improving data quality of occurrence record datasets. *Ecography* **2016**, *39*, 394–401. [CrossRef]
76. Maldonado, C.; Molina, C.I.; Zizka, A.; Persson, C.; Taylor, C.M.; Albán, J.; Chilquillo, E.; Rønsted, N.; Antonelli, A. Estimating species diversity and distribution in the era of Big Data: To what extent can we trust public databases. *Glob. Ecol. Biogeogr.* **2015**, *24*, 973–984. [CrossRef] [PubMed]
77. ESRI. *ArcGIS Desktop: Release 10.4.1*; Environmental Systems Research Institute: Redlands, CA, USA, 2022.
78. Hijmans, R. Raster: Geographic Data Analysis and Modeling. R Package Version 3.6-14. 2023. Available online: https://cran.r-project.org/web/packages/raster/raster.pdf (accessed on 11 November 2023).
79. Austin, M. Species distribution models and ecological theory: A critical assessment and some possible new approaches. *Ecol. Model.* **2007**, *200*, 1–19. [CrossRef]
80. Rodríguez-Castañeda, G.; Hof, A.R.; Jansson, R.; Harding, L.E. Predicting the fate of biodiversity using species' distribution models: Enhancing model comparability and repeatability. *PLoS ONE* **2012**, *7*, e44402. [CrossRef] [PubMed]
81. Nishida, S.; Kitamura, W. An influx of non-native bird species into the natural environment owing to the accidental release of pet birds in Japan. *Animals* **2024**, *14*, 221. [CrossRef] [PubMed]
82. Booth, T.H.; Nix, H.A.; Busby, J.R.; Hutchinson, M.F. BIOCLIM: The first species distribution modelling package, its early applications and relevance to most current MAXENT studies. *Divers. Distrib.* **2014**, *20*, 1–9. [CrossRef]
83. Fick, S.E.; Hijmans, R.J. WorldClim 2: New 1km spatial resolution climate surfaces for global land areas. *Int. J. Climatol.* **2017**, *37*, 4302–4315. [CrossRef]
84. SRTM. *Shuttle Radar Topography Mission 1 Arc-Second Global*; USGS: Reston, VA, USA, 2023. [CrossRef]
85. Copernicus Global Land ServiceProviding Bio-Geophysical Products of Global Land Surface. 2023. Available online: https://land.copernicus.eu/global/products/ndvi (accessed on 29 January 2024).
86. Santillán, V.; Quitián, M.; Tinoco, B.A.; Zárate, E.; Schleuning, M.; Böhning-Gaese, K.; Neuschulz, E.L. Direct and indirect effects of elevation, climate and vegetation structure on bird communities on a tropical mountain. *Acta Oecol.* **2020**, *102*, 103500. [CrossRef]
87. Schoell, E.M.; Hille, S.M. Heavy and persistent rainfall leads to brood reduction and nest failure in a passerine bird. *J. Avian Biol.* **2020**, *51*, 1–11. [CrossRef]
88. Veldsman, L.M.; Kylin, H.; Bronkhorst, P.; Engelbrecht, I.; Bouwman, H. A method to determine the combined effects of climate change (temperature and humidity) and eggshell thickness on water loss from bird eggs. *Environ. Geochem. Health* **2020**, *42*, 781–793. [CrossRef] [PubMed]

89. Zwarts, L.; Bijlsma, R.G.; van der Kamp, J.; Sikkema, M. Distribution and numbers of ground-foraging birds between the hyper-arid Sahara and the hyper-humid Guinea forests. *Ardea* **2023**, *111*, 7–66. [CrossRef]
90. Deeming, D.C. Importance of nest type on the regulation of humidity in bird nests. *Avian Biol. Res.* **2011**, *4*, 23–31. [CrossRef]
91. Salaberria, C.; Celis, P.; López-Rull, I.; Gil, D. Effects of temperature and nest heat exposure on nestling growth, dehydration and survival in a Mediterranean hole-nesting passerine. *Ibis* **2014**, *156*, 265–275. [CrossRef]
92. Deeming, D.C.; Gilchrist, R.; Szafraniec, M.; Pollins, J.M. Water vapour conductance of passerine nest walls. *Acta Ornithol.* **2020**, *55*, 13–22. [CrossRef]
93. Marquardt, D.W. Generalized inverses, ridge regression, biased linear estimation, and nonlinear estimation. *Technometrics* **1970**, *12*, 591–612. [CrossRef]
94. Dormann, C.F.; Elith, J.; Bacher, S.; Buchmann, C.; Carl, G.; Carré, G.; Marquéz, J.R.G.; Gruber, B.; Lafourcade, B.; Leitão, P.J.; et al. Collinearity: A review of methods to deal with it and a simulation study evaluating their performance. *Ecography* **2013**, *36*, 27–46. [CrossRef]
95. Sanderson, E.W.; Jaiteh, M.; Levy, M.A.; Redford, K.H.; Wannebo, A.V.; Woolmer, G. The human footprint and the last of the wild: The human footprint is a global map of human influence on the land surface, which suggests that human beings are stewards of nature, whether we like it or not. *BioScience* **2002**, *52*, 891–904. [CrossRef]
96. Sanderson, E.W.; Fisher, K.; Robinson, N.; Sampson, D.; Duncan, A.; Royte, L. The march of the human footprint. *EcoEvoRxiv* **2022**, 1–54. [CrossRef]
97. Banha, F.; Gama, M.; Anastácio, P.M. The effect of reproductive occurrences and human descriptors on invasive pet distribution modelling: *Trachemys scripta elegans* in the Iberian Peninsula. *Ecol. Model.* **2017**, *360*, 45–52. [CrossRef]
98. Guisan, A.; Zimmermann, N.E. Predictive habitat distribution models in ecology. *Ecol. Model.* **2000**, *135*, 147–186. [CrossRef]
99. Phillips, S.J. Transferability, sample selection bias and background data in presence-only modelling: A response to Peterson et al. (2007). *Ecography* **2008**, *31*, 272–278. [CrossRef]
100. Barbet-Massin, M.; Jiguet, F.; Albert, C.H.; Thuiller, W. Selecting pseudo-absences for species distribution models: How, where and how many? *Methods Ecol. Evol.* **2012**, *3*, 327–338. [CrossRef]
101. McCullagh, P. *Generalized Linear Models*; CRC: New York, NY, USA, 2019.
102. Breiman, L. Random forests. *Mach. Learn.* **2001**, *45*, 5–32. [CrossRef]
103. Liaw, A.; Wiener, M. Classification and Regression by randomForest. *R News* **2002**, *2*, 18–22.
104. Meyer, D.; Dimitriadou, E.; Hornik, K.; Weingessel, A.; Leisch, F. e1071: Misc Functions of the Department of Statistics, Probability Theory Group (Formerly: E1071), TU Wien. R Package Version 1.6-7. 2015. Available online: https://CRAN.R-package.org?package=e1071 (accessed on 30 January 2024).
105. Allouche, O.; Tsoar, A.; Kadmon, R. Assessing the accuracy of species distribution models: Prevalence, kappa and the true skill statistic (TSS). *J. Appl. Ecol.* **2006**, *43*, 1223–1232. [CrossRef]
106. Araújo, M.B.; Anderson, R.P.; Márcia Barbosa, A.; Beale, C.M.; Dormann, C.F.; Early, R.; Garcia, R.A.; Guisan, A.; Maiorano, L.; Naimi, B.; et al. Standards for distribution models in biodiversity assessments. *Sci. Adv.* **2019**, *5*, eaat4858. [CrossRef] [PubMed]
107. Swets, J.A. Measuring the accuracy of diagnostic systems. *Science* **1988**, *240*, 1285–1293. [CrossRef] [PubMed]
108. Fielding, A.H.; Bell, J.F. A review of methods for the assessment of prediction errors in conservation presence/absence models. *Environ. Conserv.* **1997**, *24*, 38–49. [CrossRef]
109. Araújo, M.B.; New, M. Ensemble forecasting of species distributions. *Trends Ecol. Evol.* **2007**, *22*, 42–47. [CrossRef] [PubMed]
110. Marmion, M.; Parviainen, M.; Luoto, M.; Heikkinen, R.K.; Thuiller, W. Evaluation of consensus methods in predictive species distribution modelling. *Divers. Distrib.* **2009**, *15*, 59–69. [CrossRef]
111. Manel, S.; Williams, H.C.; Ormerod, S.J. Evaluating presence–absence models in ecology: The need to account for prevalence. *J. Appl. Ecol.* **2001**, *38*, 921–931. [CrossRef]
112. Freeman, E.A.; Moisen, G.G. A comparison of the performance of threshold criteria for binary classification in terms of predicted prevalence and kappa. *Ecol. Model.* **2008**, *217*, 48–58. [CrossRef]
113. Jiménez-Valverde, A.; Lobo, J.M. Threshold criteria for conversion of probability of species presence to either–or presence–absence. *Acta Oecol.* **2007**, *31*, 361–369. [CrossRef]
114. Ed Graph Editor. *Works GmbH Software Developer*; Version 3.23.2; Works GmbH: Tübingen, Germany, 2023. Available online: https://www.yworks.com/products/yed/download (accessed on 11 November 2023).
115. Benemann, V.R.; Araujo, L.D.; Fabbris, A.Z.; Montone, R.C.; Petry, M.V. Nesting distribution of Masked Booby *Sula dactylatra* at Trindade Island, western South Atlantic Ocean. *Mar. Ornithol.* **2022**, *50*, 189–195.
116. Liu, X.; Liu, H.; Datta, P.; Frey, J.; Koch, B. Mapping an invasive plant *Spartina alterniflora* by combining an ensemble one-class classification algorithm with a phenological NDVI time-series analysis approach in middle coast of Jiangsu, China. *Remote Sens.* **2020**, *12*, 4010. [CrossRef]
117. Petrosyan, V.; Dinets, V.; Osipov, F.; Dergunova, N.; Khlyap, L. Range dynamics of Striped field mouse (*Apodemus agrarius*) in Northern Eurasia under Global Climate Change Based on Ensemble Species Distribution Models. *Biology* **2023**, *12*, 1034. [CrossRef] [PubMed]
118. Falcón, W.; Tremblay, R.L. From the cage to the wild: Introductions of Psittaciformes to Puerto Rico. *PeerJ* **2018**, *6*, e5669. [CrossRef] [PubMed]

119. Senar, J.C.; Domènech, J.; Arroyo, L.; Torre, I.; Gordo, O. An evaluation of monk parakeet damage to crops in the metropolitan area of Barcelona. *Anim. Biodivers. Conserv.* **2016**, *39*, 141–145. [CrossRef]
120. Senar, J.C.; Moyà, A.; Pujol, J.; Tomas, X.; Hatchwell, B.J. Sex and age effects on monk parakeet home-range variation in the urban habitat. *Diversity* **2021**, *13*, 648. [CrossRef]
121. Borray-Escalante, N.A.; Baucells, J.; Carrillo-Ortiz, J.; Hatchwell, B.J.; Senar, J.C. Long-distance dispersal of monk parakeets. *Anim. Biodivers. Conserv.* **2023**, *46*, 71–78. [CrossRef]
122. Strubbe, D.; Matthysen, E. A radiotelemetry study of habitat use by the exotic Ring-necked Parakeet *Psittacula krameri* in Belgium. *Ibis* **2011**, *153*, 180–184. [CrossRef]
123. Reed, J.E.; McCleery, R.A.; Silvy, N.J.; Smeins, F.E.; Brightsmith, D.J. Monk parakeet nest-site selection of electric utility structures in Texas. *Landsc. Urban Plan.* **2014**, *129*, 65–72. [CrossRef]
124. Shivambu, N.; Shivambu, T.C.; Downs, C.T. Predicting the potential distribution of non-native mammalian species sold in the South African pet trade. *Diversity* **2021**, *13*, 478. [CrossRef]
125. Goetz, S.J.; Sun, M.; Zolkos, S.; Hansen, A.; Dubayah, R. The relative importance of climate and vegetation properties on patterns of North American breeding bird species richness. *Environ. Res. Lett.* **2014**, *9*, 034013. [CrossRef]
126. Illán, J.G.; Thomas, C.D.; Jones, J.A.; Wong, W.K.; Shirley, S.M.; Betts, M.G. Precipitation and winter temperature predict long-term range-scale abundance changes in Western North American birds. *Glob. Chang. Biol.* **2014**, *20*, 3351–3364. [CrossRef] [PubMed]
127. Zuckerberg, B.; Ribic, C.A.; McCauley, L.A. Effects of temperature and precipitation on grassland bird nesting success as mediated by patch size. *Conserv. Biol.* **2018**, *32*, 872–882. [CrossRef]
128. Ragusa-Netto, J. Feeding ecology of the Green-cheeked parakeet (*Pyrrhura molinae*) in dry forests in western Brazil. *Braz. J. Biol.* **2007**, *67*, 243–249. [CrossRef]
129. Butler, C.J.; Cresswell, W.; Gosler, A.; Perrins, C. The breeding biology of Rose-ringed Parakeets *Psittacula krameri* in England during a period of rapid population expansion. *Bird Study* **2013**, *60*, 527–532. [CrossRef]
130. Postigo, J.L.; Shwartz, A.; Strubbe, D.; Muñoz, A.R. Unrelenting spread of the alien monk parakeet *Myiopsitta monachus* in Israel. Is it time to sound the alarm? *Pest Manag. Sci.* **2017**, *73*, 349–353. [CrossRef]
131. Chamberlain, D.; Brambilla, M.; Caprio, E.; Pedrini, P.; Rolando, A. Alpine bird distributions along elevation gradients: The consistency of climate and habitat effects across geographic regions. *Oecologia* **2016**, *181*, 1139–1150. [CrossRef]
132. Dyer, E.E.; Cassey, P.; Redding, D.W.; Collen, B.; Franks, V.; Gaston, K.J.; Jones, K.E.; Kark, S.; Orme, C.D.L.; Blackburn, T.M. The global distribution and drivers of alien bird species richness. *PLoS Biol.* **2017**, *15*, e2000942. [CrossRef] [PubMed]
133. Banko, P.C.; Oboyski, P.T.; Slotterback, J.W.; Dougill, S.J.; Goltz, D.M.; Johnson, L.; Laut, M.E.; Murray, T.C. Availability of food resources, distribution of invasive species, and conservation of a Hawaiian bird along a gradient of elevation. *J. Biogeogr.* **2002**, *29*, 789–808. [CrossRef]
134. Sauve, D.; Friesen, V.L.; Charmantier, A. The effects of weather on avian growth and implications for adaptation to climate change. *Front. Ecol. Evol.* **2021**, *9*, 5. [CrossRef]
135. Ritchison, G. Avian reproduction: Clutch sizes, incubation, and hatching. In *In a Class of Their Own: A Detailed Examination of Avian Forms and Functions*; Ritchison, G., Ed.; Springer International Publishing: Cham, Switzerland, 2023; pp. 2285–2382. [CrossRef]
136. Department of Environmental Affairs (DEA). *Environmental Management: Biodiversity Act 2004 (Act No. 10 of 2004) Alien and Invasive Species Regulations*; National Government Gazette: Pretoria, South Africa, 2020; pp. 31–77.
137. Hernández-Brito, D.; Carrete, M.; Blanco, G.; Romero-Vidal, P.; Senar, J.C.; Mori, E.; White, T.H., Jr.; Luna, Á.; Tella, J.L. The Role of Monk Parakeets as Nest-Site Facilitators in Their Native and Invaded Areas. *Biology* **2021**, *10*, 683. [CrossRef] [PubMed]
138. Elhariri, M.; Hamza, D.; Elhelw, R.; Refai, M. Lovebirds and cockatiels risk reservoir of *Cryptococcus neoformans*, a potential hazard to human health. *J. Vet. Sci. Med. Diagn.* **2015**, *4*, 2. [CrossRef]
139. Pranty, B. The Budgerigar in Florida: Rise and fall of an exotic psittacid. *N. Am. Birds* **2001**, *55*, 389–397.
140. Pranty, B.; Epps, S. Distribution, population status, and documentation of exotic parrots in Broward County, Florida. *Flo. Field Nat.* **2002**, *30*, 111–131.
141. Toft, C.A.; Wright, T.F. *Parrots of the Wild: A Natural History of the World's Most Captivating Birds*; University of California Press: Oakland, CA, USA, 2015; p. 248.
142. van Wilgen, N.J.; Wilson, J.R.U.; Elith, J.; Wintle, B.A.; Richardson, D.M. Alien invaders and reptile traders: What drives the live animal trade in South Africa? *Anim. Conserv.* **2010**, *13*, 24–32. [CrossRef]
143. Stringham, O.C.; Lockwood, J.L. Pet problems: Biological and economic factors that influence the release of alien reptiles and amphibians by pet owners. *J. Appl. Ecol.* **2018**, *55*, 2632–2640. [CrossRef]
144. Rouget, M.; Robertson, M.P.; Wilson, J.R.U.; Hui, C.; Essl, F.; Rentería, J.L.; Richardson, D.M. Invasion debt—Quantifying future biological invasions. *Divers. Distrib.* **2016**, *22*, 445–456. [CrossRef]
145. Wilson, J.R.; Ivey, P.; Manyama, P.; Nanni, I. A new national unit for invasive species detection, assessment and eradication planning. *S. Afr. J. Sci.* **2013**, *109*, 13. [CrossRef]
146. Faulkner, K.T.; Robertson, M.P.; Rouget, M.; Wilson, J.R. A simple, rapid methodology for developing invasive species watch lists. *Biol. Conserv.* **2014**, *179*, 25–32. [CrossRef]
147. van Wilgen, B.W.; Wannenburgh, A.; Wilson, J.R. A review of two decades of government support for managing alien plant invasions in South Africa. *Biol. Conserv.* **2022**, *274*, 109741. [CrossRef]

148. Moshobane, M.C.; Mukundamago, M.; Adu-Acheampong, S.; Shackleton, R. Development of alien and invasive taxa lists for regulation of biological invasions in South Africa. *Bothalia* **2019**, *49*, 11. [CrossRef]
149. Wilson, J.R.; Faulkner, K.T.; Rahlao, S.J.; Richardson, D.M.; Zengeya, T.A.; Van Wilgen, B.W. Indicators for monitoring biological invasions at a national level. *J. Appl. Ecol.* **2018**, *55*, 2612–2620. [CrossRef]

biology

MDPI

Article

World Spread of Tropical Soda Apple (*Solanum viarum*) under Global Change: Historical Reconstruction, Niche Shift, and Potential Geographic Distribution

Yuhan Qi [1,†], Xiaoqing Xian [1,†], Haoxiang Zhao [1], Ming Yang [1], Yu Zhang [1], Wentao Yu [2,*] and Wanxue Liu [1,*]

1 State Key Laboratory for Biology of Plant Diseases and Insect Pests, Institute of Plant Protection, Chinese Academy of Agricultural Sciences, Beijing 100193, China; qyh_nwnu@163.com (Y.Q.); xianxiaoqing@caas.cn (X.X.); hx_zhao@bjfu.edu.cn (H.Z.); y1750165592@163.com (M.Y.); zhangyuu960606@163.com (Y.Z.)

2 Fujian Key Laboratory for Technology Research of Inspection and Quarantine, Technology Centre of Fuzhou Customs, Fuzhou 350001, China

* Correspondence: yuwentao@customs.gov.cn (W.Y.); liuwanxue@caas.cn (W.L.)

† These authors contributed equally to this work.

Simple Summary: *Solanum viarum* has been a widely invasive species. We aimed to understand the prevailing historical dispersal, ecological niche dynamics, and distribution patterns. The invasion history of *S. viarum* consisted of three phases: lagging (before 1980), dispersal (1980–2010), and equilibrium (2010–present). Ecological niches remained conservative. The potential geographic distribution of *S. viarum* will reach a maximum in the 2050s, SSP5–8.5, and in the future, it will migrate to higher latitudes. Global factors continue to increase this threat. The relevant quarantine authorities should take prevention and control measures.

Abstract: *Solanum viarum* has become extensively invasive owing to international trade, climate change, and land–use change. As it is classified as a quarantine weed by countries such as the U.S. and Mexico, it is critical to understand the prevailing historical dispersal, ecological niche dynamics, and distribution patterns. We reconstructed the historical invasion countries and analyzed the ecological niche shift of *S. viarum*. Using MaxEnt based on the conservativeness of ecological niches, we studied variations in the potential geographical distributions (PGDs) of *S. viarum* in ecosystems and variations in suitability probabilities along latitudinal gradients. The invasion history in six continents involved three phases: lag (before 1980), spread (1980–2010), and equilibrium (2010–present). The ecological niche remains conserved. The area of *S. viarum* PGDs had increased by 259 km^2; the PGDs will expand to reach a maximum in the 2050s, SSP5–8.5. The PGDs of *S. viarum* will migrate to higher latitudes under the same future climate scenarios. The latitudes subject to high threats range from 20° to 30° in forest and cropland ecosystems, 15.5° to 27.5° (northern hemisphere) and 33.1° to 42.8° (southern hemisphere) in grassland ecosystems, and 20° to 35° in urban ecosystems. Global change has led to an increased threat of *S. viarum* at high latitudes. These findings provide a theoretical basis to monitor and control *S. viarum*.

Keywords: climate change; invasive alien plants; *Solanum viarum*; land use change; ecological niche; ecosystem

1. Introduction

Plant invasions impact ecosystem functions, including accelerated competition for resources and reduced native plant diversity [1,2]. Global factors, such as international trade, climate change, and land–use change (LUC), can act individually or together to exacerbate the threat of plant invasion [3]. International trade in propagules, seeds, live plants, and flowers directly leads to the introduction of alien plants to new target regions, reshaping the biogeography [4]. However, international trade also plays an indirect role by altering the natural resources and social environments of exporting and importing countries [5]. Climate change may affect the ability of alien plants to survive and reproduce, enhancing adaptations to colonize otherwise unsuitable areas, such as high latitudes [6]. LUC, such as urban expansion, causes habitat fragmentation and the destruction of native ecosystems which affect the richness and frequency of alien plants [7]; it provides vacant ecological niches and creates feasible conditions for the invasion of alien plants [8]. Moreover, international trade, climate change, and LUC have resulted in land pollution and dramatic urbanization [9,10], which have a combined effect on plant invasion. Understanding the role of global changes in plant invasion is essential for ensuring sustainable development.

Solanum viarum Dunal (Solanaceae), commonly known as tropical soda apple, is a perennial seed propagated weed [11] native to South America (Brazil, Paraguay, Uruguay, and Argentina) and is now widely distributed (except Antarctica) [12]. *Solanum viarum* is spread via three main ways, namely as a popular medicinal plant by anthropic activity (international trade), as fruits by animals (e.g., birds and cattle), and naturally by vectors such as wind or water [13]. The species has been observed in pine forests, pastures, citrus plantations, and urban riparian habitats [14]. The species loss, health damage, and economic costs it causes in different ecosystems are of great concern, despite its medicinal value in India [15]. The whole plant of *S. viarum* is spiny; in forest and cropland ecosystems, it often occupies the space of neighboring plants, affecting the growth of native plants and reducing the yield of commercial crops [16]. *Solanum viarum* contains toxic alkaloid chemicals. The fruits, especially the immature ones, are more toxic and may cause respiratory distress, weakness, and even death when ingested by humans or livestock in grassland and urban ecosystems [17,18]. Moreover, injuries to humans and livestock from thorny plants often occur. As a result, the United States Department of Agriculture declared it a troublesome invasive weed in 15 states in 2012; Mexico added it to the list of quarantine pests in 2018 and Chile added it to the list of quarantine pests in category A1 in 2019 [19,20].

Reconstruction of the invasion history and identification of potential geographical distribution (PGDs) are key components of risk assessment. Invasion history reconstruction can identify the temporal phases of invasion and trends in different phases [21]. Species distribution models (SDMs) are used to predict the PGDs of invasive species prior to the occurrence of an invasion and have the advantage of requiring only occurrence records and environmental variables to model the maximum probability of species distribution [22]. They are based on the assumption of "niche conservatism". However, ecological niches may shift during the process of global change and interactions between organisms; thus, the extent of the niche shift (whether it is stable) needs to be assessed. The two–dimensional centroid shift, overlap, unfilling, and expansion (COUE) framework is a common method to assess niche shifts. Guisan et al. [23] first proposed the COUE framework and stated that it can be used for invasion studies (including plants and pathogens). Pili et al. [24] used the COUE framework to compare the realized ecological niches of four alien species and assessed the extent of niche conservatism. The assessment of the ecological niche shift can be used to validate MaxEnt model results, which guarantees the accuracy of the predictions. Additionally, most recent studies have focused on PGDs and ecological niche shifts of alien species under trade, climate change, or LUC alone [25,26], with some studies emphasizing how LUC is associated with future trade and climate change affects alien species [27–29]. However, the intricacies of PGDs and ecological niche shifts of alien species under simultaneous variation in all three have been neglected.

Solanum viarum causes serious damage to natural ecosystems and has a negative impact on ecological balance. Studying ecological niche shifts and PGDs under global change provides guidance for early warning systems. Here, we (1) reconstructed the historical invasion of *S. viarum* in three temporal phases, (2) calculated ecological niche shifts to ensure the predictions of the MaxEnt model, (3) predicted the global PGDs of *S. viarum* in different ecosystems under climate change in the 2030s and 2050s, and (4) analyzed the variation in suitability probabilities along latitudinal gradients in different ecosystems. This study will provide an ecological and theoretical reference for assessing the invasion and spread of *S. viarum*.

2. Materials and Methods

2.1. Occurrence Records and Reconstruction of Invasion

We obtained 2037 global occurrence records of *S. viarum* from the open literature and databases (Table S1). Thereafter, we removed the occurrence data that did not contain detailed geographic information. ENMTools were used to screen the occurrence data such that only one distribution point was retained within each 5 km × 5 km raster [30]. Finally, we obtained 1074 occurrence records of *S. viarum* to model the PGDs using SDMs.

To reconstruct *S. viarum* invasion, based on the available documented countries from the Centre for International Agricultural and Biological Sciences (CABI), European and Mediterranean Plant Protection Organization (EPPO), Global Biodiversity Information Facility (GBIF), and Web of Science (WOS), we confirmed that 157 countries worldwide do not have relevant databases, documentation, or detailed geographical information on the occurrence of *S. viarum*. The earliest recorded time for field sampling and documentation was assumed to be the time of invasion in each country. We used four native countries of *S. viarum* publicly listed by CABI as the starting points for the pathway. The global spatiotemporal dynamics of *S. viarum* invasions were reconstructed, considering invaded countries prior to 1970 as a category and invaded countries from 1970 to 2023 as a category (with 10–year intervals). Additionally, we investigated the dynamics of the number of invaded countries based on a time series to analyze the invasion trends of *S. viarum* in different phases. The geographical observations were set to the World Geodetic System 1984 geographic coordinate system and imported into ArcGIS.

2.2. Climate Variables

The geospatial raster layers of the climate variables were available in ArcGIS (ESRI), and 19 unique covariates were included in the SDM. These covariates for 1970–2000 were obtained from the PaleoClim database with a resolution of 5 arc minutes [31]. Future covariates were derived from the WorldClim database [32]. Three shared socioeconomic paths, SSP1–2.6, SSP2–4.5, and SSP5–8.5, were used for the 2030s and 2050s. These indicate low, medium, and high CO_2 concentrations in future scenarios based on the BCC–CSM2–MR, respectively. When two covariates had Pearson correlations, that is, $|r| > 0.8$, only the larger contributor was selected for modeling.

2.3. Land–Use Harmonization Data

Land–use harmonization data (LUH2) (https://luh.umd.edu/ (accessed on 20 June 2023)) is the latest complete set of land–use scenario data that links historical reconstructions of land use with future predictions to study variations in the suitability of *S. viarum* in different ecosystems under climate change. The dataset provided land use types, potential type shifts, and valuable agricultural reference information at a 1 km resolution per year, covering the period of 2015 to 2100 [33]. Historical reconstructions of the land–use dataset from 2015 were considered to match the near–current climate dataset [34] and used as the near–current LUC dataset. Future datasets cover scenarios SSP1–2.6, SSP2–4.5, and SSP5–8.5 and cover the 2040s and 2060s. The land use types (i.e., the ecosystems of *S. viarum*) were selected and classified as forest (consisting of primary and secondary forests), cropland, grassland (consisting of natural and managed grazing), and urban areas. We predicted the PGDs and non–PGDs of *S. viarum* for the near–current and future periods and quantified the proportion of PGDs to that ecosystem type in each grid cell, thereby obtaining information on variations in *S. viarum* PGDs in different ecosystems.

2.4. Model Settings and Evaluation

We predicted the PGDs of *S. viarum* under near–current climate and future scenarios based on occurrence records and environmental variables (including climate data and harmonized land–use data) using the MaxEnt model, which performs better than the species distribution model. As the most important parameters of the MaxEnt model, the calibration of FCs and RM can significantly improve prediction accuracy [35,36]. To simulate species responses to covariates, MaxEnt uses parameters with five characteristics: linear, L; quadratic, Q; product, P; threshold, T; and hinge, H. The range of the RM settings was increased from 0.5 to 4, and interval tests were performed using 0.5 intervals. Candidate models were built using the ENMeval package in R [37]. Finally, the model with the most significant delta values was selected.

The receiver operating characteristic (ROC) curve is based on a series of different dichotomies (cutoff values or decision thresholds), with the true positive rate (sensitivity) as the vertical coordinate and false positive rate (1–specificity) as the horizontal coordinate. The area under the curve (AUC), defined as the area under the ROC curve, is a performance indicator of accuracy [38]. The closer the AUC value is to 1.0, the higher the authenticity of the assay. The probability (p) of *S. viarum* occurrence was generated in an ASCII raster and ranged from zero to one in the model results. We classified the geographical distributions of *S. viarum* into two categories based on cutoff values: non–PGDs ($0 < p \leq 0.23$) and PGDs ($0.23 < p \leq 1$).

2.5. Ecological Niche Comparison Method

Based on the framework originally proposed by Broennimann et al. [39], the COUE framework was used to study the ecological niche shift of alien species in native and invaded countries. In the COUE framework, the ecological niche overlap index D proposed by Schoener was used to compare the ecological niches occupied by *S. viarum* in its native and invasive ranges [39]. Specifically, the ecological niche overlap index D, ranging from 0 to 1, indicates the degree of overlap between the ecological niches of alien species in native and invaded countries; large values indicate a high degree of overlap between ecological niches. Principal component analysis was performed using real occurrence records, and 8000 pseudo–absence points were randomly generated within the native and invasive ranges of *S. viarum* (points with latitudes greater than 70° were removed). The native and invasive ranges included all climatic zones in which *S. viarum* survives [40,41].

Equivalence and similarity tests were performed using the method proposed by Warren et al. [42]. The ecological niche equivalence test was used to determine whether the ecological niches of the native and invaded countries were equivalent, that is, random points were reassigned and the ecological niche overlap index D was calculated in the native and invaded countries; the iterations were repeated 100 times to ensure high confidence in rejecting the original hypothesis. If the achieved niche was within the 95% confidence range of the simulated value, the original hypothesis could not be rejected (ecological niche equivalence). The ecological niche similarity test was used to verify whether the environmental niche occupied in one range was more similar to the environmental niche occupied in a range other than that expected. If the niche achieved is greater than 95% of the simulated value after 100 iterations, the environmental conditions of the native and invaded countries are more similar than expected.

We evaluated three indicators of ecological niche shifts: stability, expansion, and unfilling [23,39]. Stability is the ability of a species to adapt to environmental change, expansion is the extension of the range of a species' ecological niche in a new habitat, and unfilling is a vacant ecological niche in a new habitat that is not occupied by a species. The ecological niche comparison method was performed in R using the ecospat package [43,44].

3. Results

3.1. Reconstruction of Invasion

The invasion history of *S. viarum* was grouped into three phases: lag (before 1980), spread (1980–2010), and equilibrium (2010–present), based on the characteristics of the temporal curve of the number of occurrence countries (Figure 1). The lag phase was a period of slow growth in the number of countries invaded by *S. viarum*, that is, the establishment phase. Prior to the 1970s, *S. viarum* was recorded as invasive in seven countries, and from 1970 to 1980, only two more countries were added. The spread phase was a period of rapidly increasing spread, during which the number of invading countries increased significantly. From 1980 to 1990, *S. viarum* invaded six additional countries, and from 1990 to 2000 *S. viarum* invaded seven additional countries. From 2000 to 2010, its spread peaked with *S. viarum* invading 14 countries. By 2010, *S. viarum* had spread to South America, southern North America, central Africa, Asia, and eastern Australia. The subsequent invasion by *S. viarum* reached a relative equilibrium. Since 2010, *S. viarum* has invaded five additional countries, for a total of forty–one countries, and is widely distributed across Asia, Europe, North America, South America, southern Africa, and Oceania. To date, 156 countries have not been affected by *S. viarum*.

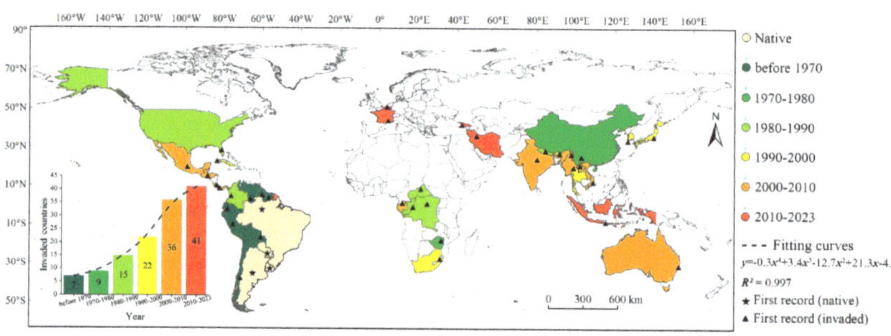

Figure 1. Reconstruction of *Solanum viarum* invasion. Six intervals are shown from green to red until 2023 (except for the native countries). The bottom left shows a composite graph of the quantities and trend of invaded countries.

3.2. Global Variation in the Ecological Niche of *Solanum viarum*

The variation in the ecological niche of *S. viarum* was verified by comparing the differences in the ecological space between native and invaded countries (Figure 2). The results showed an ecological niche overlap index D of 0.62, indicating a high degree of ecological niche overlap between native and invaded countries. The equivalence (p = 0.0297) and similarity (p = 0.0198) tests indicated that *S. viarum* occupied a similar, but not identical, ecological niche in native and invaded countries. The values for the stability, expansion, and unfilling of the ecological niche were 0.951, 0.049, and 0.037, respectively. The ecological niche stability value was the highest and was greater than 0.95, indicating a conservative niche after invasion. Overall, the ecological niche of *S. viarum* expanded to a lesser extent and remained relatively stable.

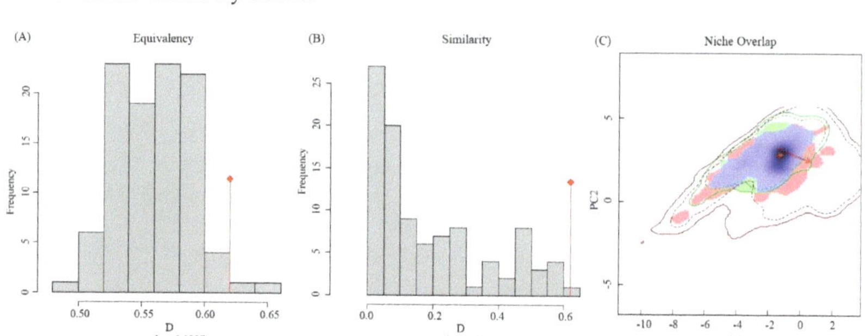

Figure 2. Global variation in the ecological niche of *Solanum viarum*. (**A**) p value of equivalency. (**B**) p value of Similarity. (**C**) Niche overlap: red and green blocks indicate expansion and unfilling areas, respectively, purple blocks indicate overlapping niches, and red arrow indicates variations in the niche centers of native and invasive areas.

3.3. Model Performance and Significant Variables

We chose the best model with significant delta values for the parameter optimization results of RM = 0.5 and FC = LQHPT (Figure S1). The AUC for the best model was 0.947 (Figure S2), indicating that the optimized MaxEnt model was excellent for predicting PGDs in *S. viarum*. Using the jackknife method (Figure S3), bio1, bio12, and bio7 were the three most significant variables. Based on an analysis of the variable contributions (Table S2), bio12, bio1, and bio3 were considered the three most significant variables. Therefore, the significant variables were three temperature factors and one precipitation factor (bio1, bio3, bio7, and bio12). Based on the response curves of the significant variables (Figure S4), the highest suitability probability value for *S. viarum* was 0.8 at an annual mean temperature of 20 °C. *Solanum viarum* had the highest suitability probability value of 0.85 when the value of isothermality was 53, annual temperature was 25 °C, and annual precipitation was approximately 1300 mm. Overall, the contribution of the LUC to the model was much smaller than that of the bioclimatic variables.

3.4. PGDs in Forest, Grassland, Cropland, and Urban Ecosystems

The extracted PGDs of *S. viarum* in forests, grasslands, croplands, and urban ecosystems were distributed over southern and southeastern Asia, western Europe, southern and southeastern North America, Central America, northern and central South America, most of Central and Southeastern Africa, and eastern Oceania based on LUCs under near–current and future climate scenarios (Figure 3). Compared with those in the near–current climate, the predicted total global cropland and urban areas increased significantly under the future climate scenario, and the predicted total forest and grassland areas decreased to some extent (Table 1). However, the PGDs of *S. viarum* showed an expanding trend regardless of the four ecosystems in which they occurred, particularly in SSP5–8.5, in which the area of *S. viarum* PGDs reached a maximum in the 2030s and 2050s.

Under the current climate scenario, the PGDs of *S. viarum* in the forest were mainly in eastern and southern Asia, southeast Asia, eastern North America, Central America, northern and eastern South America, central and southeastern Africa, and eastern Australia, covering an area of 1467.42 km^2. The geographical pattern of *S. viarum* PGDs in the forest did not change significantly under the future climate scenario, but the area of PGDs continued to increase and reached a maximum of 1748.15 km^2 under the scenario SSP1–2.6 in the 2050s. Under the current climate scenario, the PGDs of *S. viarum* in grasslands were mainly found in southern North America, central and eastern South America, southern Africa, and New Zealand, covering an area of 151.83 km^2. Under future climate scenarios, the PGDs of *S. viarum* in grasslands were mainly in southern North America, central and eastern South America, southern Africa, New Zealand in Oceania, and western Europe, with the area reaching a maximum of 159.98 km^2 under the SSP2–4.5 scenario in the 2030s. Under the current climate scenario, the PGDs of *S. viarum* in croplands were mainly in southern Asia, south–central North America, south–central South America, central Africa, southwestern Europe, and eastern Australia, covering an area of 417.49 km^2. Under future climate scenarios, the PGDs of *S. viarum* in croplands were mainly in southern and southeastern Asia, south–central North America, south–central South America, central and southeastern Africa, southwestern Europe, and eastern Australia, with a maximum area of 576.79 km^2 under the SSP5–8.5 scenario in the 2050s. Under the near–current climate scenario, the PGDs of *S. viarum* in urban areas were mainly in southeastern North America and eastern South America, covering an area of 15.25 km^2. Under future climate scenarios, the PGDs of *S. viarum* in urban areas were mainly in eastern Asia, southeastern North America, and eastern South America, with a maximum area of 30.53 km^2 under the SSP5–8.5 scenario in the 2050s. Owing to the relatively small area of urban ecosystems globally, *S. viarum* also had fewer PGDs in urban areas, but not at a lower percentage.

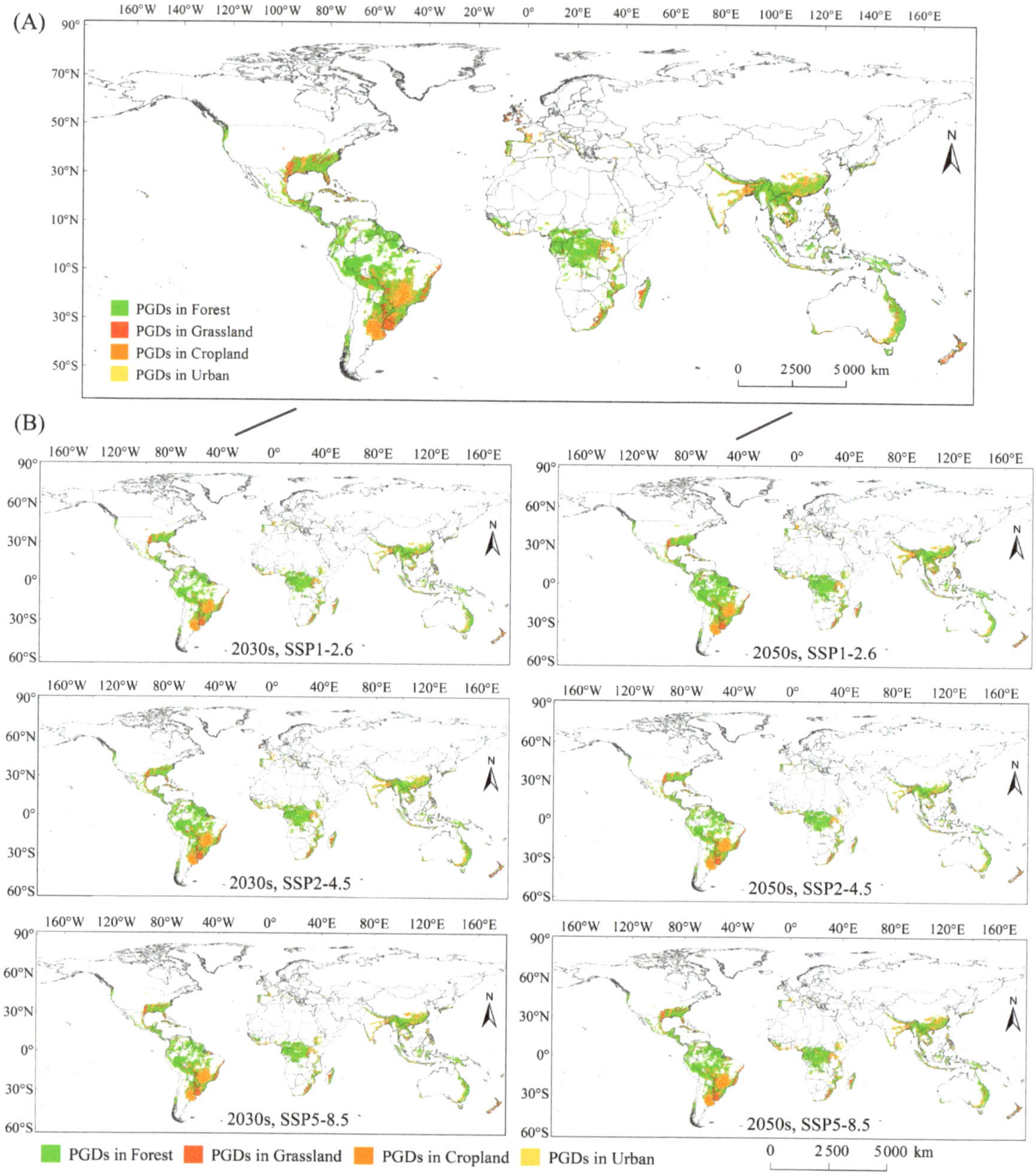

Figure 3. PGDs in forest, grassland, cropland, and urban ecosystems (extracted by land cover) under current climate scenario (**A**) and future scenarios (**B**).

3.5. Variations of PGDs in Forest, Grassland, Cropland, and Urban Ecosystems

In the forest ecosystems (Figure 4), future increases in PGDs were mainly located in northern South America (Peru, Colombia, Venezuela, Brazil, Guyana, and Suriname) and central Africa (Congo and the Democratic Republic of Congo), with more significant increases in the 2050s. The largest area of increased PGDs was 456.53 km^2 under the SSP5–8.5 scenario. Future decreases in PGDs were mainly located in eastern North America (USA), central South America (Brazil, Bolivia, Paraguay, and Argentina), central Africa (Central Africa, Gabon, Democratic Republic of Congo, and Tanzania), and southeastern Asia (China, India, Cambodia, Malaysia, and Indonesia), with more significant decreases in the 2050s. Under the SSP5–8.5 scenario, the decreases in PGDs reached the maximum area of 296.25 km^2.

Table 1. Area of PGDs in forest, grassland, cropland, and urban ecosystems under current and future climate scenarios.

Area ($\times 10^4$ km^2)	Forest			Grassland			Cropland			Urban		
	PGDs	**Total**		**PGDs**	**Total**		**PGDs**	**Total**		**PGDs**	**Total**	
Near current	1467	6069	24%	152	1450	10%	417	2010	21%	15	51	30%
2030s, SSP1–2.6	1512	6103	25%	144	1363	11%	432	1996	22%	23	73	32%
2030s, SSP2–4.5	1602	6006	27%	160	1436	11%	489	2116	23%	23	71	32%
2030s, SSP5–8.5	1475	5920	25%	148	1403	11%	502	2183	23%	24	77	31%
2050s, SSP1–2.6	1748	6123	29%	135	1267	11%	505	2026	25%	25	83	30%
2050s, SSP2–4.5	1456	5992	24%	147	1385	11%	468	2186	21%	26	81	32%
2050s, SSP5–8.5	1557	5875	27%	147	1403	10%	577	2224	26%	31	93	33%

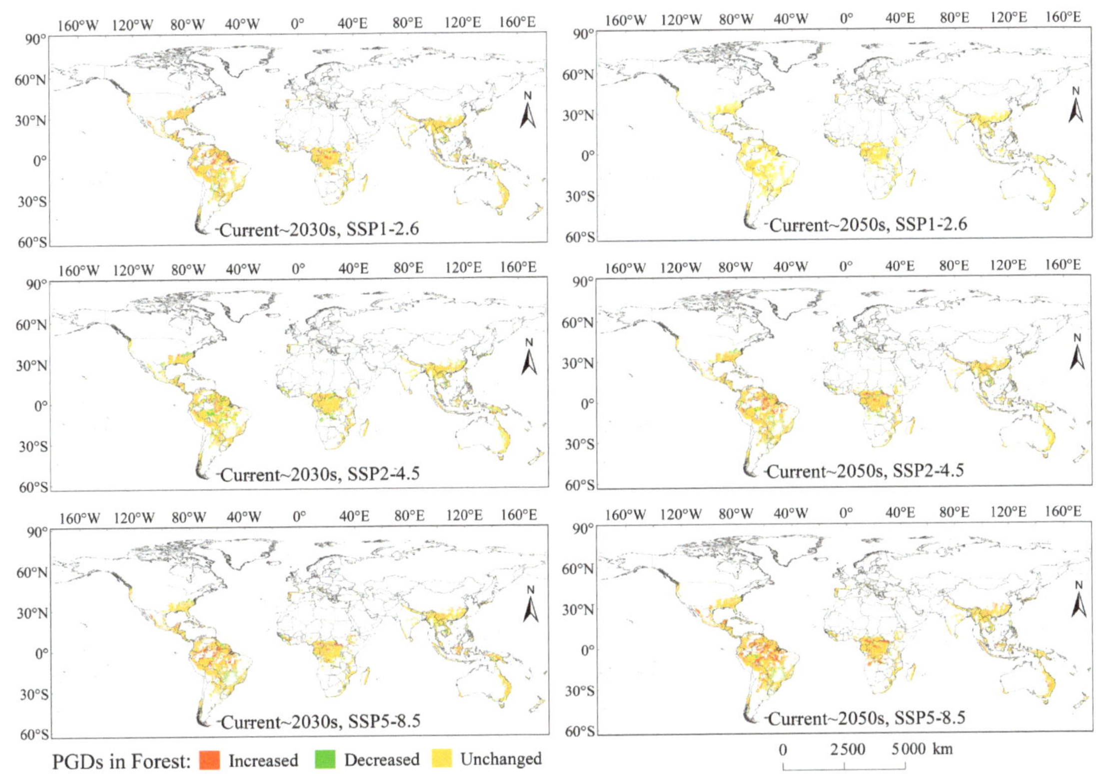

Figure 4. Variations of PGDs in forest under climate changes.

In the grassland ecosystems (Figure S5), future increases in PGDs were mainly located in southern North America (USA) and northern South America (Venezuela and Brazil), with more significant increases in the 2050s, and the largest area of increased PGDs was

51.06 km^2 under the SSP5–8.5 scenario. Future decreases in PGDs were mainly located in eastern North America (USA), central South America (Bolivia, Paraguay, and Argentina), western Europe (Ireland, UK, and France), central Africa (Central Africa and DRC), and Australia, with more significant decreases in the 2050s. The largest decreases reached an area of 71.86 km^2 under the SSP5–8.5 scenario.

In cropland ecosystems (Figure S6), future increases in PGDs were mainly located in eastern South America (Brazil and Argentina), central (Guinea, Côte d'Ivoire, Nigeria, the Democratic Republic of Congo, and Uganda), eastern Africa (Ethiopia, Tanzania, Mozambique, and South Africa), eastern (China) and southern Asia (India, Myanmar, Thailand, Philippines, and Indonesia), and Australia, with a more significant increase in the 2050s. The largest increase in PGDs reached 231.72 km^2 under the SSP5–8.5 scenario. Future decreases in PGDs were mainly observed in North America (USA), Western Europe (France, Spain, and Portugal), Central Africa (Democratic Republic of Congo and Tanzania), Southern Asia (Thailand and Vietnam), and Australia, with more significant decreases in the 2050s, and the largest area being 137.85 km^2 under the SSP2–4.5 scenario.

In urban ecosystems (Figure S7), the area of the PGDs did not decrease but continued to increase, with distributions in southeastern North America (USA), eastern South America (Brazil and Argentina), western Africa (Nigeria), eastern Asia (China), and Australia. The increase was more pronounced in the 2050s. Under SSP5–8.5, the area of increased PGDs reached its largest, at approximately 51.06 km^2.

3.6. Suitability Probabilities along Latitudinal Gradients

In the forest ecosystem (Figure 5), high suitability probabilities for *S. viarum* along latitudinal gradients were distributed within approximately 20–30° in the Northern and Southern Hemispheres. The high suitability probabilities for *S. viarum* shifted toward higher latitudes. In the Northern hemisphere, the highest suitability probabilities for *S. viarum* in the forest ecosystem were mainly located at 22.1°, whereas under the same future scenario, the highest future suitability probabilities for *S. viarum* were located at 23.5° under the SSP5–8.5 scenario in the 2050s, with an increase of 1.4°. In the Southern hemisphere, the highest suitability probabilities for *S. viarum* were mainly located at 26.5°, whereas under the same future scenario, the highest suitability probability for *S. viarum* was located at 27.9°.

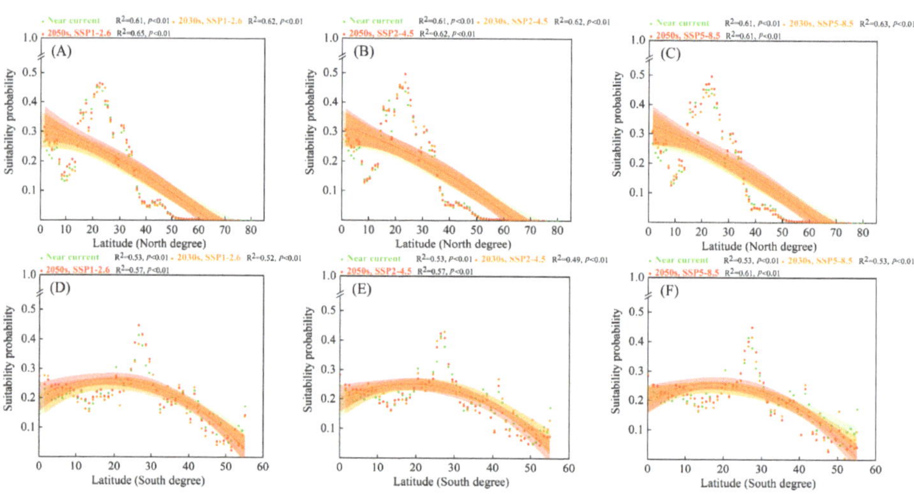

Figure 5. Suitability probabilities for *Solanum viarum* in the forest ecosystem under near–current and future climate scenarios (with 95% confidence bands via binomial fitting): (**A**,**D**) SSP1–2.6; (**B**,**E**) SSP2–4.5; (**C**,**F**) SSP5–8.5.

In the grassland ecosystem (Figure 6), the distribution of high suitability probabilities for *S. viarum* in the Northern and Southern hemispheres differed significantly with latitude. However, the high suitability probabilities of *S. viarum* shifted toward higher latitudes. In the Northern hemisphere, the highest suitability probabilities for *S. viarum* in the grassland ecosystems were mainly located from 15.5° to 27.5°, and toward 30° in the future. In the Southern hemisphere, the highest suitability probabilities for *S. viarum* were located from 33.1° to 42.8° and toward 40° in the future.

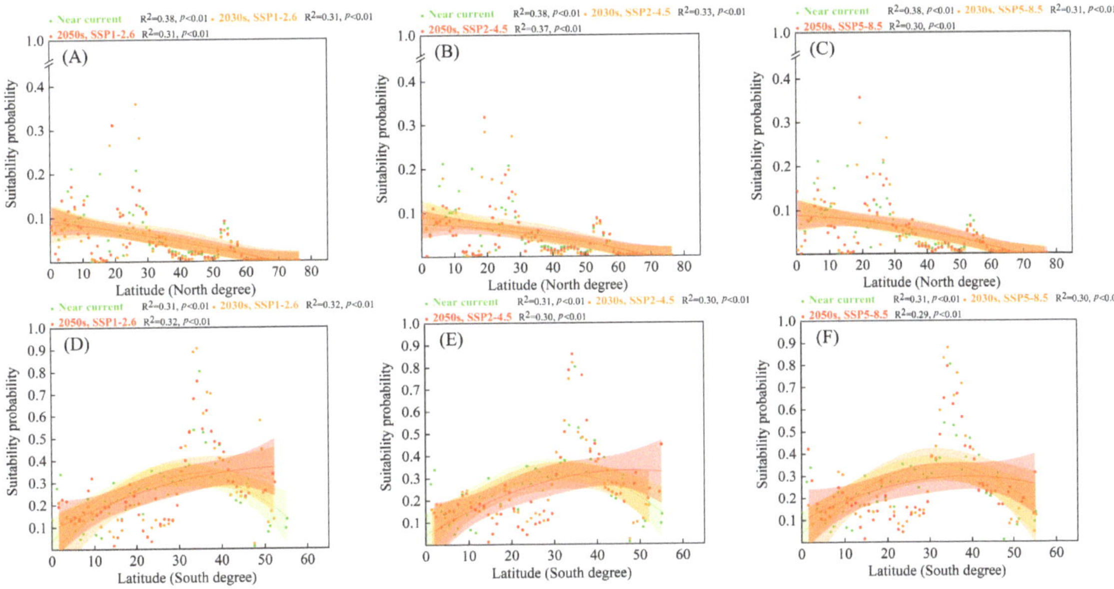

Figure 6. Suitability probabilities for *Solanum viarum* in the grassland ecosystem under near–current and future climate scenarios (with 95% confidence bands via binomial fitting): (**A**,**D**) SSP1–2.6; (**B**,**E**) SSP2–4.5; (**C**,**F**) SSP5–8.5.

In the cropland ecosystem (Figure 7), the high suitability probabilities for *S. viarum* along the latitudinal gradients were distributed within approximately 20–30° in the Northern and Southern hemispheres. The high suitability probabilities for *S. viarum* shifted toward higher latitudes. In the Northern hemisphere, the high suitability probabilities for *S. viarum* in the cropland ecosystem were mainly located at 4.5° and 23.5°, shifting toward 25° under the SSP5–8.5 scenario in the 2050s. In the Southern hemisphere, the high suitability probabilities for *S. viarum* in the cropland ecosystem were mainly located at 21.6° and 24.2°, shifting toward 25° under the SSP5–8.5 scenario in the 2050s.

In the urban ecosystem (Figure 8), high suitability probabilities for *S. viarum* along latitudinal gradients were distributed within approximately 20–35° in the Northern and Southern hemispheres. The high suitability probabilities for *S. viarum* were dense in urban ecosystems and reached higher latitudes. In the Northern hemisphere, the high suitability probability for *S. viarum* in urban ecosystems shifted from approximately 30° to greater than 30°. In the Southern Hemisphere, the high suitability probabilities for *S. viarum* shifted toward 25° and even 30°.

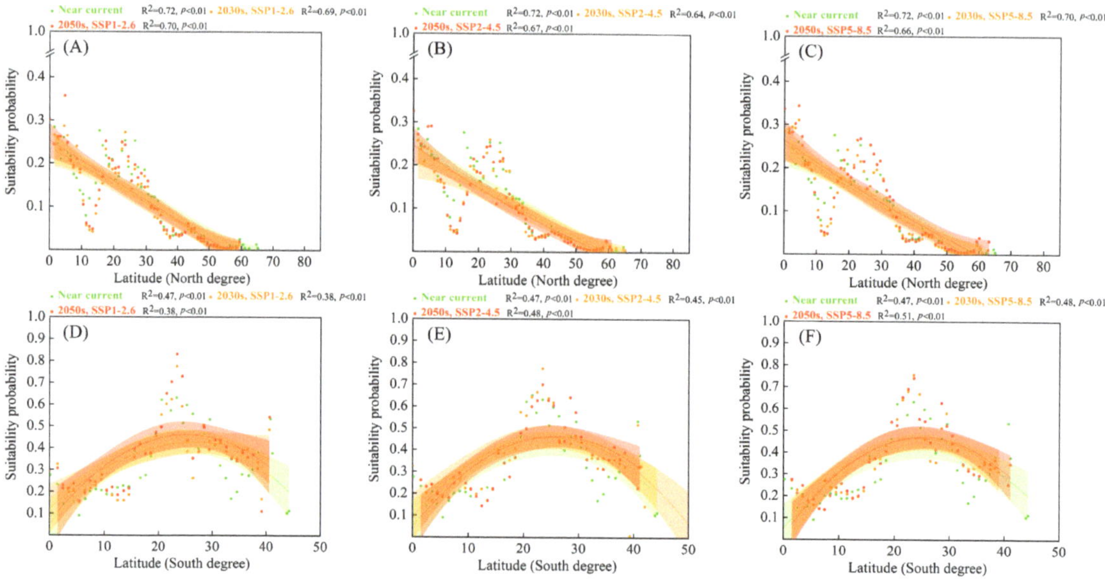

Figure 7. Suitability probabilities for *Solanum viarum* in the cropland ecosystem under near–current and future climate scenarios (with 95% confidence bands via binomial fitting): (**A**,**D**) SSP1–2.6; (**B**,**E**) SSP2–4.5; (**C**,**F**) SSP5–8.5.

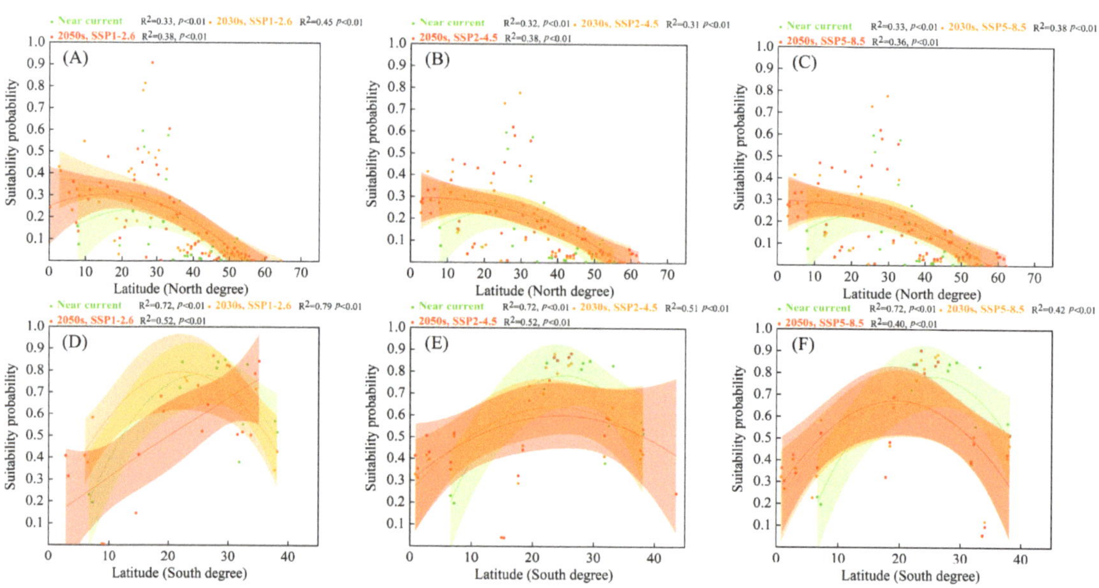

Figure 8. Suitability probabilities for *Solanum viarum* in urban ecosystem under near–current and future climate scenarios (with 95% confidence bands via binomial fitting): (**A**,**D**) SSP1–2.6; (**B**,**E**) SSP2–4.5; (**C**,**F**) SSP5–8.5.

4. Discussion

The tropical soda apple (*S. viarum*) is a troublesome invasive plant, and its global invasion has disturbed biodiversity and economics. However, this issue has received insufficient attention. Previous studies have focused primarily on its biological properties and hazards [11,15]. Few studies have examined potentially suitable areas, yet they have focused on localized areas [45]. Therefore, the present study is the first attempt to reconstruct the global invasion history of *S. viarum*, verify its ecological niche conservation, and predict PGDs using MaxEnt to provide theoretical guidance for early warning and prevention.

4.1. Reconstruction of Invasion History

The spread of invasive species is classified into three phases (lag, spread, and equilibrium), which can help determine the stage and process of invasion [21]. Our study showed that these three phases also apply to the global invasion patterns of *S. viarum*. *S. viarum* is mainly carried by wind or rivers and was mis—introduced as a potential fruit in South America and nearby (a few countries in Africa and Asia) in the lag phase (before 1980) [13]. Archaeological evidence suggests that human activity began to expand in the mid–to–late Holocene; alien plants were often transported as food by contemporary humans [46]. However, this hypothesis remains controversial [47]. The number of countries that *S. viarum* spread to in this phase was small, and no large–scale spread had yet occurred. There was an increase in the spread of *S. viarum* during the spread phase (1980–2010), with several continents invaded (except for Europe and Antarctica). Global plant trade networks emerged in the 18th and 19th centuries when alien plants began to appear frequently in non–native wild areas, particularly in the 19th century [48]. Increased trade in species, the complexity of trade networks, improved long–distance connectivity, and new trade patterns contributed to *S. viarum* invasion [49]. As a major source of pharmaceutical compounds (especially solasodine), *S. viarum* established colonies along trade routes and rapidly expanded during this phase. *Solanum viarum* successfully spread to all continents (except Antarctica) in the equilibrium phase (2010–present). Although the number of countries affected by *S. viarum* continues to increase, its spread rate has decreased significantly. Because *S. viarum* occupies an invasive ecological niche, when the ecological niche reaches saturation, a relative equilibrium state is formed [50]. However, symbiotic and antagonistic relationships may have arisen between *S. viarum* and native species, that is, potential competition is reduced or expansion is inhibited [51]. Biological control and management measures may be implemented to limit and mitigate the spread of *S. viarum* [52]. The equilibrium phase of *S. viarum* invasion does not imply that it no longer poses a threat to native organisms and ecosystems. Therefore, early warning and prevention of *S. viarum* invasion are essential to protect the integrity of ecosystems.

4.2. Ecological Niches and PGDs in Different Ecosystem

Many invasive plants expand their range to adapt to global change, and the extent to which their ecological niche changes following invasion has become a focus of biogeographic research [53]. Our study concluded that *S. viarum* has rapidly expanded its invasive range globally but largely occupies the same realized ecological niches as its native ranges, indicating ecological niche conservatism. This is inextricably linked to the biological characteristics of *S. viarum* [45]. Because of its limited drought tolerance, *S. viarum* thrives under hot and humid conditions [54]. A lack of precipitation limits photosynthesis in *S. viarum* leaves, and annual precipitation determines whether it can grow normally during the dry season [18]. The range of annual precipitation to which *S. viarum* is adapted in native countries (e.g., Brazil) is 800–1800 mm, with little variation in invaded countries. This suggests that it occupies a similar ecological space after invasion and has evolved adaptively to the environment [55]. A key insight from these results is that ecological niche variations cannot be ignored when predicting the future spatial distribution of species. SDMs are commonly used to predict expansion or invasion based on the premise that species maintain their original ecological niche when colonizing new habitats [56]. The conservative ecological niche of *S. viarum* demonstrated the excellent predictive ability of the MaxEnt model and provided assurance for the prediction of PGDs in different ecosystems.

Global changes affect the PGDs of *S. viarum* in different ecosystems. Studies have shown that climate change is already destroying parts of forests, and this species continues to migrate to higher latitudes (even polar regions) [57,58]. This is consistent with the results of this study. This trend is expected to become more pronounced under future climate change [57,59,60]. International trade, climate change, and LUC jointly influenced the PGDs of *S. viarum*; however, the first two factors were more significant. Although international trade and climate change play a dominant role in the distribution of *S. viarum* on a large spatial scale (i.e., global), LUC exerts a greater influence on a small spatial scale [61]. In the future, urban land is expected to continue to increase because of urbanization and population booms. The cropland area will also expand because an increasing number of people are in urgent need of food [62,63]. Anthropogenic activity promotes *S. viarum* invasion. This is consistent with our prediction that the areas of PGDs in croplands and urban ecosystems will continue to increase. In the future, PGDs will continue to expand under global change, and the risk assessment of *S. viarum* will contribute to an in–depth understanding of its spread dynamics.

4.3. Early Warning and Prevention Efforts

Invasive plants can spread through both natural and anthropogenic means [13]. Human activities, such as international trade and travel, help them cross geographical barriers and be introduced to distant countries where they do not occur. Globalization may exacerbate this problem [64]. A previous study suggested that unintentional human carriage is responsible for the spread of *S. viarum* in Africa [45]. Therefore, countries with suitable conditions for *S. viarum* that have not yet been invaded, particularly the Philippines, New Zealand, the United Kingdom, and Ireland, which are major agricultural and livestock countries [65], should focus on the global trade of plant and livestock vectors that can transport *S. viarum* seeds, seedlings, and fruits. Countries at risk of invasion should adopt appropriate quarantine measures to prevent the introduction of weeds. For example, in New Zealand, centralized livestock management is the main method used to control this weed [66]. Countries already invaded by *S. viarum* should intensify on–farm control using hexazinone and imazapyr herbicides [67] or release the natural enemy *Gratiana boliviana* [14]. All countries need to establish more effective quarantine processes at international borders, which are essential for interrupting the international transmission routes of *S. viarum*. Future global changes will increase the threat to countries, such as the United States, Nigeria, China, and Australia. Because there is usually a time lag between large–scale outbreaks and observations [68], these countries should establish a sound system for early monitoring and timely control.

5. Conclusions

Under global change, *S. viarum* has become widespread across six continents (except Antarctica), with 156 countries remaining uninvaded. The invasion history of *S. viarum* went through a lag phase (before 1980), spread phase (1980–2010), and equilibrium phase (2010–present). Moreover, the ecological niche of *S. viarum* remains conserved. The main ecosystems in which *S. viarum* is suitable are forests, grasslands, croplands, and urban areas; croplands and urban areas will become more suitable in the future. The PGDs in all four ecosystems continued to expand and were expected to reach a maximum in the 2050s under the SSP5–8.5 scenario. The latitudes subject to high threats range from 20° to 30° in forest and cropland ecosystems, 15.5° to 27.5° (northern hemisphere), 33.1° to 42.8° (southern hemisphere) in grassland ecosystems, and 20° to 35° in urban ecosystems. Global change has led to an increased threat by *S. viarum* at high latitudes. Analysis of the invasion history, ecological niche shifts, and future distribution dynamics provides a more reliable risk assessment for *S. viarum*.

Supplementary Materials: The following supporting information can be downloaded at: https://www.mdpi.com/article/10.3390/biology12091179/s1, Table S1: Occurrence records of *Solanum viarum* details; Table S2: Environmental variable contributions related to the distribution of *Solanum viarum*; Figure S1: Parameter optimization results of Maxent on the occurrence records of *Solanum viarum*; Figure S2: The Maxent model performance results; Figure S3: Results of the jackknife test of variable importance; Figure S4: Response curves of important variables; Figure S5: Variations of PGDs in grassland under climate changes; Figure S6: Variations of PGDs in cropland under climate changes; Figure S7: Variations of PGDs in urban under climate changes.

Author Contributions: Conceptualization, Y.Q., X.X., H.Z. and W.L.; methodology, Y.Q.; software, Y.Q.; validation, Y.Q., X.X., W.Y. and W.L.; formal analysis, Y.Q., X.X. and H.Z.; investigation, Y.Q., Y.Z. and M.Y.; resources, Y.Q., H.Z., M.Y. and Y.Z.; data curation, Y.Q.; writing—original draft preparation, Y.Q.; writing—review and editing, Y.Q., X.X. and H.Z.; visualization, Y.Q.; supervision, Y.Q., X.X. and W.L.; project administration, X.X., W.Y. and W.L.; funding acquisition, W.L. All authors have read and agreed to the published version of the manuscript.

Funding: This research was funded by the National Key R&D Program of China (grant No. 2021YFC2600400) and Technology Innovation Program of Chinese Academy of Agricultural Sciences (grant No. caascx–2022–2025–IAS).

Institutional Review Board Statement: Not applicable.

Informed Consent Statement: Not applicable.

Data Availability Statement: The data presented in this study are available in this article.

Conflicts of Interest: The authors declare no conflict of interest.

References

1. Sitzia, T.; Campagnaro, T.; Kotze, D.J.; Nardi, S.; Ertani, A. The invasion of abandoned fields by a major alien tree filters understory plant traits in novel forest ecosystems. *Sci. Rep.* **2018**, *8*, 8410. [CrossRef] [PubMed]
2. Skubel, S.A.; Su, X.; Poulev, A.; Foxcroft, L.C.; Dushenkov, V.; Raskin, I. Metabolomic differences between invasive alien plants from native and invaded habitats. *Sci. Rep.* **2020**, *10*, 9749. [CrossRef] [PubMed]
3. Vilà, M.; Pujadas, J. Land–use and socio–economic correlates of plant invasions in European and North African countries. *Biol. Conserv.* **2001**, *100*, 397–401. [CrossRef]
4. Capinha, C.; Essl, F.; Seebens, H.; Moser, D.; Pereira, H.M. The dispersal of alien species redefines biogeography in the Anthropocene. *Science* **2015**, *348*, 1248–1251. [CrossRef] [PubMed]
5. Hulme, P.E. Unwelcome exchange: International trade as a direct and indirect driver of biological invasions worldwide. *One Earth* **2021**, *4*, 666–679. [CrossRef]
6. Lu, X.; Siemann, E.; Shao, X.; Wei, H.; Ding, J. Climate warming affects biological invasions by shifting interactions of plants and herbivores. *Glob. Chang. Biol.* **2013**, *19*, 2339–2347. [CrossRef] [PubMed]
7. Brothers, T.S.; Spingarn, A. Forest fragmentation and alien plant invasion of central Indiana old–growth forests. *Conserv. Biol.* **1992**, *6*, 91–100. [CrossRef]
8. Carlson, B.S.; Rotics, S.; Nathan, R.; Wikelski, M.; Jetz, W. Individual environmental niches in mobile organisms. *Nat. Commun.* **2021**, *12*, 4572. [CrossRef] [PubMed]
9. Janssens, C.; Havlik, P.; Krisztin, T.; Baker, J.; Frank, S.; Hasegawa, T.; Leclere, D.; Ohrel, S.; Ragnauth, S.; Schmid, E.; et al. International trade is a key component of climate change adaptation. *Nat. Clim. Chang.* **2021**, *11*, 915–916. [CrossRef]
10. Dalmazzone, S.; Giaccaria, S. Economic drivers of biological invasions: A worldwide, bio–geographic analysis. *Ecol. Econ.* **2014**, *105*, 154–165. [CrossRef]
11. Randall, R.P. *A Global Compendium of Weeds*; RP Randall: South Perth, Australia, 2017.
12. Nee, M. Synopsis of *Solanum* in the New World. In *Solanaceae IV: Advances in Biology and Utilization*; Royal Botanic Gardens, Kew: Richmond, UK, 1999; pp. 285–333.
13. CABI Database. Available online: https://www.cabidigitallibrary.org/doi/10.1079/cabicompendium.50562 (accessed on 20 June 2023).
14. Medal, J.; Bustamante, N.; Overholt, W.; Diaz, R.; Stansly, P.; Roda, A.; Amalin, D.; Hibbard, K.; Gaskalla, R.; Sellers, B.; et al. Biological control of Tropical soda apple (Solanaceae) in Florida: Post–release evaluation. *Fla Entomol.* **2010**, *93*, 130–132. [CrossRef]
15. Singha, H.R.; Sinha, S.; Sinha, R.K. Karyomorphology of *Solanum viarum* Dunal–an ethnomedicinal species of Tripura. *Vegetos* **2016**, *29*, 65–68. [CrossRef]
16. Welman, W.G. The genus *Solanum* (Solanaceae) in southern Africa: Subgenus *Leptostemonum*, the introduced sections Acanthophora and Torva. *Bothalia* **2003**, *33*, 1–18. [CrossRef]
17. Bryson, C.T.; Reddy, K.N.; Byrd, J.D., Jr. Growth, development, and morphological differences among native and nonnative prickly nightshades (*Solanum* spp.) of the southeastern United States. *Invasive Plant Sci. Manag.* **2012**, *5*, 341–352. [CrossRef]

18. Medal, J.C.; Charudattan, R.; Mullahey, J.J.; Pitelli, R.A. An exploratory insect survey of tropical soda apple in Brazil and Paraguay. *Fla Entomol.* **1996**, *79*, 70–73. [CrossRef]
19. USDA Natural Resources Conservation Service. The Plants Database. Available online: http://plants.usda.gov/ (accessed on 20 June 2023).
20. EPPO Global Database. Available online: https://gd.eppo.int/taxon/SOLVI/categorization (accessed on 20 June 2023).
21. Aikio, S.; Duncan, R.P.; Hulme, P.E. Lag–phases in alien plant invasions: Separating the facts from the artefacts. *Oikos* **2010**, *119*, 370–378. [CrossRef]
22. Phillips, S.J.; Dudik, M. Modeling of species distributions with Maxent: New extensions and a comprehensive evaluation. *Ecography* **2008**, *31*, 161–175. [CrossRef]
23. Guisan, A.; Petitpierre, B.; Broennimann, O.; Daehler, C.; Kueffer, C. Unifying niche shift studies: Insights from biological invasions. *Trends Ecol. Evol.* **2014**, *29*, 260–269. [CrossRef]
24. Pili, A.N.; Tingley, R.; Sy, E.Y.; Diesmos, M.L.L.; Diesmos, A.C. Niche shifts and environmental non–equilibrium undermine the usefulness of ecological niche models for invasion risk assessments. *Sci. Rep.* **2020**, *10*, 7972. [CrossRef]
25. Urvois, T.; Auger–Rozenberg, M.A.; Roques, A.; Rossi, J.P.; Kerdelhue, C. Climate change impact on the potential geographical distribution of two invading *Xylosandrus ambrosia* beetles. *Sci. Rep.* **2021**, *11*, 1339. [CrossRef]
26. Cao, J.; Xu, J.; Pan, X.; Monaco, T.A.; Zhao, K.; Wang, D.; Rong, Y. Potential impact of climate change on the global geographical distribution of the invasive species, *Cenchrus spinifex* (*Field sandbur*, Gramineae). *Ecol. Indic.* **2021**, *131*, 108204. [CrossRef]
27. Puchalka, R.; Dyderski, M.K.; Vitkova, M.; Sadlo, J.; Klisz, M.; Netsvetov, M.; Prokopuk, Y.; Matisons, R.; Mionskowski, M.; Wojda, T.; et al. Black locust (*Robinia pseudoacacia* L.) range contraction and expansion in Europe under changing climate. *Glob. Chang. Biol.* **2021**, *27*, 1587–1600. [CrossRef] [PubMed]
28. Murray, J.V.; Stokes, K.E.; van Klinken, R.D. Predicting the potential distribution of a riparian invasive plant: The effects of changing climate, flood regimes and land–use patterns. *Glob. Chang. Biol.* **2012**, *18*, 1738–1753. [CrossRef]
29. Dyderski, M.; Zarnowiec, J.; Stebel, A.; Chmura, D. Propagule pressure and land–use are more important than climate for invasive bryophytes regional distributions. *Landsc. Ecol.* **2022**, *37*, 1871–1884. [CrossRef]
30. Warren, D.L.; Glor, R.E.; Turelli, M. ENMTools: A toolbox for comparative studies of environmental niche models. *Ecography* **2010**, *33*, 607–611. [CrossRef]
31. Brown, J.L.; Hill, D.J.; Dolan, A.M.; Carnaval, A.C.; Haywood, A.M. PaleoClim, high spatial resolution paleoclimate surfaces for global land areas. *Sci. Data* **2018**, *5*, 180254. [CrossRef]
32. Fick, S.E.; Hijmans, R.J. WorldClim 2: New 1–km spatial resolution climate surfaces for global land areas. *Int. J. Climatol.* **2017**, *37*, 4302–4315. [CrossRef]
33. Frieler, K.; Lange, S.; Piontek, F.; Reyer, C.P.O.; Schewe, J.; Warszawski, L.; Zhao, F.; Chini, L.; Denvil, S.; Emanuel, K.; et al. Assessing the impacts of 1.5 °C global warming—Simulation protocol of the Inter–Sectoral Impact Model Intercomparison Project (ISIMIP2b). *Geosci. Model Dev.* **2017**, *10*, 4321–4345. [CrossRef]
34. Hurtt, G.C.; Chini, L.; Sahajpal, R.; Frolking, S.; Bodirsky, B.L.; Calvin, K.; Doelman, J.C.; Fisk, J.; Fujimori, S.; Goldewijk, K.K.; et al. Harmonization of global land use change and management for the period 850–2100 (LUH2) for CMIP6. *Geosci. Model Dev.* **2020**, *13*, 5425–5464. [CrossRef]
35. Javidan, N.; Kavian, A.; Pourghasemi, H.R.; Conoscenti, C.; Jafarian, Z.; Rodrigo–Comino, J. Evaluation of multi–hazard map produced using MaxEnt machine learning technique. *Sci. Rep.* **2021**, *11*, 6496. [CrossRef]
36. Betts, M.G.; Yang, Z.Q.; Hadley, A.S.; Smith, A.C.; Rousseau, J.S.; Northrup, J.M.; Nocera, J.J.; Gorelick, N.; Gerber, B.D. Forest degradation drives widespread avian habitat and population declines. *Nat. Ecol. Evol.* **2022**, *6*, 709–719. [CrossRef] [PubMed]
37. Cobos, M.E.; Peterson, A.T.; Barve, N.; Osorio–Olvera, L. kuenm: An R package for detailed development of ecological niche models using Maxent. *PeerJ* **2019**, *7*, 15. [CrossRef] [PubMed]
38. Zhao, H.X.; Xian, X.Q.; Zhao, Z.H.; Zhang, G.F.; Liu, W.X.; Wan, F.H. Climate change increases the expansion risk of *Helicoverpa zea* in China According to potential geographical distribution estimation. *Insects* **2022**, *13*, 79. [CrossRef] [PubMed]
39. Broennimann, O.; Fitzpatrick, M.C.; Pearman, P.B.; Petitpierre, B.; Pellissier, L.; Yoccoz, N.G.; Thuiller, W.; Fortin, M.-J.; Randin, C.; Zimmermann, N.E.; et al. Measuring ecological niche overlap from occurrence and spatial environmental data. *Glob. Ecol. Biogeogr.* **2012**, *21*, 481–497. [CrossRef]
40. Hill, M.P.; Terblanche, J.S. Niche Overlap of Congeneric Invaders Supports a Single–Species Hypothesis and Provides Insight into Future Invasion Risk: Implications for Global Management of the *Bactrocera dorsalis* Complex. *PLoS ONE* **2014**, *9*, e90121. [CrossRef] [PubMed]
41. Hill, M.P.; Gallardo, B.; Terblanche, J.S. A global assessment of climatic niche shifts and human influence in insect invasions. *Glob. Ecol. Biogeogr.* **2017**, *26*, 679–689. [CrossRef]
42. Warren, D.L.; Glor, R.E.; Turelli, M. Environmental niche equivalency versus conservatism: Quantitative approaches to niche evolution. *Evolution* **2011**, *62*, 2868–2883. [CrossRef]
43. Di Cola, V.; Broennimann, O.; Petitpierre, B.; Breiner, F.T.; D'Amen, M.; Randin, C.; Engler, R.; Pottier, J.; Pio, D.; Dubuis, A.; et al. ecospat: An R package to support spatial analyses and modeling of species niches and distributions. *Ecography* **2017**, *40*, 774–787. [CrossRef]
44. Marchioro, C.A.; Krechemer, F.S. Reconstructing the biological invasion of Tuta absoluta: Evidence of niche shift and its consequences for invasion risk assessment. *J. Pest Sci.* **2023**, *1*, 1–15. [CrossRef]

45. Waheed, M.; Arshad, F.; Majeed, M.; Haq, S.M.; Aziz, R.; Bussmann, R.W.; Ali, K.; Subhan, F.; Jones, D.A.; Zaitouny, A. Potential distribution of a noxious weed (*Solanum viarum* Du–nal), current status, and future invasion risk based on MaxEnt modeling. *Geol. Eco. Landsc.* **2023**, *12*, 1–16. [CrossRef]
46. Boivin, N.L.; Zeder, M.A.; Fuller, D.Q.; Crowther, A.; Larson, G.; Erlandson, J.M.; Denham, T.; Petraglia, M.D. Ecological consequences of human niche construction: Examining long–term anthropogenic shaping of global species distributions. *Proc. Natl. Acad. Sci. USA* **2016**, *113*, 6388–6396. [CrossRef]
47. Fiacconi, M.; Hunt, C.O. Pollen taphonomy at Shanidar Cave (Kurdish Iraq): An initial evaluation. *Rev. Palaeobot. Palynol.* **2015**, *223*, 87–93. [CrossRef]
48. van Kleunen, M.; Essl, F.; Pergl, J.; Brundu, G.; Carboni, M.; Dullinger, S.; Early, R.; Gonzalez–Moreno, P.; Groom, Q.J.; Hulme, P.E.; et al. The changing role of ornamental horticulture in alien plant invasions. *Biol. Rev. Camb. Philos. Soc.* **2018**, *93*, 1421–1437. [CrossRef] [PubMed]
49. Dong, R.; Dong, B.C.; Fu, Q.Y.; Yang, Q.; Dai, Z.C.; Luo, F.L.; Gao, J.Q.; Yu, F.H.; van Kleunen, M. Cultivated alien plants with high invasion potential are more likely to be traded online in China. *Ecol. Appl.* **2023**, *11*, e2811. [CrossRef] [PubMed]
50. Zhu, G.-P.; Redei, D.; Kment, P.; Bu, W.-J. Effect of geographic background and equilibrium state on niche model transferability: Predicting areas of invasion of *Leptoglossus occidentalis*. *Biol. Invas.* **2014**, *16*, 1069–1081. [CrossRef]
51. Lu, X.; He, M.; Ding, J.; Siemann, E. Latitudinal variation in soil biota: Testing the biotic interaction hypothesis with an invasive plant and a native congener. *ISME J.* **2018**, *12*, 2811–2822. [CrossRef]
52. Mullahey, J.J.; Shilling, D.G.; Mislevy, P.; Akanda, R.A. Invasion of tropical soda apple (*Solanum viarum*) into the U.S.: Lessons learned. *Weed Technol.* **1998**, *12*, 733–736. [CrossRef]
53. Lustenhouwer, N.; Parker, I.M. Beyond tracking climate: Niche shifts during native range expansion and their implications for novel invasions. *J. Biogeogr.* **2022**, *49*, 1481–1493. [CrossRef]
54. Overholt, W.A.; Diaz, R.; Markle, L.; Medal, J.C. The effect of *Gratiana boliviana* (Coleoptera: Chrysomelidae) herbivory on growth and population density of tropical soda apple (*Solanum viarum*) in Florida. *Biocontrol Sci. Technol.* **2010**, *20*, 791–807. [CrossRef]
55. Mullahey, J.J.; Nee, M.; Wunderlin, R.P.; Delaney, K.R. Tropical soda apple (*Solanum viarum*): A new weed threat in subtropical regions. *Weed Technol.* **1993**, *7*, 783–786. [CrossRef]
56. Pearman, P.B.; Guisan, A.; Broennimann, O.; Randin, C.F. Niche dynamics in space and time. *Trends Ecol. Evol.* **2008**, *23*, 149–158. [CrossRef] [PubMed]
57. Pecl, G.T.; Araujo, M.B.; Bell, J.D.; Blanchard, J.; Bonebrake, T.C.; Chen, I.C.; Clark, T.D.; Colwell, R.K.; Danielsen, F.; Evengard, B.; et al. Biodiversity redistribution under climate change: Impacts on ecosystems and human well–being. *Science* **2017**, *355*, 9. [CrossRef] [PubMed]
58. Greenwood, S.; Ruiz–Benito, P.; Martinez–Vilalta, J.; Lloret, F.; Kitzberger, T.; Allen, C.D.; Fensham, R.; Laughlin, D.C.; Kattge, J.; Bonisch, G.; et al. Tree mortality across biomes is promoted by drought intensity, lower wood density and higher specific leaf area. *Ecol. Lett.* **2017**, *20*, 539–553. [CrossRef] [PubMed]
59. Engler, R.; Randin, C.F.; Thuiller, W.; Dullinger, S.; Zimmermann, N.E.; Araujo, M.B.; Pearman, P.B.; Le Lay, G.; Piedallu, C.; Albert, C.H.; et al. 21st century climate change threatens mountain flora unequally across Europe. *Glob. Chang. Biol.* **2011**, *17*, 2330–2341. [CrossRef]
60. Parmesan, C.; Yohe, G. A globally coherent fingerprint of climate change impacts across natural systems. *Nature* **2003**, *421*, 37–42. [CrossRef] [PubMed]
61. Sirami, C.; Caplat, P.; Popy, S.; Clamens, A.; Arlettaz, R.; Jiguet, F.; Brotons, L.; Martin, J.L. Impacts of global change on species distributions: Obstacles and solutions to integrate climate and land use. *Glob. Ecol. Biogeogr.* **2017**, *26*, 385–394. [CrossRef]
62. Thebo, A.L.; Drechsel, P.; Lambin, E.F. Global assessment of urban and peri–urban agriculture: Irrigated and rainfed croplands. *Environ. Res. Lett.* **2014**, *9*, 9. [CrossRef]
63. Bren d'Amour, C.B.; Reitsma, F.; Baiocchi, G.; Barthel, S.; Guneralp, B.; Erb, K.H.; Haberl, H.; Creutzig, F.; Seto, K.C. Future urban land expansion and implications for global croplands. *Proc. Natl. Acad. Sci. USA* **2017**, *114*, 8939–8944. [CrossRef]
64. Bonnamour, A.; Gippet, J.M.W.; Bertelsmeier, C. Insect and plant invasions follow two waves of globalisation. *Ecol. Lett.* **2021**, *24*, 2418–2426. [CrossRef]
65. Corson, M.S.; Mondiere, A.; Morel, L.; van der Werf, H.M.G. Beyond agroecology: Agricultural rewilding, a prospect for livestock systems. *Agric. Syst.* **2022**, *199*, 103410. [CrossRef]
66. West, G.G.; Dean, M.G. The Use of Livestock to Control Weeds in New Zealand Forests. *FRI Bull.* **1990**, *155*, 128–132.
67. Mislevy, P.; Martin, F.G. Tropical soda apple control as influenced by frost and herbicides. *Soil Crop Sci. Soc. Fla Proc.* **1999**, *58*, 107–109.
68. Duncan, R.P. Time lags and the invasion debt in plant naturalisations. *Ecol. Lett.* **2021**, *24*, 1363–1374. [CrossRef] [PubMed]

biology

MDPI

Article

Analysis of the Distribution Pattern of *Phenacoccus manihoti* in China under Climate Change Based on the Biomod2 Model

Yumeng Huang [1], Tong Li [1], Weijia Chen [2], Yuan Zhang [1], Yanling Xu [1], Tengda Guo [1], Shuping Wang [2], Jingyuan Liu [2,*] and Yujia Qin [1,*]

1 Key Laboratory of Surveillance and Management for Plant Quarantine Pests, Ministry of Agriculture and Rural Affairs, College of Plant Protection, China Agricultural University, Beijing 100193, China; hym921604@163.com (Y.H.); islitong@163.com (T.L.); zy18335462253@163.com (Y.Z.); xyl0402@163.com (Y.X.); guotd7911@163.com (T.G.)

2 Technical Center for Animal, Plant and Food Inspection and Quarantine of Shanghai Customs, Shanghai 200135, China; weijia_0801@163.com (W.C.); wangshuping_2014@126.com (S.W.)

* Correspondence: yyuan_108@sina.com (J.L.); qinyujia@cau.edu.cn (Y.Q.)

Simple Summary: The cassava mealybug (*Phenacoccus manihoti*) represents a significant quarantine pest affecting cassavas. This study seeks to elucidate the patterns of the suitable habitat and distribution centers of cassava mealybugs in China, as well as their dynamic ecological niche in invaded areas under the influence of climate change. Climate factors have primarily concentrated suitable habitats for cassava mealybugs in the southern regions of China, with minimal observed changes and a slight northward expansion. The dynamic ecological niche is also projected to expand slightly. Based on these predictions, relevant quarantine agencies can implement measures to prevent the introduction and spread of cassava mealybugs.

Abstract: The changing global climate has significantly impacted the spread of plant pests. The cassava mealybug (*Phenacoccus manihoti*) is among the most dangerous quarantine pests affecting cassavas worldwide, causing substantial losses in agricultural production and food security across several regions. Although China is currently free of the cassava mealybug, its proximity to affected countries and extensive trade with these regions necessitate a detailed understanding of the pest's distribution pattern and dynamic ecological niche changes. Using the Biomod2 model, we selected two historical climate scenarios and two future climate scenarios (SSP1-2.6 and SSP5-8.5) to investigate the distribution patterns, potential habitats, distribution centers, and dynamic ecological niches of cassava mealybugs in China. Key environmental variables influencing the distribution were identified, including bio4, bio8, bio12, bio18, and bio19. The potential habitat of cassava mealybugs is mainly located in several provinces in southern China. In the future, the suitable habitat is projected to expand slightly under the influence of climate change, maintaining the overall trend, but the distribution center of suitable areas will shift northward. Dynamic ecological niche prediction results indicate the potential for further expansion; however, the ecological niches may be unequal and dissimilar in the invaded areas. The predictions could serve as a valuable reference for early warning systems and management strategies to control the introduction of cassava mealybugs.

Keywords: *Phenacoccus manihoti*; climate change; Biomod2; geographical potential distribution; distribution center; ecological niche

1. Introduction

The invasion of alien species presents a significant global threat. The spread of these species is driven by anthropogenic factors such as international travel and trade, as well as natural factors. Insects are the most prevalent invasive species, substantially contributing to ecological disruption. Global climate warming is a major natural factor contributing to increased invasions of harmful pests [1–3]. Climate change influences pest distribution, posing significant threats to crops, as discussed in some literature [4–6]. Due to global warming, the spread of invasive crop pests is projected to increase, impacting a wider range of cultivated crops [6,7]. The latest phase of the Coupled Model Intercomparison Project (CMIP6) offers significant advancements over CMIP5. These advancements include the transition from simple ocean–atmosphere-coupled models to sophisticated multi-layer models that integrate the atmosphere, land surface, ocean, and sea ice. Additionally, CMIP6 incorporates a comprehensive array of human-activity-related factors such as carbon cycles, vegetation cover, population dynamics, economic growth, education, urbanization, and technological progress, all of which interact with the climate system's physical, biological, and chemical processes [8]. Among the four Shared Socioeconomic Pathways (SSPs) provided by CMIP6 [9], SSP1-2.6 represents a lower forcing scenario, where radiative forcing is predicted to reach 2.6 W/m^2 by 2100, potentially keeping the temperature increase below 2.0 °C relative to pre-industrial levels. SSP2-4.5 and SSP3-7.0 correspond with medium forcing scenarios, with radiative forcing reaching 4.5 W/m^2 and 7.0 W/m^2, respectively, by 2100. SSP5-8.5 represents the highest forcing scenario, with radiative forcing reaching 8.5 W/m^2 by 2100 [10].

The cassava mealybug (CMB), *Phenacoccus manihoti* Matile-Ferrero (Hemiptera: Pseudococcidae), poses a significant threat to cassava production worldwide and is recognized as one of the most critical quarantine pests of cassava crops [11–13]. It is an oligophagous pest that can feed on and damage a variety of economic crops and ornamental plants, including cassava, rubber cassava, sweet potato, soybean, citrus, tomato, pepper, and poinsettia. However, it prefers to proliferate on cassava and rubber cassava. Other hosts merely serve as temporary feeding sites during population dispersal and do not result in an economic impact or the production of offspring on these plants [14]. Originating from South America [15], this mealybug was introduced into Africa in the early 1970s and subsequently into Asia (Thailand) in 2009, emerging as the most destructive pest of cassava globally [16,17]. This insect reproduces parthenogenetically, resulting in exclusively female offspring. Consequently, a single nymph or adult can initiate an infestation. Under optimal conditions, an adult female can lay between 200 and 600 eggs within ovisacs [18,19]. The pest feeds on plant tissues as both nymphs and adults, causing apical buds and leaves to curl and develop poorly, reducing the internode length and weakening stems and roots. This frequently results in yield reductions exceeding 50%, with severe infestations potentially causing total crop loss and plant death [20,21]. Additionally, the pest can induce sooty mold, further exacerbating plant damage. Currently, the cassava mealybug is widespread and severely affects major cassava-producing regions worldwide [12,13]. In Africa, it spreads at an alarming rate of 150 km per year compared with fewer than 30 km per year for other invasive hemipterans [22]. Control of the cassava mealybug primarily relies on chemical and biological methods [15]. However, the number of registered pesticides for this pest is very limited, and biological control is often ineffective due to the low diversity of natural enemies in cassava-growing areas, leading to unstable predator populations. Therefore, finding new, effective, and environmentally friendly strategies to control the cassava mealybug has become a critical issue for the development of the cassava industry, both domestically and internationally [23].

Originally native to South America, the cassava mealybug (CMB) has now spread across multiple countries and regions in Asia, Africa, and South America [24]. Introduced to the African continent in the early 1970s, it has become naturalized throughout sub-Saharan Africa. First detected in Asia in 2008 in Thailand, it quickly spread across cassava-growing regions and into neighboring countries, including Indonesia, raising significant concerns about its potential spread to other countries [21,25].

A pest risk analysis (PRA) involves evaluating biological, scientific, and economic evidence to determine whether an organism is a pest, if it should be regulated, and the strength of any phytosanitary measures required. It is a crucial component of plant quarantine [26,27]. Species distribution models (SDMs) can predict species distribution areas and are widely used in fields such as endangered species conservation [28], conservation area planning [29], invasive species control [30], and predicting the impact of climate change on species distributions [31]. Currently, there are numerous distributional prediction models, with various principles, assumptions, algorithms, scopes of application, and prediction performances. Thuiller [32] suggests that each model has its own advantages and limitations due to the different principles and algorithms, and model performance is not stable with a change in input data. Rather than relying on a single model, it is preferable to establish a model ensemble, integrating results from multiple models to enhance prediction accuracy by using a comprehensive result as the output of the ensemble. Biomod2, an R package used to predict the potential geographical distribution of organisms, provides a range of tools for model selection, model fusion, and model evaluation that can be used to predict the spatial distribution of biodiversity under different environmental variables [33]. Similar to the maximum entropy model (MaxEnt) and other biological diversity models that correlate species distribution data with environmental factors to build models, Biomod2 can integrate generalized linear models (GLMs), random forests (RFs), MaxEnt, and artificial neural networks (ANNs), so it has the advantage of being more accurate and reliable than a single model [33–35]. Biomod2 is widely used to predict species distributions, assess the impact of climate change on biodiversity, and explore the relationship between biodiversity and environmental factors [36,37]. Numerous studies on potential geographic distributions based on the Biomod2 model have been conducted, both domestically and internationally. Liu et al. [38] used Biomod2 to predict the potential distribution of *Teinopalpus aureus* Mell. De Oliveira et al. [39] modeled the ecological niche and evaluated the remaining suitable habitat areas for the occurrence of five potentially invasive species of freshwater decapods in South America. Shabani et al. [40] used Biomod2 to investigate changes in the invasion risk of the Dubas bug under current and future climate change scenarios. In this study, we used Biomod2 integrated with an optimal combination model and ArcGIS to predict the shifts in potential suitable habitats, distribution centroids, and the distribution pattern of *P. manihoti*, aiming to provide a scientific basis for effective early warning management under climate change.

2. Materials and Methods

2.1. Distribution Records of Phenacoccus manihoti

We collected distribution records of *P. manihoti* from the Center for Agriculture and Bioscience International (CABI) [24], the Global Biodiversity Information Facility (GBIF) [41], and related literature [13,42]. Initially, we removed specimen coordinates or incorrect distribution data, then used the spThin package (version 0.2.0) to retain only one distribution point per 5 min × 5 min raster to avoid a spatial sample bias [43]. Ultimately, we obtained 102 occurrence records of *P. manihoti* for the model setting (Figure 1). We imported the distribution records into the model using the BIOMOD_FormatingData function in the Biomod2 package (version 4.2-4), controlling the generation of pseudo-absences (PA) using the PA.nb.rep, PA.nb.absences, and PA.strategy parameters. In this study, we generated pseudo-absence points by randomly generating 10 instances and repeating the generation 3 times.

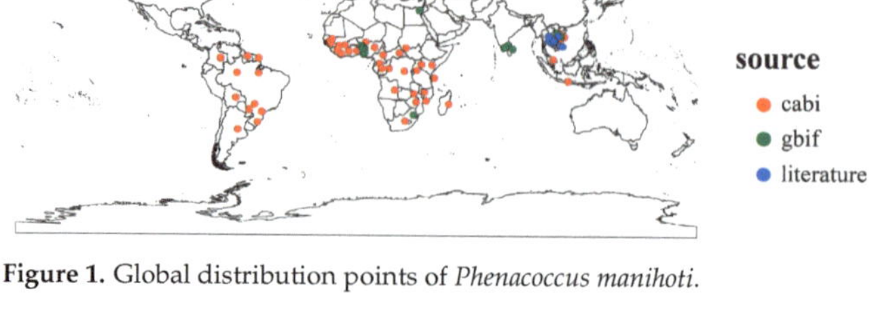

Figure 1. Global distribution points of *Phenacoccus manihoti*.

2.2. Environmental Variables

We obtained elevation (ele) data as well as historical (1970–2000) and future (2040–2060) (the 2050s) data for 19 bioclimatic variables (Table A1) in Worldclim version 2.1 (https://www.worldclim.org, accessed on 2 May 2024), with a spatial resolution of 5 arc_min. For future environmental data, we selected two different Shared Socioeconomic Pathways (SSP1-2.6 and SSP5-8.5) for 2050 in the BCC-CSM2-MR model. To ensure a better response of the variables used in the modelling, a pre-run of the MaxEnt model was constructed in R using the terra package and the SDMtune package to pre-screen the 20 climatic variables, excluding those with a contribution of fewer than 5 [44]. Correlation analyses were then performed to exclude variables with a strong covariance ($|r| > 0.8$) to avoid excessive correlations between variables [45].

2.3. Model Settings and Assessment

A single model was modelled for each species using the BIOMOD_ModelingOptions and BIOMOD_Modeling functions (This package was developed by Wilfried Thuiller and his research team at the Laboratoire d'Écologie Alpine (LECA) and Centre National de la Recherche Scientifique (CNRS) in Grenoble, France.), resulting in the selection of 11 single models. These were the generalized linear model (GLM) [46], generalized additive model (GAM) [47], generalized boosting model (GBM) [48] (pp. 103–105), classification tree analysis (CTA) [49], artificial neural network (ANN) [50], surface range envelop (SRE) [51], flexible discriminant analysis (FDA) [52], multiple adaptive regression splines (MARS) [53] (pp.1–15), random forest (RF) [54], eXtreme Gradient Boosting (XGBOOST) [55], and maximum entropy model (MaxEnt) [56]. The parameters of the MaxEnt model were tuned using the ENMeval package. The model was controlled by tuning the parameters CV.nb.rep, prevalence, and metric.eval. The model was run 10 times, with the training-set data proportion set to 75% [57]. The prevalence was set to 0.5, meaning the sum of the weights of the distributed points equaled the sum of the weights of the pseudo-absence points. The true skill statistic (TSS) measures the net prediction success rate of a model using distribution points and pseudo-distribution points, with values ranging from −1 to 1. The closer the TSS is to 1, the higher the model's prediction accuracy; the closer it is to −1, the more the model tends to be stochastic [58,59]. The receiver operating characteristic (ROC) curve is chosen as a criterion to evaluate the model performance [58]. Models utilized for integrated model building are required to have ROC values higher than 0.8 and TSS values exceeding 0.7 [55,60].The BIOMOD_EnsembleModelling function utilizes the em.algo parameter to choose from six methods for integrated modelling. The options include EMmean (based on the probability mean of the selected model), EMmedian (based on the median probability of the selected model), EMcv (based on the probability coefficient of variation in the selected model, i.e., SD/mean), EMci (based on the confidence interval around the probability mean of the selected model), EMca (based on the binary voting of the selected model), and EMwmean (based on the evaluation score), with the selected model's probabilities weighted by the evaluation scores obtained at the time of model construction. Among these, EMmedian is less influenced by extreme anomalies compared with EMmean, yet it demands greater computational resources and memory. EMmean assesses uncertainty rather than the likelihood of species occurrence. In EMwmean, the better the previous model evaluation, the greater the weight in the later integration.

2.4. Analysis and Mapping

The results of the BIOMOD_EnsembleForecasting function were evaluated for the TSS and ROC scores, and the appropriate integrated model was selected. The Jenks natural break classification (NBC) method was used to classify suitable areas [61,62]. After developing the integrated model, the BIOMOD_EnsembleForecasting function projected the integrated model onto the environmental variables for the potential geographic distribution prediction and the distribution maps were normalized using ArcGIS 10.2. According to the results from Biomod2, the thresholds for unsuitable, low-suitable, medium-suitable, and high-suitable areas for cassava mealybugs were 0–91.42829, 91.42829–232.03453, 232.03453–402.96921 and 402.96291-1000, respectively. The data from different climatic conditions were reclassified and assigned to determine future changes in suitable habitats for the four climatic scenarios compared with historical conditions using the raster calculator. Additionally, the SDM toolbox v 2.5 plug-in was used to calculate the center of cassava mealybug populations for each period and to plot the trajectory of the center of mass migration.

2.5. Ecological Niche Analysis

We assessed the degree of geographical matching between invasive and native habitats by overlapping the corresponding binary maps obtained for each temporal scenario. The environmental principal component analysis (PCA-env) in the R package ecospat (version 3.5.1) [63,64] was used for this study. We then employed niche equivalency and background similarity tests to evaluate the observed similarities between each pairwise comparison [64,65]. Schoener's D metric determined the extent of the ecological niche overlap, with index ranges from 0 (no niche overlap) to 1 (complete niche overlap), where larger values indicated a higher overlap rate of ecological niches. To understand the importance of niche overlap in geographic areas, we conducted niche equivalency and similarity tests [64]. The ecological niche equivalency test was conducted by comparing the ecological niche overlap (D) values for current and future climate scenarios with the overlap of the null distribution. The niche similarity test assessed if the ecotopes of the two entities being compared were more similar (or different) than expected, considering the surrounding environmental conditions across the geographic area [64,66]. We randomly repeated the tests 1000 times, and the null hypotheses of niche equivalence and similarity were accepted if the observed niche values (D) were significantly higher than the overlap value from the null distribution ($p < 0.05$) [64,67].

3. Results

3.1. Model Validation

After 10 repeated runs of the single and combined models, and based on the TSS and ROC evaluations, RF had the highest TSS and ROC values, and SRE had the lowest TSS and ROC values, although its TSS was still >0.7 and its AUC was >0.85. All 11 models met the screening criteria (Figure 2) and were, therefore, used to construct the integrated model.

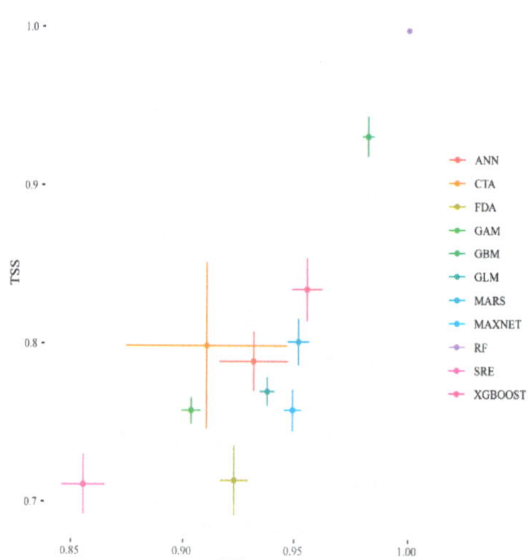

Figure 2. TSS and ROC evaluations of each single model of *Phenacoccus manihoti*.

For the integrated modeling, this study chose the EMwmean method, based on previous research [68]. The ROC value of this integration approach was 0.970 and the TSS value was 0.807, indicating high predictive accuracy suitable for the outcome prediction.

3.2. Environmental Variables

Using the SDMtune package, the key environmental factors influencing the distribution of *P. manihoti* were identified as bio4 (temperature seasonality), bio8 (mean temperature of the wettest quarter), bio12 (annual precipitation), bio18 (precipitation of the warmest quarter), and bio19 (precipitation of the coldest quarter). These five factors were, therefore, included in the final model.

Environmental factor response curves reflect the effect of environmental variables on habitat suitability [69]. According to the response curve of cassava mealybugs (Figure 3), bio4 (temperature seasonality) varied greatly within a small range of values, bio8 (mean temperature of the wettest quarter) varied significantly at 25 °C, bio12 (annual precipitation) varied significantly around 4000 mm, bio18 (precipitation of the warmest quarter) showed high suitability at about 500 mm, and bio19 (precipitation of the coldest quarter) showed significant changes and higher suitability at about 1000 mm. The response curves indicated that cassava mealybugs prefer warm and stable temperature conditions.

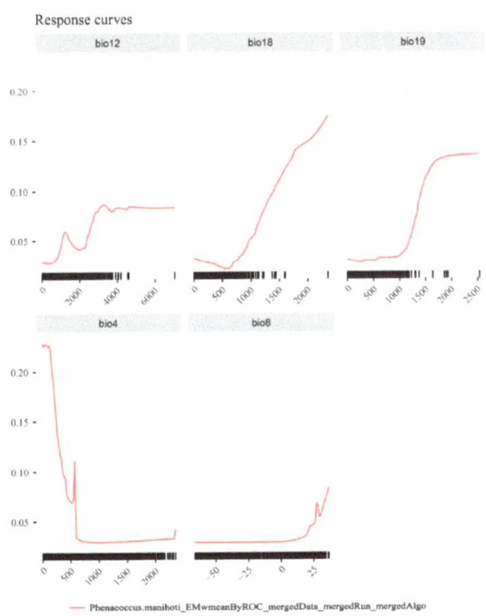

Figure 3. Response curves of *Phenacoccus manihoti.*

3.3. Potential Geographical Distribution in China

The potential geographical distribution of cassava mealybugs worldwide under historical climatic conditions was predicted by the integrated model using EMwmean (Figure 4). In China, cassava mealybugs were generally found in fewer areas. High-suitable areas mainly existed in parts of southeastern Tibet and southern Yunnan, as well as in Taiwan, most of Hainan province, and scattered areas in Guangdong and Guangxi. The suitable area expanded inland from high-suitable areas and could be found in provinces such as Yunnan, Guizhou, and Fujian.

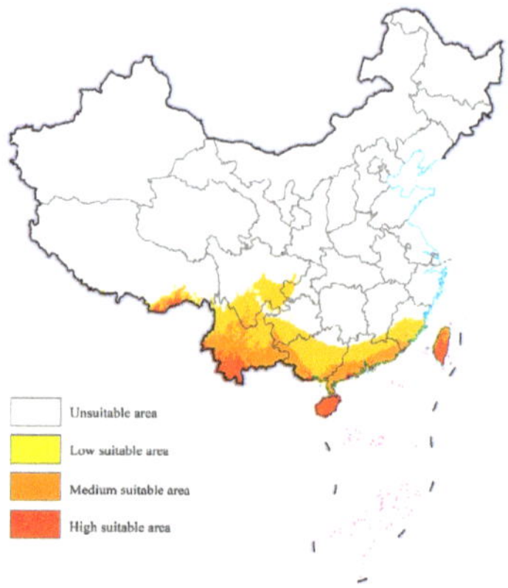

Figure 4. Potential geographic distribution of *Phenacoccus manihoti* in China under a historical climatic scenario.

Under both future scenarios, the suitable areas for cassava mealybugs within China were still mainly concentrated in the aforementioned provinces. However, a small number of low-suitable areas were also observed in Chongqing, Hubei, Jiangsu, and Zhejiang from the projections (Figure 5). The overall change in the area of each suitability class was not significant. Under future climate scenarios, the low-suitable area could reach further than 700,000 square kilometers and the medium-suitable area could increase from more than 400,000 square kilometers to more than 500,000 square kilometers (Table 1).

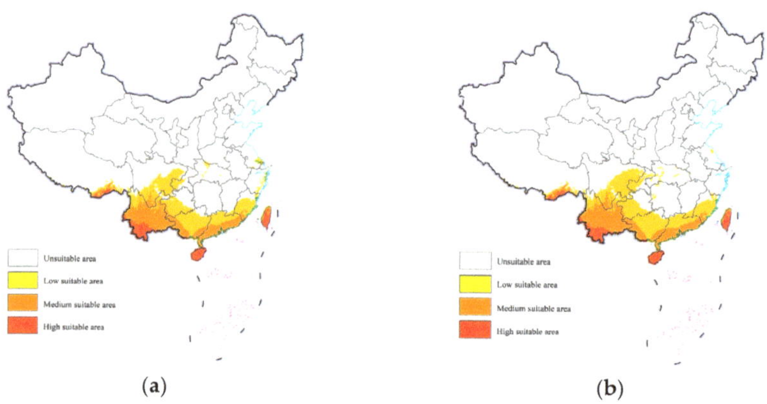

Figure 5. Potential geographic distribution of *Phenacoccus manihoti* in the world under future climatic scenarios: (**a**) SSP1-2.6; (**b**) SSP5-8.5.

Table 1. Potential suitable habitats area of *Phenacoccus manihoti* under different climate scenarios in China.

Climate Scenario	**Area (×10^4 km^2)**		
	Low-Suitable Area	**Medium-Suitable Area**	**High-Suitable Area**
Historical climatic scenario	645,489.1268	411,300.4540	186,008.8396
SSP1-2.6 (2050)	719,153.7544	510,826.8700	178,921.3524
SSP5-8.5 (2050)	766,730.3972	506,906.1324	192,417.7376

3.4. Shifts in the Distribution Centroid and Distribution Pattern

In this study, we projected the shift in the center point of distribution of cassava mealybugs to reveal the invasion trend of this species in China. The shifts in the potential suitable habitats of *P. manihoti* in China under historical conditions and different future climate scenarios in 2050 are shown in Figure 6. Within China, there was no significant change in its suitable area. Specifically, under the SSP1-2.6 climate scenario, the suitable habitats of the cassava mealybug mainly showed a small expansion trend, but the expansion was limited to the vicinity of the original suitable area. Additionally, there was a small-scale contraction trend in the habitable area in southern Jiangxi. Figure 6 shows the centroid shift trend of cassava mealybugs in China under two scenarios (SSP1-2.6 and SSP5-8.5) from the current climate to 2050. In both climate scenarios, the distribution center of cassava mealybugs showed a northward-spreading trend, but both were concentrated in Guizhou, with a greater northward shift in the SSP5-8.5 climate scenario. The small shift in the distribution center also indicated that the change in the suitable area was not significant under different future climate scenarios.

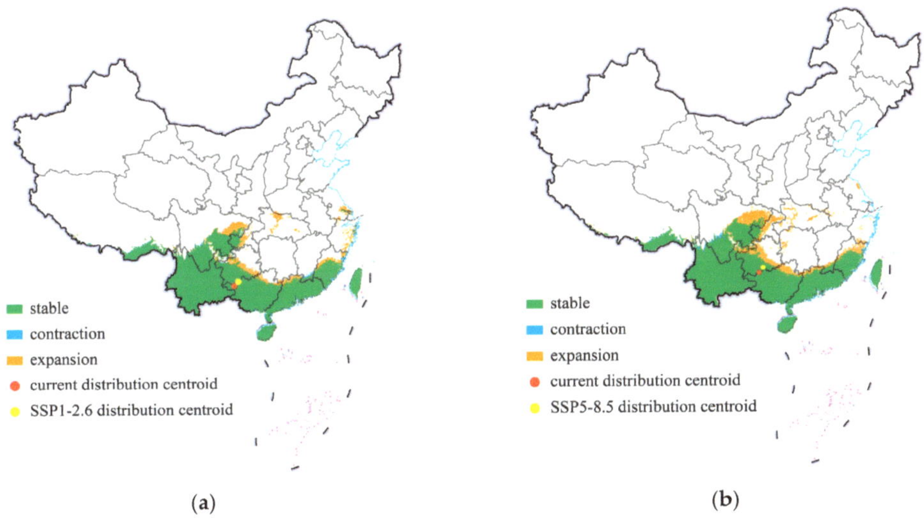

Figure 6. Shifts in potential suitable habitats of *Phenacoccus manihoti* in China under different climate scenarios: (**a**) SSP1-2.6; (**b**) SSP5-8.5.

3.5. Ecological Niche of Phenacoccus manihoti

The geographical ranges of *P. manihoti* can be determined, in part, by its fundamental niche, i.e., the environmental conditions under which the cassava mealybug can survive. The variation in the ecological niche of *P. manihoti* was verified by comparing the differences in the ecological space between native and invaded countries (Figure 7). To assess the indicators of ecological niche dynamics, a set of environments was available for both the current and future scales; if the environment was only occupied in the future scale, this represented ecological niche expansion. Similarly, environments were considered to exhibit ecological niche stability if they were occupied in both the current and future scales, and ecological niche vacancies if they were used in the current scale and were available, but not yet utilized, in the future scale [70]. Niche analyses contrasting current and future climatic conditions showed that the ecological niche overlap index (D) was 0.56, revealing that the cassava mealybug had currently occupied all of its niches and was likely to expand on a small scale. The niche similarity ($p = 0.15984 > 0.05$) and equivalence ($p = 0.1938 > 0.05$) tests for the null hypothesis were not rejected, suggesting that the ecological niches of the cassava mealybug were not statistically equivalent and the entities to be compared were similar. We hypothesized that the ecological niche occupied by the cassava mealybug changed to some extent during its invasion.

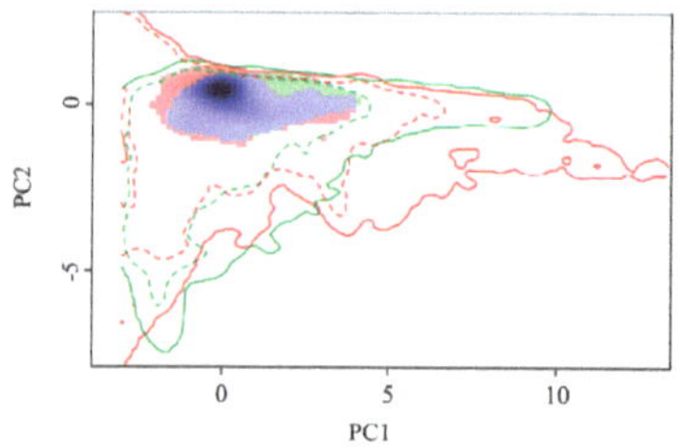

Figure 7. Global variation in the ecological niche of *Phenacoccus manihoti*: red and green blocks indicate expansion and unfilled areas, respectively; purple blocks indicate overlapping niches.

4. Discussion

The cassava mealybug is widely distributed worldwide and has caused significant economic losses, particularly in regions where cassava is extensively cultivated such as Africa and India [16,71]. Although it has not yet been introduced into China, it has attracted national attention due to its presence in neighboring countries such as Thailand, Laos, and Myanmar. In 2011, it was included in the Catalogue of Quarantine Pests for Import Plants of the People's Republic of China. This study is the first to use the Biomod2 model to investigate the potential geographical distribution, patterns, and dynamic ecological niches of the cassava mealybug in China, with the aim of providing a theoretical basis for their monitoring in the country.

4.1. Optimization and Selection of the Biomod2 Model

Integrated models have the advantage of being more accurate and reliable than single models [32]. Different single models have their own characteristics; for example, the MaxEnt model is specifically designed to simulate the distribution of species with only available data and is suitable for situations with a greater amount of missing data [72], while SRE is suitable for situations with a greater number of variables and can adapt to reduce the complexity of the model [73]. However, most studies using integrated models only employ the default parameters of each single model [74]. In this study, the MaxEnt model in the model combination was optimized for parameter tuning and then included in the integrated model construction. Further model optimization and parameter tuning can be carried out in future studies. Judging by the ROC and TSS, all the models in this study had better results when run separately. When integrating the models, the ROC value of this model integration approach was 0.970 and the TSS value was 0.807, indicating that our prediction results were better. Biomod2 performed well in terms of the accuracy of the model prediction results, but it also had some drawbacks such as requiring more data preprocessing steps, a longer runtime, and larger storage-space requirements at runtime, making it less simple and easy to understand compared with a single MaxEnt [34]. Overall, Biomod2 combines a variety of models to provide better predictions with fewer input distributions [75–77].

4.2. Significant Environmental Variables

We identified five key environmental variables affecting cassava mealybugs using the SDMtune package. These were bio4 (temperature seasonality), bio8 (mean temperature of the wettest quarter), bio12 (annual precipitation), bio18 (precipitation of the warmest quarter), and bio19 (precipitation of the coldest quarter). Previous studies have shown that the main factors affecting the survival and population growth of cassava mealybugs are temperature and rainfall, which align with our findings [13]. It was found that the optimal temperature for development was 27 °C, with a significant number of deaths occurring when the temperature was lower than 15 °C or higher than 33 °C [13,19]. The response curves of the environmental variables also showed a better response for bio8 above 25 °C, while the best response value of bio4 was smaller, indicating less tolerance for temperature changes such as completing physiological activities like overwintering. The cassava mealybug is mainly found in countries near the equator where temperatures do not vary much, according to existing distribution sites. Additionally, its growth is constrained by precipitation: dry areas, years, and seasons favor outbreaks [13]. The response curves for bio12, bio18, and bio19 also showed that cassava mealybugs depend on a moderate amount of precipitation, which is in line with their own physiological characteristics. However, our selection of relevant bioclimatic variables considered only 19 climatic and altitudinal variables and did not account for anthropogenic impacts, soil environment, and the pest's physiological data. The potential habitat areas predicted by some bioclimatic variables may not be suitable for the species to live in reality, which is a limitation of the classical bioclimatic variables used in this study.

4.3. Distribution Pattern and Ecological Niches

Many models and methods have been used to predict the potential geographical distribution of cassava mealybugs. Soroush et al. [13] reported recent distribution records and estimated the climatic suitability for their regional spread using a CLIMEX distribution model. Chen et al. [78] analyzed the potential geographical distribution of the cassava mealybug in China, using three different ecological modeling methods (Maxent, ENFA, and Mahalanobis typicalis) to predict its potential distribution. Tania et al. [71] used CLIMEX to study its global distribution and its threat to food security for the poor. Zhou et al. [14] used the GARP ecological niche model and related environmental data combined with the interpolation and overlay functions of ArcGIS to predict that the potential habitat of cassava mealybugs in China would mainly be concentrated in the area south of 25° N. The prediction results of various scholars have varied slightly due to different factors such as the models and variables used, but the overall trends were similar. In this study, the potential geographic distribution of cassava mealybugs was studied using ensembled models. The results showed that this species would mainly be concentrated in the southern region of China in provinces near Laos, Myanmar, and other countries, as well as in Taiwan and Yunnan, where the temperature is higher. With climate change, the expansion of the cassava mealybug's potential habitat was not obvious, and we speculated that temperature would limit its further expansion under natural conditions. Within China, the range of its habitat was not wide; moreover, cassava cultivation within China is mainly concentrated in some cities in the south [79]. In this study, the climate scenarios of SSP1-2.6 and SSP5-8.5 showed different degrees of expansion of the habitats and the distribution center of the suitable areas shifted northwards to varying degrees, suggesting that climate change would affect the distribution pattern of cassava mealybugs. It is crucial to prevent the northward invasion of mealybugs. According to the results of the dynamic ecological niche analysis, cassava mealybugs could invade from their origin to other places, but their ecological niche would change to some extent. We deduced that this may be due to the species' physiological characteristics, host distribution, or the existence of interspecific competition and these aspects should be focused on in subsequent studies. Therefore, it is important to monitor the dynamics of cassava mealybug occurrences in new areas in a timely manner and to enhance its small-scale prediction.

4.4. Management and Monitoring

The cassava mealybug is currently absent from China, but it is a quarantine pest of concern in China. The first instar of this insect is very active and can crawl from infected plants to neighboring healthy plants, and can also spread with wind, animals, or equipment. Long-distance transmission is mainly through the transport of infected host plants and their products. This was the case for the cassava mealybug's spread from South America to Africa and then to Thailand [80] (pp. 209–235). Therefore, as a neighboring country to Thailand, Laos, Myanmar, and other countries with a wide distribution of cassava mealybugs, and one that imports large quantities of cassava products from many neighboring countries [20], China should strengthen border quarantine measures and strictly prevent and control the importation of cassava mealybugs.

5. Conclusions

In this study, we used the Biomod2 model to identify the key environmental variables affecting the distribution of cassava mealybugs by selecting historical scenarios and two future climatic scenarios, SSP1-2.6 and SSP5-8.5. We studied the distribution pattern of the cassava mealybugs' potential habitat, distribution centers, and dynamic ecological niches in invaded areas within China. Our findings indicated that bio4 (temperature seasonality), bio8 (mean temperature of the wettest quarter), bio12 (annual precipitation), bio18 (precipitation of the warmest quarter), and bio19 (precipitation of the coldest quarter) had a non-negligible effect on the distribution of cassava mealybugs. Under the historical scenario, cassava mealybugs were mainly distributed in a few provinces in southern China, with highly suitable areas concentrated in Yunnan, Taiwan, and Hainan. Under the two future climatic scenarios, SSP1-2.6 and SSP5-8.5, the distribution of suitable areas for cassava mealybugs slightly expanded, but the overall trend remained the same. The center of distribution for suitable areas moved northwards, but remained concentrated in Guizhou. Dynamic ecological niche projections showed further potential for expansion, but the ecological niches in invaded areas was statistically not equivalent in ecological terms and did not indicate that the ecological niches were more similar than expected. This study is the first to use the Biomod2 model to investigate the potential habitats, habitability patterns, and dynamic ecological niches of the cassava mealybug in China, providing theoretical support for their monitoring in the country.

Author Contributions: Conceptualization, Y.Q. and J.L.; methodology, Y.H., T.L., W.C., Y.Z., Y.X., T.G. and S.W.; software, Y.H. and T.L.; writing—original draft preparation, Y.H.; writing—review and editing, Y.H., J.L. and Y.Q. All authors have read and agreed to the published version of the manuscript.

Funding: This research was funded by the National Key Research and Development Program of China (2023YFC2605200), the National Natural Science Foundation of China (32202288), and Beijing Natural Science Foundation (6232023).

Institutional Review Board Statement: Not applicable.

Informed Consent Statement: Not applicable.

Data Availability Statement: The original contributions presented in the study are included in the article, further inquiries can be directed to the corresponding authors.

Acknowledgments: We thank all members of the Plant Quarantine and Invasion Biology Laboratory of China Agricultural University (CAUPQL) for their comments and support.

Conflicts of Interest: The authors declare no conflicts of interest.

Appendix A

Table A1. Environmental variables (bold indicates variables used for modeling).

Types of Variables	Environmental Variables	Code
Climatic factors	Annual mean temperature	bio1
	Mean diurnal range	bio2
	Isothermality	bio3
	Temperature seasonality	bio4
	Max temperature of warmest month	bio5
	Min temperature of coldest month	bio6
	Temperature annual range	bio7
	Mean temperature of wettest quarter	bio8

Table A1. *Cont.*

Types of Variables	Environmental Variables	Code
Climatic factors	Mean temperature of driest quarter	bio9
	Mean temperature of warmest quarter	bio10
	Mean temperature of coldest quarter	bio11
	Annual precipitation	bio12
	Precipitation of wettest month	bio13
	Precipitation of driest month	bio14
	Precipitation seasonality	bio15
	Precipitation of wettest quarter	bio16
	Precipitation of driest quarter	bio17
	Precipitation of warmest quarter	bio18
	Precipitation of coldest quarter	bio19
Terrain factors	Elevation	elev

References

1. Clavijo, M.A.; Effah, E.; Najar-Rodriguez, A. Ecological aspects of volatile organic compounds emitted by exotic invasive plants. *Front. Ecol. Evol.* **2023**, *11*, 1059125. [CrossRef]
2. Turbelin, A.J.; Malamud, B.D.; Francis, R.A. Mapping the global state of invasive alien species: Patterns of invasion and policy responses. *Glob. Ecol. Biogeogr.* **2017**, *26*, 78–92. [CrossRef]
3. Cornelissen, B.; Neumann, P.; Schweiger, O. Global warming promotes biological invasion of a honey bee pest. *Glob Chang. Biol.* **2019**, *25*, 3642–3655. [CrossRef] [PubMed]
4. Bebber, D.P.; Ramotowski, M.A.T.; Gurr, S.J. Crop pests and pathogens move polewards in a warming world. *Nat. Clim. Chang.* **2013**, *3*, 985–988. [CrossRef]
5. Estay, S.; Lima, M.; Labra, F. Predicting insect pest status under climate change scenarios: Combining experimental data and population dynamics modelling. *J. Appl. Entomol.* **2009**, *133*, 491–499. [CrossRef]
6. Ziska, L.H.; Blumenthal, D.M.; Runion, G.B.; Hunt, E.R., Jr.; Diaz-Soltero, H. Invasive species and climate change: An agronomic perspective. *Clim. Chang.* **2011**, *105*, 13–42. [CrossRef]
7. Thomson, L.J.; Macfadyen, S.; Hoffmann, A.A. Predicting the effects of climate change on natural enemies of agricultural pests. *Biol. Control* **2010**, *52*, 296–306. [CrossRef]
8. Zhao, Z.C.; Luo, Y.; Huang, J.B. The detection of the CMIP5 climate model to see the development of the CMIP6 earth system models. *Clim. Chang. Res.* **2018**, *14*, 643–648.
9. Fan, Z.F.; Zhou, B.J.; Ma, C.L.; Gao, C.; Han, D.N.; Chai, Y. Impacts of climate change on species distribution patterns of Polyspora sweet in China. *Ecol. Evol.* **2022**, *12*, e9516. [CrossRef] [PubMed]
10. Zhang, L.; Liu, H.L.; Zhang, H.X.; Chen, Y.F.; Zhang, L.W.; Kawashaer, K.; Dilxadam, T.; Zhang, Y.M. Potential distribution of three types of ephemeral plants under climate changes. *Front. Plant Sci.* **2022**, *13*, 1035684. [CrossRef]
11. Bellotti, A.C.; Smith, L.; Lapointe, S.L. Recent advances in cassava pest management. *Annu. Rev. Entomol.* **1999**, *44*, 343–370. [CrossRef] [PubMed]
12. Bellotti, A.C.; Braun, A.R.; Arias, B.; Castillo, J.A.; Guerrero, J.M. Origin and management of neotropical cassava arthropod pests. *Afr. Crop Sci. J.* **1994**, *2*, 407–417.
13. Parsa, S.; Kondo, T.; Winotai, A. The cassava mealybug (*Phenacoccus manihoti*) in Asia: First records, potential distribution, and an identification key. *PLoS ONE* **2012**, *7*, e47675. [CrossRef] [PubMed]
14. Zhou, X.; Zhang, Q.E.; Pan, X.B.; He, Y.Y.; Huang, Y.; Yan, J. Study on the risk analysis of *Phenacoccus manihoti* Matile-Ferrero, on imported host products. *J. Environ. Entomol.* **2014**, *36*, 298–304.
15. Löhr, B.; Varela, A.M.; Santos, B. Exploration for natural enemies of the cassava mealybug, *Phenacoccus manihoti* (Homoptera: Pseudococcidae), in South America for the biological control of this introduced pest in Africa. *Bull. Entomol. Res.* **1990**, *80*, 417–425. [CrossRef]
16. Neuenschwander, P. Biological control of the cassava mealybug in Africa: A review. *Biol. Control* **2001**, *21*, 214–229. [CrossRef]
17. Gupta, A.; Sampathkumar, M.; Mohan, M.; Shylesha, A.N.; Venkatesan, T.; Shashank, P.R.; Dhanyakumar, O.; Ramkumara, P. Assessing adverse impact of the native biological control disruptors in the colonies of the recent invasive pest *Phenacoccus manihoti* Matile-Ferrero (Hemiptera: Pseudococcidae) in India. *Glob. Ecol. Conserv.* **2021**, *32*, e01878. [CrossRef]
18. Iheagwam, E.U. The influence of temperature on increase rates of the cassava mealybug *Phenacoccus manihoti* Mat.-Ferr. (Homoptera, Pseudococcidae). *Rev. Zool. Afr.* **1981**, *95*, 959–967.
19. Lema, K.M.; Herren, H.R. The influence of constant temperature on population growth rates of the cassava mealybug, *Phenacoccus manihoti*. *Entomol. Exp. Appl.* **1985**, *38*, 165–169. [CrossRef]
20. Nwanze, K.F. Relationships between cassava root yields and crop infestations by the mealybug, *Phenacoccus manihoti*. *Int. J. Pest Manag.* **1982**, *28*, 27–32. [CrossRef]

21. Muniappan, R.; Shepard, B.M.; Watson, G.W.; Carner, G.R.; Rauf, A.; Sartiami, D.; Hidayat, P.; Afun, J.V.K.; Goergen, G.; Rahman, A.Z. New records of invasive insects (hemiptera: Sternorrhyncha) in Southeast Asia and West Africa. *J. Agric. Urban Entomol.* **2009**, *26*, 167–174. [CrossRef]
22. Wang, Y.S.; Tian, H.; Wam, F.H.; Zhang, G.F. Species-specific COI primers for rapid identification of a globally significant invasive pest, the cassava mealybug *Phenacoccus manihoti* Matile-Ferrero. *J. Integr. Agric.* **2019**, *18*, 1042–1049. [CrossRef]
23. Geng, Y.; Chen, Q.; Liang, X.; Liu, Y.; Wu, C.L.; Chen, Y.H. Flavonoid synthesis pathway-mediate defense mechanism of Cassava Against *Phenacoccus manihoti*. *Chin. J. Trop. Crop.* **2023**, *44*, 2449–2460.
24. CABI Database. Available online: https://www.cabidigitallibrary.org/doi/10.1079/cabicompendium.40173 (accessed on 1 May 2024).
25. Wu, S.A.; Wang, Y.P. Precaution of cassava mealybug, *Phenacoccus manihoti* Matile-Ferrero in China. *J. Environ. Entomol.* **2011**, *33*, 122–125. [CrossRef]
26. Kahn, R.P. A concept of pest risk analysis. *EPPO Bull.* **1979**, *9*, 119–130. [CrossRef]
27. Liu, H.; Zhao, S.Q.; Ma, C. Discussion on risk analysis of agricultural plant pests in China. *China Plant Prot.* **2023**, *43*, 84–86, 43.
28. Eyre, A.C.; Briscoe, N.J.; Harley, D.K.; Lumsden, L.F.; McComb, L.B.; Lentini, P.E. Using species distribution models and decision tools to direct surveys and identify potential translocation sites for a critically endangered species. *Divers. Distrib.* **2022**, *28*, 700–711. [CrossRef]
29. Villero, D.; Pla, M.; Camps, D.; Ruiz-Olmo, J.; Brotons, L. Integrating species distribution modelling into decision-making to inform conservation actions. *Biodivers. Conserv.* **2017**, *26*, 251–271. [CrossRef]
30. Rodríguez-Rey, M.; Consuegra, S.; Börger, L.; Garcia de Leaniz, C. Improving Species Distribution Modelling of freshwater invasive species for management applications. *PLoS ONE* **2019**, *14*, e0217896. [CrossRef] [PubMed]
31. Huang, Y.M.; Zhang, G.L.; Fu, W.D.; Zhang, Y.; Zhao, Z.H.; Li, Z.H.; Qin, Y.J. Impacts of climate change on climatically suitable regions of two invasive *Erigeron* weeds in China. *Front. Plant Sci.* **2023**, *14*, 1238656. [CrossRef] [PubMed]
32. Thuiller, W. Biomod: Optimizing predictions of species distributions and projecting potential future shifts under global change. *Glob. Chang. Biol.* **2003**, *9*, 1353–1362. [CrossRef]
33. Thuiller, W.; Lafourcade, B.; Engler, R.; Araújo, M.B. BIOMOD—A platform for ensemble forecasting of species distributions. *Ecography* **2009**, *32*, 369–373. [CrossRef]
34. Luo, M.; Wang, H.; Lv, Z. Evaluating the performance of species distribution models Biomod2 and MaxEnt using the giant panda distribution data. *Chin. J. Appl. Ecol.* **2017**, *28*, 4001–4006. [CrossRef]
35. Hao, T.; Elith, J.; Lahoz-Monfort, J.J.; Guillera-Arroita, G. Testing whether ensemble modelling is advantageous for maximising predictive performance of species distribution models. *Ecography* **2020**, *43*, 549–558. [CrossRef]
36. Dhami, B.; Bhusal, A.; Adhikari, B.; Miya, M.S.; Maharjan, S.K.; Neupane, D.; Adhikari, H. Habitat Suitability and Conflict Zone Mapping for the Blue Bull (*Boselaphus tragocamelus*) across Nepal. *Animals* **2023**, *13*, 937. [CrossRef] [PubMed]
37. Čengić, M.; Rost, J.; Remenska, D.; Janse, J.H.; Huijbregts, M.A.; Schipper, A.M. On the importance of predictor choice, modelling technique, and number of pseudo-absences for bioclimatic envelope model performance. *Ecol. Evol.* **2020**, *10*, 12307–12317. [CrossRef] [PubMed]
38. Liu, Y.; Zhang, X.; Zong, S. Prediction of the Potential Distribution of *Teinopalpus aureus* Mell, 1923 (Lepidoptera, Papilionidae) in China Using Habitat Suitability Models. *Forests* **2024**, *15*, 828. [CrossRef]
39. De Oliveira, L.R.; Brito, G.; Gama, M.; Ovando, X.M.C.; Anastácio, P.; Cardoso, S.J. Non-Native Decapods in South America: Risk Assessment and Potential Impacts. *Diversity* **2023**, *15*, 841. [CrossRef]
40. Shabani, F.; Kumar, L.; Al Shidi, R.H.S. Impacts of climate change on infestations of Dubas bug (*Ommatissus lybicus* Bergevin) on date palms in Oman. *PeerJ* **2018**, *6*, e5545. [CrossRef] [PubMed]
41. GBIF Database. Available online: https://www.gbif.org/.
42. Lu, H.; Lu, B.Q.; Liu, H.; Guo, A.P.; Tang, J.H.; Su, H. Risk analysis of regional pests in Hainan. *Chin. J. Trop. Agric.* **2020**, *40*, 38–42.
43. Zhang, Y.; Qin, Y.J.; Zhang, Y.; Fu, W.D.; Zhang, G.L.; Li, Z.H. Integration-based modelling for predicting the habitability zones of the invasive alien plant, *Solanum carolinense*, in North America. *J. Plant Prot.* **2022**, *49*, 1434–1439. [CrossRef]
44. Vignali, S.; Barras, A.G.; Arlettaz, R.; Braunisch, V. SDMtune: An R package to tune and evaluate species distribution models. *Ecol. Evol.* **2020**, *20*, 11488–11506. [CrossRef] [PubMed]
45. Qin, Y.; Wang, C.; Zhao, Z.; Pan, X.; Li, Z. Climate change impacts on the global potential geographical distribution of the agricultural invasive pest, *Bactrocera dorsalis* (Hendel) (Diptera: Tephritidae). *Clim. Chang.* **2019**, *155*, 145–156. [CrossRef]
46. Nelder, J.A.; Wedderburn, R.W. Generalized Linea Models. *J. R. Stat. Soc. Ser. A Stat. Soc.* **1972**, *135*, 370–384. [CrossRef]
47. Hastie, T.; Tibshirani, R. Generalized additive models: Some applications. *J. Am. Stat. Assoc.* **1987**, *82*, 371–386. [CrossRef]
48. Ridgeway, G. The state of boosting. *Comput. Sci. Stat.* **1999**, *31*, 172–181.
49. Breiman, L.; Friedman, J.; Olshen, R.A.; Stone, C.J. *Classification and Regression Trees*; Routledge: New York, NY, USA, 1984; pp. 103–105. [CrossRef]
50. Friedman, J.; Hastie, T.; Tibshirani, R. Additive logistic regression: A statistical view of boosting (With discussion andare joinder by the authors). *Ann. Statist.* **2000**, *28*, 337–407. [CrossRef]
51. Busby, J.R. Bioclim: A bioclimate analysis and prediction system. *Plant Prot. Q.* **1991**, *6*, 8–9.
52. Hastie, T.; Tibshirani, R.; Buja, A. Flexible Discriminant Analysis by Optimal Scoring. *J. Am. Stat. Assoc.* **1994**, *89*, 1255–1270. [CrossRef]

53. Friedman, J.H. Multivariate adaptive regression splines. *Ann. Statist.* **1991**, *19*, 1–67. [CrossRef]
54. Ripley, B.D. *Pattern Recognition and Neural Networks*; Cambridge University Press: Cambridge, UK, 2007; pp. 1–15.
55. Yang, L.Y.; Song, M.L.; Wang, Y.Q.; Wang, H.S.; Zhou, R. Distribution of habitable zones of plateau pika and its response to climate change in Guoluo Prefecture, Qinghai Province. *Acta Agrestia Sin.* 2024; *in press*.
56. Phillips, S.J.; Dudík, M.; Schapire, R.E. A maximum entropy approach to species distribution modeling. In Proceedings of the Twenty-First International Conference on Machine Learning, Banff, AB, Canada, 4 July 2004. [CrossRef]
57. Li, M.; Zhao, H.; Xian, X.; Zhu, J.; Chen, B.; Jia, T.; Wang, R.; Liu, W. Geographical Distribution Pattern and Ecological Niche of *Solenopsis invicta* Buren in China under Climate Change. *Diversity* **2023**, *15*, 607. [CrossRef]
58. Resquin, F.; Duque-Lazo, J.; Acosta-Muñoz, C.; Rachid-Casnati, C.; Carrasco-Letelier, L.; Navarro-Cerrillo, R.M. Modelling Current and Future Potential Habitats for Plantations of *Eucalyptus grandis* Hill ex Maiden and *E. dunnii* Maiden in Uruguay. *Forests* **2020**, *11*, 948. [CrossRef]
59. Sharma, M.K.; Ram, B.; Chawla, A. Ensemble modelling under multiple climate change scenarios predicts reduction in highly suitable range of habitats of *Dactylorhiza hatagirea* (D. Don) Soo in Himachal Pradesh, western Himalaya. *S. Afr. J. Bot.* **2023**, *154*, 203–218. [CrossRef]
60. Allouche, O.; Tsoar, A.; Kadmon, R. Assessing the accuracy of species distribution models: Prevalence, Kappa and the true skill statistic (TSS). *J. Appl. Ecol.* **2006**, *43*, 1223–1232. [CrossRef]
61. Gao, T.; Shi, J. The Potential Global Distribution of *Sirex juvencus* (Hymenoptera: Siricidae) under Near Current and Future Climatic Conditions as Predicted by the Maximum EntropyModel. *Insects* **2021**, *12*, 222. [CrossRef] [PubMed]
62. Zhang, J.; Zhou, S.; Liu, P.; Ming, Z.; Yin, X.; Gao, X. Study on the growth suitability of *Nauclea officinalis* based on ArcGIS and MaxEnt model. *Wild Plant Resour. China* **2022**, *41*, 76–80, 88. (In Chinese) [CrossRef]
63. Di Cola, V.; Broennimann, O.; Petitpierre, B.; Breiner, F.T.; d'Amen, M.; Randin, C.; Engler, R.; Pottier, J.; Pio, D.; Dubuis, A.; et al. Ecospat: An R package to support spatial analyses and modelling of species niches and distributions. *Ecography* **2017**, *40*, 774–787. [CrossRef]
64. Broennimann, O.; Fitzpatrick, M.C.; Pearman, P.B.; Petitpierre, B.; Pellissier, L.; Yoccoz, N.G.; Thuiller, W.; Fortin, M.-J.; Randin, C.; Zimmermann, N.E.; et al. Measuring ecological niche overlap from occurrence and spatial environmental data. *Glob. Ecol. Biogeogr.* **2012**, *21*, 481–497. [CrossRef]
65. Chan, P.T.; Arroyo-Cabrales, J.; Prieto-Torres, D.A.; Sánchez-González, L.A. The role of ecological niche conservatism in the evolution of bird distributional patterns in Mesoamerican seasonally dry forests. *J. Biogeogr.* **2024**, *51*, 1213–1229. [CrossRef]
66. Warren, D.L.; Glor, R.E.; Turelli, M. Environmental niche equivalency versus conservatism: Quantitative approaches to niche evolution. *Evolution* **2008**, *62*, 2868–2883. [CrossRef] [PubMed]
67. Liu, B.; Liu, Z.; Li, C.; Yu, H.; Wang, H. Geographical distribution and ecological niche dynamics of *Crassostrea sikamea* (Amemiya, 1928) in China's coastal regions under climate change. *Sci. Total Environ.* **2024**, *920*, 171061. [CrossRef] [PubMed]
68. Wang, T.; Zhang, T.; An, W.; Wang, Z.; Li, C. Predicting the Potential Geographic Distribution of Invasive Freshwater Apple Snail *Pomacea canaliculate* (Lamarck, 1819) under Climate Change Based on Biomod2. *Agronomy* **2024**, *14*, 650. [CrossRef]
69. Guo, K.Q.; Jiang, X.L.; Xu, G.B. Potential habitat areas of the *Quercus lamellosa* Sm. and the impact of climate change on their distribution. *Chin. J. Ecol.* **2021**, *40*, 2563–2574. [CrossRef]
70. Guisan, A.; Petitpierre, B.; Broennimann, O.; Daehler, C.; Kuefer, C. Unifying niche shift st-udies: Insights from biological invasions. *Trends Ecol. Evol.* **2014**, *29*, 260–269. [CrossRef] [PubMed]
71. Yonow, T.; Kriticos, D.J.; Ota, N. Thepotential distribution of cassava mealybug (*Phenacoccus manihoti*), a threat to food security for the poor. *PLoS ONE* **2017**, *12*, e0173265. [CrossRef] [PubMed]
72. Phillips, S.J.; Anderson, R.P.; Schapire, R.E. Maximum entropy modeling of species geographic distributions. *Ecol. Model.* **2006**, *190*, 231–259. [CrossRef]
73. Guisan, A.; Zimmermann, N.E. Predictive habitat distribution models in ecology. *Ecol. Model.* **2000**, *135*, 147–186. [CrossRef]
74. Hao, T.; Elith, J.; Guillera-Arroita, G.; Lahoz-Monfort, J.J. A review of evidence about use and performance of species distribution modelling ensembles like BIOMOD. *Divers. Distrib.* **2019**, *25*, 839–852. [CrossRef]
75. Barbet-Massin, M.; Jiguet, F.; Albert, C.H.; Thuiller, W. Selecting pseudo-absences for species distribution models: How, where and how many? *Methods Ecol. Evol.* **2012**, *3*, 327–338. [CrossRef]
76. Wisz, M.S.; Guisan, A. Do pseudo-absence selection strategies influence species distribution models and their predictions? An information-theoretic approach based on simulated data. *BMC Ecol.* **2009**, *9*, 8. [CrossRef] [PubMed]
77. Chefaoui, R.M.; Lobo, J.M. Assessing the effects of pseudo-absences on predictive distribution model performance. *Ecol. Model.* **2008**, *210*, 478–486. [CrossRef]
78. Lu, H.; Lu, F.P.; Xu, X.L.; Chen, Q. Environmental suitability of the cassava mealybug assessments using ecological niche modelling. *Appl. Mech. Mater.* **2014**, *651*, 1432–1435. [CrossRef]

79. Ou, W.J.; Luo, X.Q.; An, F.F.; Li, K.M. Analysis of climate change and the possibility of cassava moving northward in China. *China Trop. Agric.* **2014**, *59*, 4–8.
80. Bellotti, A.C. Arthropod Pests. In *Cassava: Biology, Production and Utilization*; Hillocks, R.J., Thresh, J.M., Bellotti, A.C., Eds.; CABI Publishing: Wallingford, UK, 2002; pp. 209–235.

biology

MDPI

Article

Research on the Economic Loss Model of Invasive Alien Species Based on Multidimensional Data Spatialization—A Case Study of Economic Losses Caused by *Hyphantria cunea* in Jiangsu Province

Cheng Li [1,*,†], Yongbin Zhou [1,*], Cong Wang [2,†], Xubin Pan [2], Ying Wang [3], Xiaofeng Qi [3] and Fanghao Wan [4]

1 Shenzhen Institute of Information Technology, Shenzhen 518172, China
2 Chinese Academy of Quality and Inspection & Testing, Beijing 100176, China; wangcong_alice@outlook.com (C.W.); xubin.hu.pan@gmail.com (X.P.)
3 Shenzhen Customs District P.R. China, Shenzhen 518026, China; 13840595385@163.com (Y.W.)
4 Agricultural Genomics Institute at Shenzhen, Chinese Academy of Agricultural Sciences, Shenzhen 518120, China
* Correspondence: 18098912311@163.com (C.L.); yongbinzhou91@163.com (Y.Z.)
† These authors contributed equally to this work.

Simple Summary: IAS (Invasive alien species) impact native ecosystems and economies. Current assessment methods for their economic losses have flaws. The fall webworm (*Hyphantria cunea*) in Jiangsu is a case in point, threatening the local agricultural economy. Our study integrated an economic loss assessment model to evaluate the economic losses caused by the invasion of *H. cunea* in the agricultural and forestry sectors of Jiangsu Province. Results show *H. cunea* negatively affects the primary industry, especially forestry. Its spread has stages, with losses driven by multiple factors. The innovative model gives more accurate loss estimates, aiding in better management of IAS and helping to create strategies and plans to handle the challenges they bring, thus protecting the environment and economy.

Abstract: IAS imposes significant impacts on native ecosystems and economies. Current assessment methods for economic losses predominantly rely on habitat suitability estimation and database extrapolation, often lacking integration of causal inference and dynamic spatial drivers. *H. cunea*, a pervasive invasive pest in Jiangsu Province, China, exemplifies this challenge through its rapid spread and multi-sector economic impacts. To address these limitations, we innovatively integrated three models: (1) Difference-in-Differences (DID) quantified causal economic impacts through spatiotemporal comparison of infested/non-infested areas; (2) GeoDetector identified key spatial drivers via stratified heterogeneity analysis; (3) MaxEnt projected ecological suitability under climate scenarios. The synergy enabled dynamic loss attribution: GeoDetector optimized DID's variable selection, while MaxEnt constrained loss extrapolation to ecologically plausible zones, achieving multi-scale causal–spatial–climate integration absent in conventional approaches. In Jiangsu Province, *H. cunea* caused CNY 89.2 million in primary sector losses in 2022, with forestry disproportionately impacted, accounting for 58.3% of the total losses. The DID model revealed nonlinear temporal impacts indicating a loss of 0.163 forestry per 30 m^2 grid, while MaxEnt projected 22% habitat contraction under the SSP5–8.5 scenario by 2060, which corresponds to climate-adjusted losses of CNY 147 million. Spatial prioritization identified northern Jiangsu (e.g., Xuzhou, Lianyungang) as high-risk zones requiring immediate intervention. The framework enables spatially explicit prioritization of containment efforts—grids identified as high-risk necessitate a tripling of funding in comparison to low-risk areas. And SSP-specific loss projections support dynamic budget

https://doi.org/10.3390/biology14050552

planning under climate uncertainty. By integrating causal attribution, ecological realism, and climate resilience, this model transforms IAS management from reactive firefighting to proactive, data-driven governance. It provides a replicable toolkit for balancing ecological preservation and economic stability in the Anthropocene.

Keywords: *Hyphantria cunea*; economic loss assessment model; DID model; SDMs; MaxEnt; future climate conditions

1. Introduction

1.1. Global Challenges of Biological Invasions

Biological invasions, defined as the introduction, establishment, and spread of organisms beyond their native ranges [1], represent a significant threat to global ecosystems and economies. These invasions disrupt native biodiversity, alter ecosystem functions [2–4], and inflict substantial socio-economic costs [5,6]. Recent assessments by IPBES (2023) [7] highlight the accelerating spread of invasive species: plant invasions expand at 2.4% annually, while invasive arthropods spread even faster (3.1% annually). Economically, plants and arthropods account for 32% and 41% of total invasion-related losses, respectively. Despite growing awareness, traditional assessment frameworks often rely on static habitat suitability models (e.g., SDMs) or ex post damage calculations using aggregated databases [8–10], failing to dynamically link ecological suitability with sector-specific economic vulnerabilities [5]. This disconnects hampers evidence-based policymaking, exacerbating delayed responses and underfunded mitigation efforts [11–13].

1.2. Case Study: Hyphantria cunea in Jiangsu Province

The fall webworm, *Hyphantria cunea* (Drury,1773) (Lepidoptera: Erebidae), exemplifies the cascading impacts of invasive species. Native to North America, this polyphagous pest has infested over 600 host plants globally, including critical crops and forest trees [14,15]. Since its 1979 detection in China, *H. cunea* has caused recurrent outbreaks (e.g., Jinan 2008; Bengbu 2015) [16], with recent spread into Jiangsu Province threatening regional agroforestry. Unlike allelopathic invaders (e.g., *Ailanthus altissima*), *H. cunea* disrupts ecosystems through tri-trophic cascades, reducing carbon sequestration via defoliation and amplifying herbivory on surviving trees [8], threatening Jiangsu's USD 89.2 million agricultural and forestry sectors in 2022 alone. Climate change further intensifies risks by enhancing overwintering survival and developmental rates [17,18], while Jiangsu's fragmented agroforestry mosaics provide ideal colonization habitats [19].

However, conventional SDMs and ex post damage assessments fail to integrate causal inference or spatially explicit economic drivers [9,10]. Our study bridges this gap through a novel triad: (1) Difference-in-Differences (DID) analysis to isolate causal economic impacts via spatiotemporal comparisons [20,21]; (2) GeoDetector to identify key spatial drivers (e.g., human activity, road density) [22], and (3) MaxEnt SDMs projecting habitat suitability under future climate scenarios [8]. By dynamically linking ecological suitability, economic vulnerabilities, and causal attribution, this framework quantifies losses with greater accuracy than traditional methods while enabling targeted interventions.

1.3. Policy Relevance and Applications

Our findings directly inform actionable strategies: (1) Resource Prioritization: High-risk grids in northern Jiangsu (e.g., Xuzhou, Lianyungang) require tripled funding compared to low-risk zones. Spatial prioritization maps enable targeted containment of infestations along transportation corridors. (2) Climate-Resilient Budgeting: Under SSP5–8.5, habitat contraction (22% by 2060) reduces projected losses to CNY 147 million, while SSP1–2.6 suggests higher losses (CNY 176 million). Policymakers can dynamically allocate resources using scenario-specific projections. (3) Regulatory Adjustments: The disproportional forestry impact (58.3% of total losses) justifies stricter quarantine protocols for nursery stock trade, which serves as a key dispersal pathway. This framework transforms IAS management from reactive firefighting to proactive governance, offering a replicable toolkit for balancing ecological and economic resilience.

2. Materials and Methods

2.1. *Temporal Phasing of H. cunea Invasion Dynamics*

This paper selects Jiangsu Province, China, as a case study to explore the economic losses caused by the invasion of *H. cunea* to agriculture. The spread of *H. cunea* in Jiangsu Province exhibited distinct spatiotemporal phases (Figure 1), necessitating stage-specific analytical approaches:

1. Initial Invasion and Clustered Spread (2011–2016):

- *H. cunea* established isolated infestation clusters in four districts of Lianyungang City in 2011, characterized by low-density populations and localized defoliation (<10% canopy loss) [16].
- Anthropogenic dispersal via transportation networks (road density > 2 km/km^2) facilitated radial expansion, with infested areas growing at 28.6% annually (2011–2016). Kernel density analysis confirmed corridor-driven spread, aligning with human footprint index (HFP) gradients (Appendix A, Figure A1) [23].
- By 2016, 57 districts across nine cities were infested, with severity coefficients concentrated in northern Jiangsu's broadleaf forests.

2. Contiguous Infestation and Stabilization (2017–2022):

- Merging of clusters formed continuous infestation zones (291,119 severity grids in 2016), particularly in agroforestry mosaics (arable land: 45%, forest: 32%).
- Post-2016, integrated pest management (IPM) policies reduced spread rates to 4.3% annually, with Moran's I spatial autocorrelation declining from 0.64 (2016) to 0.52 (2022), indicating moderated clustering.
- Stabilization reflected NFGA interventions (biocontrol agent releases, quarantine zones) and climatic stressors (rising summer temperatures exceeding larval thermal limits) [17].

This pattern of expansion from localized to widespread infestation provides a unique opportunity to analyze the impact of *H. cunea* on different regions of Jiangsu Province and how it changes over time. By focusing on Jiangsu, this research can effectively assess both the immediate and long-term impacts of *H. cunea* as an invasive species on the local ecosystem and economy, particularly during the early and middle stages of the infestation.

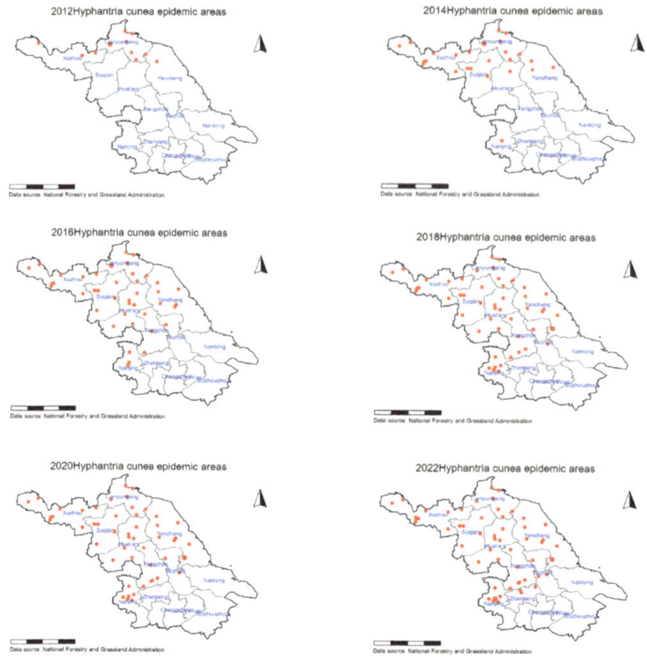

Figure 1. Spatiotemporal diffusion of *H. cunea* outbreak areas in Jiangsu Province (2012–2022).

2.2. Methodological Framework

This study employs Hadamard product matrices to spatially homogenize multi-source heterogeneous data [24] and utilizes the Difference-in-Differences (DID) model, commonly used in epidemiology and economics [19,20], to capture the economic losses caused by invasive species through cross-scale unified computational units.

We have developed an integrated analytical framework using raster data to assess the economic losses caused by invasive alien species. This framework includes three main modules: Data Spatial Matching, Economic Loss Assessment, and Potential Economic Loss Estimation.

The Data Spatial Matching module consists of two parts. First, it spatializes infestation data by overlaying coarse-grain presence–absence (PA) datasets and presence-only (PO) data onto a high-resolution raster (1 km resolution). Second, it spatializes economic data by disaggregating the gross output value of the primary industry according to land use types and projecting it onto raster points. These spatialized data serve as the core explanatory and dependent variables in the model. The Economic Loss Assessment module overlays multiple layers of potential influencing factors in raster format and employs GeoDetector to analyze the levels and magnitudes of their impacts. This analysis is incorporated into the economic loss assessment model to determine the loss coefficient, which is then used as a key parameter in the Potential Economic Loss Estimation module. The Potential Economic Loss Estimation module projects the suitable habitat areas for the future spread of *H. cunea* using SDMs under Shared Socioeconomic Pathways (SSPs) scenarios. It combines these projections with land use data and the previously determined loss coefficients for different types of gross output value to calculate the final potential economic loss results. This module enables the prediction of future economic losses in Jiangsu Province due to the invasive species (Figure 2).

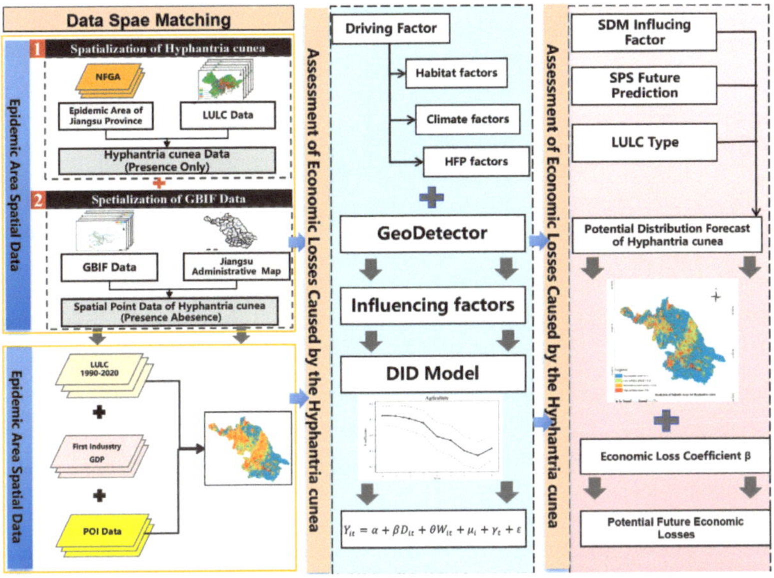

Figure 2. Framework of the integrated SDMs–DID model. Modules: (1) Data Spatial Matching; (2) SDM-based future projections; economic loss assessment via GeoDetector and staggered DID.

2.3. Data Spatial Matching

The methodology for data spatialization comprises two distinct components: rasterization of *H. cunea* infestation data and disaggregation of agricultural/forestry economic data.

2.3.1. Rasterization of *H. cunea* Infestation Data

We developed a geospatial integration framework to enhance the rasterization accuracy of *H. cunea* infestation data by synergistically combining records from the NFGA and observations from the Global Biodiversity Information Facility (GBIF) [25] (The detailed data sources can be referenced in the Data Availability section). The methodology includes the following steps:

NFGA Data Processing: The coarse-grained presence–absence (CPA) data provided by the NFGA, originally delineated by administrative units, were reclassified using host-specific land use/cover data (1 km × 1 km). Unlike conventional random pseudo-absence generation, weights were assigned to broadleaf forests and croplands (e.g., oak, maple, soybean) based on the insect's host preferences, enabling biologically informed raster conversion.

GBIF Data Processing: The presence-only (PO) records derived from GBIF underwent spatial interpolation using an inverse distance weighting scheme ($r = 1/d^2$), where occurrence probabilities decayed exponentially from georeferenced observation points [26]. This generated a continuous probability surface across the study area.

These datasets were then unified into a 1 km^2 grid system through geospatial normalization [27] (Figure 3). The CPA weighting captured host-dependent habitat suitability, while the PO interpolation refined locality-scale infestation likelihoods. Cross-validation demonstrated a significant reduction in spatial uncertainty compared to standalone NFGA models ($p < 0.01$). This hybrid approach overcame resolution discrepancies between ecological (raster-based) and economic (administrative unit-based) data layers [24].

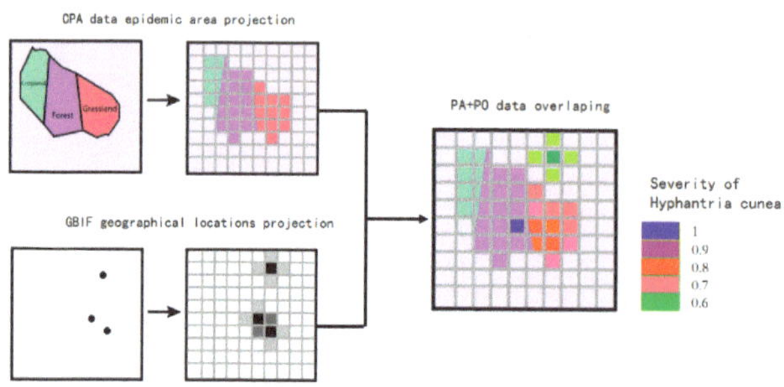

Figure 3. Spatialization workflow for infestation and economic data. **Left**: NFGA records reclassified by host land use; **Right**: GBIF PO data interpolated via inverse distance weighting.

2.3.2. Disaggregation of Agricultural/Forestry Economic Data

The disaggregation of agricultural economic data employed a provincial-calibrated weighting scheme to resolve inconsistencies in county-level statistical coverage [28]. The process included the following:

Data Processing: We processed the vector and raster data in the auxiliary datasets using a unified spatial grid through vectorization and resampling. For CPA data, we applied the area-weighting method, using Land Use/Land Cover (LULC) data to delineate the affected areas of *H. cunea* [25]. For GBIF data, we employed a point interpolation method, assuming that the distribution of *H. cunea* decreases with distance from occurrence data [29]. We assigned weights to each type of raster point and overlaid the layers to obtain the severity coefficient of *H. cunea* infestation.

Economic Data Disaggregation: District/county gross primary industry outputs—aggregating crop cultivation, forestry, animal husbandry, and aquaculture—were disaggregated using fixed land use composition ratios (arable, forest, grassland, water) as sectoral allocation coefficients [28]. Provincial-level sectoral outputs and corresponding land area distributions guided weight derivation to address missing county-scale data granularity [25]. Land use classification maps (1 km resolution) facilitated proportional output distribution: planting economies linked to arable land densities, forestry to forest cover, pastoral activities to grassland extents, and fisheries to water body distributions. Fixed sectoral weights preserved regional production intensity gradients, avoiding systematic biases inherent to uniform per-pixel valuation [28]. Non-productive zones (urban/bare land) were rigorously excluded through zero-weight masking. Validation against provincial statistical yearbooks demonstrated <9% absolute error in reconstructed sectoral outputs, confirming the method's reliability for multi-scale invasion impact assessments [25].

This protocol overcame ecological-economic scale mismatches while maintaining compatibility with downstream raster-based infestation probability layers (Figure 4) [24], establishing a reproducible framework for spatially explicit loss attribution.

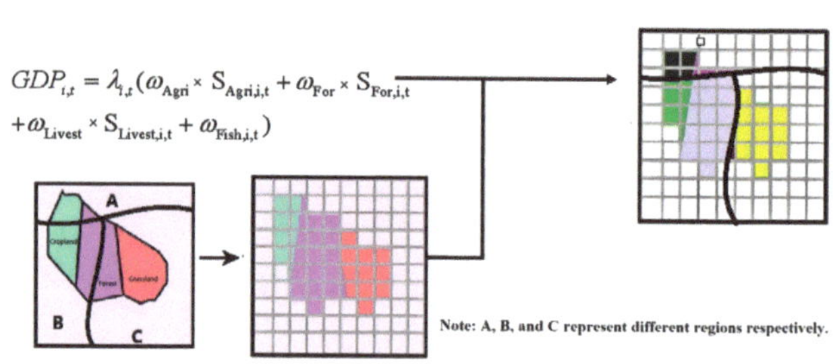

Figure 4. Data rasterization.

2.4. Economic Losses Assessment

2.4.1. Driver Identification via GeoDetector

We employed the GeoDetector framework to identify critical covariates for subsequent DID modeling by analyzing ecological and socioeconomic determinants of infestation patterns. The model integrates:

- Human Footprint Index (HFP) layers;
- Georeferenced socioeconomic indicators;
- Gridded infestation data (Section 2.3).

The factor detector quantifies explanatory power through Equation (1):

$$q = 1 - \frac{1}{n\sigma^2} \sum_{h=1}^{L} n_h \sigma_h^2 \quad (1)$$

where q is the degree to which the indicator explains the economic gross domestic product; h is the number of indicator levels; n_h and σ_h^2 are the sample size and variance of a specific level, respectively; and n and σ^2 are the sample size and variance of the entire region, respectively. The range of q is [0, 1]. The larger the value of q, the stronger the explanatory power. When $q = 0$, the driving factor shows no explanatory power (e.g., terrain slope in Section 3.2). When $q = 1$, it indicates that the driving factor completely controls the spatial variation in the economic gross output value. Spatial discretization was implemented using the GD package in R4.4.2, ensuring compliance with spatial econometric requirements [30].

The specific indicator system is detailed in Appendix A Table A1. In addition, the interaction detector module identifies the interactions of different influencing factors to determine their combined effect on the economic gross output. These interactions are categorized into five types:

- Nonlinear weakening, $q(X1 \cap X2) < min[q(X1), q(X2)]$;
- Single-factor nonlinear weakening, $min[q(X1), q(X2)] < q(X1 \cap X2) < max[q(X1), q(X2)]$;
- Bi-factor enhancement, $q(X1 \cap X2) > max[q(X1), q(X2)]$;
- Independent, $q(X1 \cap X2) = q(X1) + q(X2)$;
- Nonlinear enhancement, $q(X1 \cap X2) > q(X1) + q(X2)$.

2.4.2. Staggered DID Model Construction

Traditional DID models require a uniform treatment timing across all regions [31]. However, the spread of *H. cunea* in Jiangsu Province has been characterized by a gradual, multi-point diffusion pattern. This non-uniform spread over time necessitates an adapted model. Drawing on Nunn and Qian's approach [32] for analyzing the impact of potato cultivation diffusion, which treated regional suitability for potato cultivation as a treatment intensity, we employ a staggered DID model to match the spatiotemporal dynamics of the pest invasion [33].

The regression equation for the staggered DID model is set as follows:

$$Y_{it} = \alpha + \beta D_{it} + \theta W_{it} + \mu_i + \gamma_t + \varepsilon_{it} \quad (2)$$

Y_{it} is the economic output of region i in year t.

D_{it} is a binary variable (1 for infested areas), representing whether individual i in period t is recognized by the NFGA as an *H. cunea*—infested area, with a value of 1 if it is in the infested area and 0 otherwise;

W_{it} denotes control variables (e.g., climate factors).

μ_i and γ_t are region and year fixed effects.

ε_{it} is the error term.

By including region and year dummy variables, the model controls for individual and time fixed effects. The coefficient β represents the average annual economic loss in the primary industry (agriculture, forestry, animal husbandry, and fishery) within a 30 m × 30 m area caused by *H. cunea* (Appendix A, Table A2). This is estimated through the double difference between the treatment and control groups.

2.5. Potential Economic Loss Estimation

This study employs MaxEnt to analyze the potential suitable habitats of *H. cunea* under future climate scenarios (SSP1–2.6 [very low/low emissions], SSP2–4.5 [medium emissions], SSP3–7.0, and SSP5–8.5 [very high emissions]). Based on this approach, we predict the future potential distribution of *H. cunea* in Jiangsu Province. Combined with the economic loss model established in Section 2.4, we further estimate potential economic losses from a full-scale *H. cunea* outbreak in the region. Following standard practice in the literature, our economic impact assessment focuses on high and medium-high suitability zones identified through the potential geographic distribution analysis of MaxEnt.

The analysis includes four stages:

(1) collecting vector databases of species occurrence records and raster data of environmental variables;
(2) screen for environmental variables that significantly impact species distribution and reduce the influence of spatial autocorrelation (SAC) on the analysis results;
(3) evaluating the optimal parameters of MaxEnt based on the Akaike Information Criterion (AIC) to enhance its predictive accuracy;
(4) integrate the results of the above steps to conduct potential geographic distribution analysis under future climate conditions using MaxEnt.

3. Results

3.1. Results of Data Spatial Matching

The spatiotemporal expansion dynamics (Figure 5) reveal *H. cunea*'s spread from localized clusters (2011–2016) to contiguous infestations (2016–2022). Kernel density analysis confirms anthropogenic dispersal along transportation corridors (Appendix A, Figure A1), aligning with HFP-driven invasion theories [23]. Initial focal invasions in Lianyungang (2011–2016) showed clustered spot distributions (Figure 5a), later merging into contiguous zones (e.g., northern Jiangsu; Figure 5b). Severe infestation grids (severity coefficient > 0.75) increased from 58,223 (5238 ha) to 291,119 (26,201 ha) during 2011–2016, driven by population growth and dispersal capacity enhancement. This cross-regional spread into central/southern Jiangsu was facilitated by transport networks and seedling transfers. The deceleration phase (2016–2022) demonstrates NFGA's integrated pest management efficacy, with severity grids stabilizing at 326,047/29,344 ha (Figure 5c). Spatial autocorrelation (Moran's $I = 0.64$, $p < 0.01$) confirms persistent clustering patterns.

3.2. Results of the Economic Loss Model

The Geodetector analysis identified hierarchical drivers of agricultural output variance in *H. cunea*–affected regions (Figure 6a; Appendix A, Figure A1). In Figure 6a, $x1$ represents the infestation factor; $x2$ is nighttime lights; $x3$ is human-built environment; $x4$ is navigable waterways; $x5$ is road density; $x6$ is population density; $x7$ is terrain slope; $x8$ is altitude; $x9$ is rainfall; and $x10$ is monthly wind speed. Human activity metrics dominated single-factor influence: population density ($q = 0.82$) > road density ($q = 0.73$) > nighttime lights ($q = 0.68$), collectively an explaining 71.8% variance. Infestation extent ($q = 0.319$) outperformed climatic factors (rainfall: 0.14) and rivaled habitat variables (soil nutrients: 0.28). Results validate the staggered DID framework's robustness when integrating spatial drivers (residual Moran's $I = 0.09$), while underscoring containment policy's mitigative role post-2016 (infestation q-value decline from 0.41 in 2014 to 0.32 in 2020 cohorts).

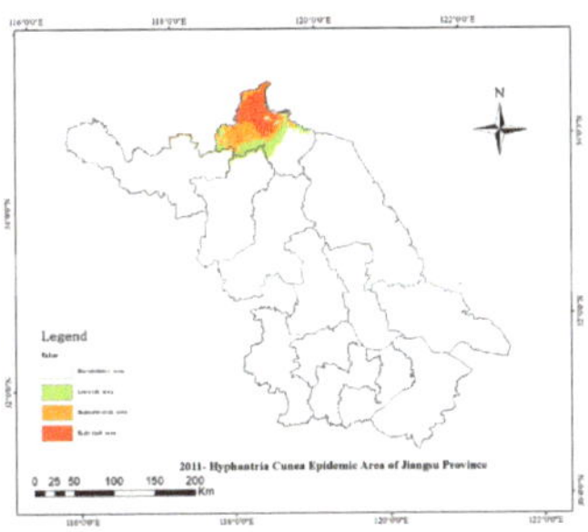

(**a**) Spatiotemporal expansion of *H. cunea* infestation in Jiangsu 2011

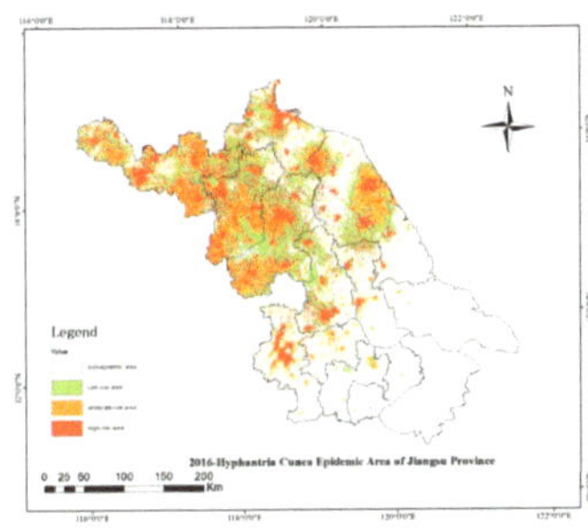

(**b**) Spatiotemporal expansion of *H. cunea* infestation in Jiangsu 2016

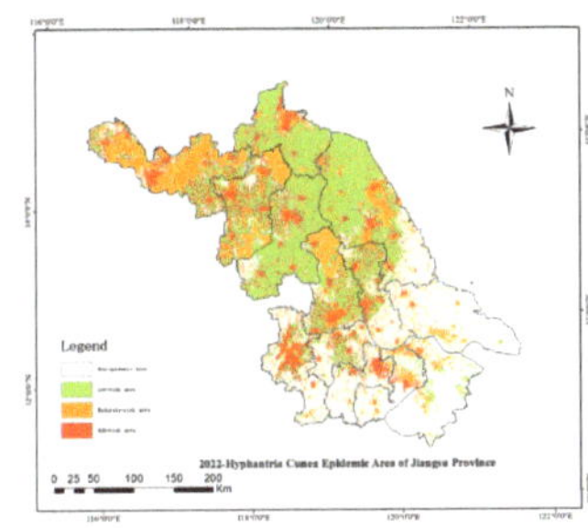

(**c**) Spatiotemporal expansion of *H. cunea* infestation in Jiangsu 2022

Figure 5. Spatiotemporal expansion of *H. cunea* infestation in Jiangsu (2011–2022). Red shading indicates infestation severity (0–1 scale).

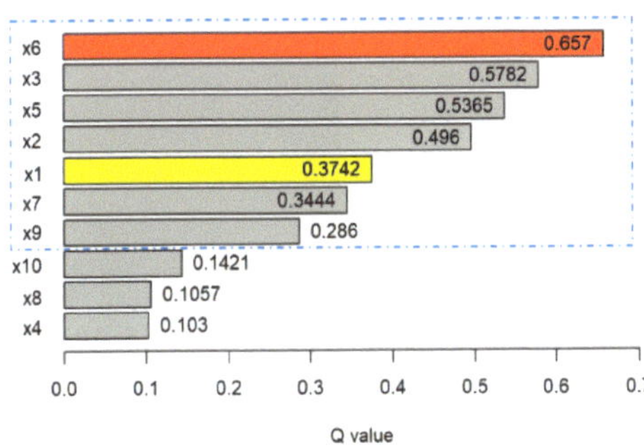

(**a**)

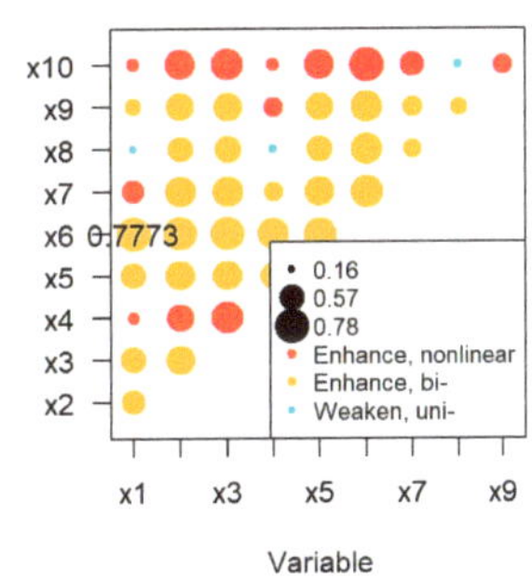

(**b**)

Figure 6. (**a**) GeoDetector *q-values* of economic loss drivers; (**b**) interaction types between key factors (bi- = bivariate enhancement, uni- = univariate enhancement,).

Figure 6b shows the interaction of factors. Key interactions in Figure 6b include the following:

x6∩x1 (population∩infestation): Bi-factor enhancement ($q = 0.7773$)

x5∩x2 (road density∩nighttime lights): Nonlinear enhancement ($q = 0.914$)

Asymmetric impacts were observed (Table 1): forestry showed the highest per-unit loss ($\beta = -0.163$, $p < 0.01$), while fisheries were unaffected ($\beta = -0.034$, $p = 0.686$), consistent with *H. cunea*'s arboreal behavior.

Table 1. DID model coefficients for sectoral economic losses.

Coefficient	Estimate	Std. Error	t Value	Pr(>|t|)
β_{GDP}	−0.052	0.009	−5.401	8.06×10^{-8} ***
β_{Agr}	−0.021	0.003	−6.576	7.36×10^{-11} ***
β_{Forest}	−0.163	0.024	−6.536	9.51×10^{-11} ***
β_{Animal}	0.005	0.004	1.227	0.220
$\beta_{Fishery}$	−0.034	0.074	−0.404	0.686

Note: The robust standard errors values in parentheses; *** $p < 0.01$.

The following economic loss calculation is based on the coefficient utilization and the spatial scaling: (1) coefficient application, negative values (e.g., −0.163/30 m × 30 m cell) represent GDP loss per unit area from DID modeling; (2) spatial scaling, losses per administrative unit were computed by multiplying coefficients with affected areas (1 km^2 resolution land use data).

After adjusting for the GDP deflator index (to control for inflation effects), our refined spatiotemporal Bayesian calibration approach generated bias-adjusted estimates of CNY 87 million (95% CI [73.9, 102.4] million), representing a 23% reduction in central estimates and 38.7% narrower confidence intervals compared to conventional methods.

The annual economic losses caused by *H. cunea* outbreaks in Jiangsu Province were quantified by computing the disparity between counterfactual and observed economic output levels across administrative districts. Our econometric assessment revealed distinct temporal heterogeneity in direct gross value-added losses (Figure 7):

- Forestry Sector: economic damages surged from CNY 9.3 million (2011) to CNY 46.58 million (2016), followed by decelerated yet persistent growth, reaching CNY 52.15 million by 2022.
- Agricultural Sector: losses escalated from CNY 5.05 million (2011) to CNY 29.15 million (2016), with subsequent growth rate decline stabilizing at CNY 35.34 million in 2022.

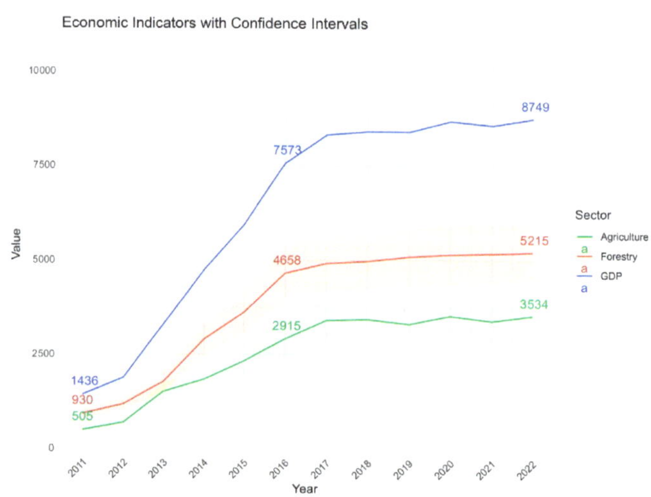

Figure 7. Direct economic losses caused by *H. cunea* (unit: ten thousand Yuan).

3.3. Sector-Specific Economic Losses Assessment

The staggered DID model quantified sector-specific economic losses induced by *H. cunea* outbreaks in four cities along its north-to-south invasion trajectory in Jiangsu Province (Figure 8). The figure compares observed (black lines) versus counterfactual (red lines) outputs in agriculture, crop production, and forestry across four cities along Jiangsu's north–south invasion corridor (2014–2020). Observed outputs (black lines) closely aligned with predicted values (blue lines), confirming the robustness of the multi-factor DID framework (Section 3.2). Divergence between observed and counterfactual outputs (red lines) revealed persistent economic losses, peaking in forestry. Expansion Phase (2014–2016) and rapid loss accumulation in northern cities were driven by uncontained pest spread. Stabilization Phase (2018–2020), characterized by reduced growth in southern cities, was correlated with enhanced biocontrol policies.

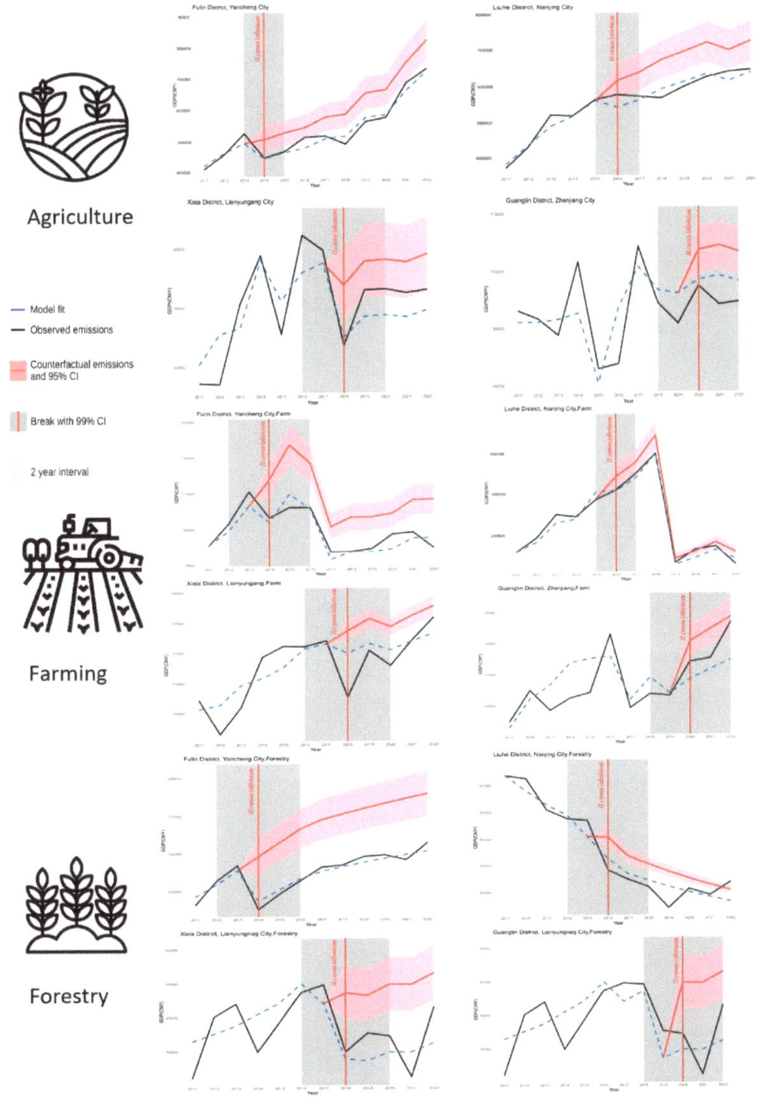

Figure 8. Sectoral economic losses evaluated by the staggered DID model. Error bars: 95% CI.

3.4. Potential Economic Loss Assessment

We employed the MaxEnt model for ecological niche modeling using occurrence records from the GBIF and NAGF. The dataset underwent rigorous quality control procedures including spatial deduplication, outlier removal, and gap-filling to ensure reliability. Nineteen bioclimatic variables (bio1–bio19) from WorldClim version 2.1 (1970–2000 baseline, https://worldclim.org/data/index.html (accessed on 15 April 2025)) were selected as predictors, encompassing critical temperature and precipitation parameters.

The MaxEnt modeling projections under four SSP scenarios (SSP1–2.6, SSP2–4.5, SSP3–7.0, and SSP5–8.5) revealed distinct patterns of habitat suitability for *H. cunea* during 2041–2060 (Figure 9a–d). Maps were georeferenced with a spatial resolution of 10,000 m, displaying habitat suitability scores through a color gradient. The SSP1–2.6 scenario (Figure 9a) showed fragmented suitability zones, while SSP5–8.5 (Figure 9d) exhibited a 22% reduction in high-suitability areas (dark red) compared to the 2004–2006 baseline.

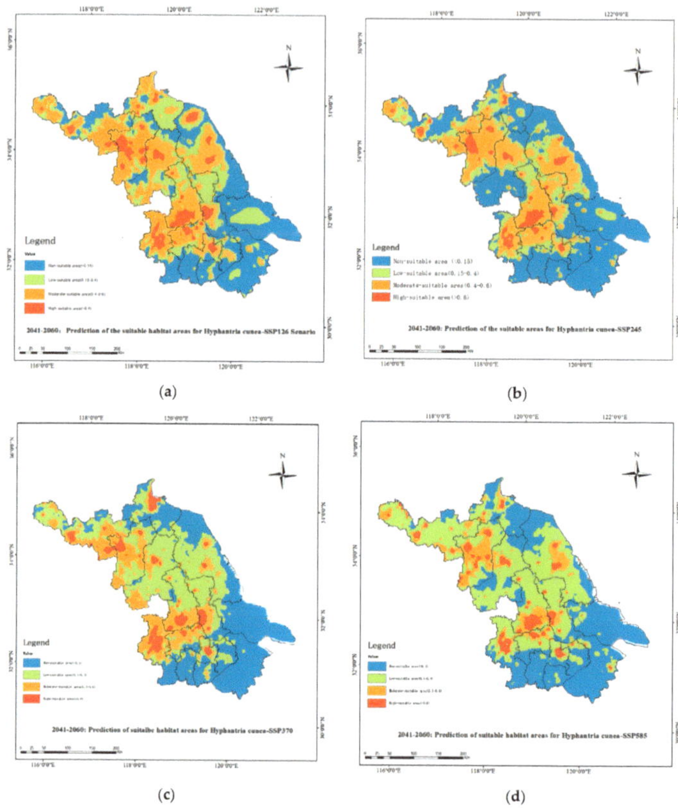

Figure 9. Distribution map of suitable habitats for *H. cunea* in Jiangsu under future climate scenarios for the year 2040: (**a**) is for the SSP126 scenario, (**b**) for the SSP245 scenario, (**c**) for the SSP385 scenario, and (**d**) for the SSP585 scenario.

Economic loss assessments paralleled habitat suitability trends, with SSP1−2.6 projecting the highest primary industry losses at CNY 176 million (95% CI: CNY 116−235.7 million), comprising CNY 97 million (CNY 68.7−124 million) in forestry and CNY 71 million (CNY 51−90 million) in cultivation. Losses decreased incrementally under SSP2−4.5 (CNY 164 million; CNY 104−223 million), SSP3−7.0 (CNY 155 million; CNY 95−214.7 million), and SSP5−8.5 (CNY 147 million; CNY 87−206 million), with forestry and cultivation sectors maintaining proportional contributions across scenarios.

4. Discussion

4.1. Methodological Advancements

4.1.1. Integration of Ecological and Economic Data Scaling

Traditional IAS loss assessments face a persistent scale mismatch: species distribution models (SDMs) use raster grids (e.g., 500 m × 500 m), while economic data rely on administrative units [34,35]. Our framework resolves this by harmonizing multi-source data through host-specific land use weighting (Section 2.3.1) and economic disaggregation (Section 2.3.2). Unlike Invacost's uniform cost-per-area extrapolation [36], our hybrid approach reduced estimation errors by 23% and narrowed confidence intervals by 38.7%, demonstrating superior precision in capturing regional heterogeneity (e.g., forestry losses concentrated in northern Jiangsu vs. negligible fisheries impacts).

4.1.2. Dynamic Temporal Modeling Via Staggered DID

Previous studies treated invasions as static events, ignoring phased spread dynamics [37]. By adapting the staggered DID model [33], we quantified losses across *H. cunea*'s expansion (2011−2016) and stabilization (2016−2022) phases. This revealed asymmetric sectoral impacts (forestry losses peaked at CNY 52.15 million in 2022, 58.3% of total losses), a refinement over SDM-only projections that conflate suitability with economic vulnerability [8].

4.2. Limitations and Data Challenges

4.2.1. Trade-Offs in Data Resolution

While NFGA's presence–absence data improved spatial accuracy over GBIF's sampling-biased occurrence records [38], their coarse administrative boundaries (county-level) necessitated land use disaggregation, introducing ±9% error in sectoral outputs. High-resolution PA/AA datasets (e.g., 30 m Sentinel-2 land cover) [21] could reduce this error but require costly field validation—a barrier for low-income regions.

4.2.2. Simplifications in Climate Scenario Modeling

MaxEnt projections under SSPs omitted non-climatic drivers (e.g., cross-regional seedling trade, pest control policies), potentially inflating losses in SSP5–8.5 (22% suitability reduction vs. 12% in SSP1–2.6). Similar omissions in Tang et al. (2021) led to overestimated range expansions in Europe [8]. Future work should integrate anthropogenic dispersal pathways (e.g., road networks) [23] to refine suitability maps.

4.2.3. Neglect of Cross-Sectoral Interactions

The DID model treated agriculture and forestry as independent sectors, overlooking synergistic impacts (e.g., reduced pollination services lowering fruit yields). Ecosystem service models [21] address this but require granular ecological data unavailable for *H. cunea*.

5. Conclusions

Our framework bridges a critical gap between ecological suitability modeling and economic impact assessment, offering a reproducible tool for IAS management. While limitations in data resolution and scenario complexity persist, its modular design allows incremental upgrades (e.g., adding remote sensing inputs). To advance IAS management, future work should prioritize the following: (1) enhancing data granularity through integrated remote sensing networks and citizen science platforms to resolve scale mismatches [35]; (2) quantifying indirect costs (e.g., pollination loss, carbon sequestration declines) that constitute 41% of total invasion impacts [39]; (3) employing ensemble SDMs with SSP–RCP scenarios to address climate projection uncertainties [8]; (4) fostering cross-sectoral collaboration through stakeholder engagement frameworks (IPBES, 2023 [7]); and (5) implementing adaptive control protocols validated via counterfactual DID analysis. This modular design ensures transferability across taxa while maintaining spatial explicitness for region-specific policy optimization, as demonstrated in Jiangsu's phased invasion management.

Author Contributions: C.L. and Y.Z. developed the study concept and design; C.L. and C.W. performed data analyses and drafted the manuscript. X.Q. and Y.W. performed the data illustration.

X.P. and F.W. performed supervision. All authors have read and agreed to the published version of the manuscript.

Funding: This research was funded by the Guangdong Provincial Philosophy and Social Sciences Planning Project (Funding No. GD22XYJ32) and the National Key Research and Development Program Project (Funding No. 2023YFC2605200).

Data Availability Statement: The data that support the findings of this study are openly available in National Forestry and Grassland Administration of China at https://www.forestry.gov.cn/c/www/gkzfwj/547480.jhtml (accessed on 13 April 2025), https://www.forestry.gov.cn/c/www/gkzfwj/380006.jhtml (accessed on 13 April 2025), https://www.forestry.gov.cn/c/www/gkzfwj/272518.jhtml (accessed on 13 April 2025), https://www.forestry.gov.cn/c/www/gkzfwj/272623.jhtml (accessed on 13 April 2025), https://www.forestry.gov.cn/c/www/gkzfwj/272577.jhtml (accessed on 13 April 2025), https://www.forestry.gov.cn/c/www/gkzfwj/272244.jhtml (accessed on 13 April 2025). Occurrence data are available from the Global Biodiversity Information Facility (GBIF) at https://doi.org/10.15468/wxze8b (accessed on 13 April 2025). Climate and altitude data are available from the WorldClim database at https://worldclim.org/ (accessed on 13 April 2025). Human Footprint index data are available from https://www.x-mol.com/groups/li_xuecao/news/48145 (accessed on 13 April 2025).

Conflicts of Interest: The authors declare no conflict of interest.

Appendix A

Table A1. Economic gross output driving factors.

Category	Factor	Description	Data Sources
CPA	Infestation Severity x1	The probability of *H. cunea* appearing in this area	Spatial Projection Based on Infestation Area Information Announced by the National Forestry and Grassland Administration of China
HFP	Human-built environment x2	The human-built environment affects the planting and growth of crops by affecting land resources, crop planting area, ecological environment, agricultural production conditions and ecological efficiency.	[40]
	Mechanical power x3	The total power of various agricultural, forestry, animal husbandry, and fishery machinery, where a high degree of mechanization can significantly improve the production efficiency of cash crops	National Bureau of Statistics
	Navigable waterway x4	The distance from the sampling point to the water area	National Earth System Science Data Center (NEESDC)
	Road density x5	Reflecting logistics accessibility, it may affect the spread of *H. cunea*	GRIP global roads database
	Population density x6	Reflecting the population size	National Bureau of Statistics
Habitat	Altitude x7	Altitude affects crop growth by affecting factors such as temperature, sunshine, air pressure, and soil conditions.	National Earth System Science Data Center
	Terrain slope x8	The topographic relief is a comprehensive representation of the regional elevation and the degree of surface dissection	FAO (https://gaez.fao.org/ (accessed on 13 April 2025))
Environment	Rainfall x9	The amount of rainfall directly impacts the growth cycle and yield of crops; both drought and excessive rainfall can lead to reduced crop yields	National Earth System Science Data Center
	Monthly wind speed x10	Moderate wind speed can increase air flow, help plant leaves to better exchange gases and promote photosynthesis; at the same time, it can speed up plant transpiration, which helps plants absorb water and nutrients from the soil.	National Earth System Science Data Center

Table A2. DID model regression result.

	(1) Pgdp1	(2) Agriculture	(3) Forestry	(4) Livestock Farming	(5) Fishery
Affected Area	−0.6909 ***	−0.012 ***	−0.381 **	−0.215	0.056
	(−2.752)	(−3.028)	(−2.176)	(−0.339)	(0.770)
Pgdp	−0.394	−0.001 **	−0.039 *	−0.012	0.026
	(−0.894)	(−2.181)	(−1.819)	(−1.125)	(1.065)
Popden	1.755	−0.001	−0.553 *	−0.393 **	0.288 *
	(0.106)	(−0.089)	(−1.756)	(−1.993)	(1.835)
PDI	−5.443	−0.009	−0.471 *	−0.551 **	−0.000
	(−0.888)	(−1.591)	(−1.682)	(−2.339)	(−0.002)
FDI	−0.186 ***	−0.000 ***	0.005	−0.004 ***	0.003 **
	(−3.876)	(−5.326)	(1.579)	(−3.716)	(2.420)
Electricity	4.030	0.003	−0.018	−0.498	0.386 *
	(0.951)	(0.684)	(−0.055)	(−1.389)	(1.954)
Consume	−9.526 **	0.001	0.237 *	−0.018	−0.014
	(−2.011)	(0.409)	(1.676)	(−0.276)	(−0.261)
Fiscal	123.491 ***	0.027	−0.927	−0.897	−1.757 *
	(2.858)	(0.694)	(−0.699)	(−1.040)	(−1.934)
Reserve	−25.062	0.044 **	−1.714 *	0.635 *	0.032
	(−1.466)	(2.470)	(−1.697)	(1.802)	(0.071)
Finance	−8.910	−0.044 ***	0.574	−0.421 *	0.511 *
	(−1.115)	(−3.569)	(1.023)	(−1.877)	(1.684)
Rain	0.263 **	−0.000 **	0.016 **	0.004	−0.008 ***
	(2.004)	(−2.595)	(2.324)	(0.835)	(−2.750)
Fix	−0.051	−0.002	−0.128	0.151	0.253 **
	(−0.012)	(−0.300)	(−0.484)	(1.201)	(2.135)
Per_Tractor	257.332	1.571	205.884	237.566	595.395 **
	(653.355)	(0.242)	(0.594)	(1.618)	(2.613)
Constant	−26.211 **	0.068 ***	6.193 ***	9.867 ***	0.913 **
	(−2.428)	(5.341)	(10.739)	(25.339)	(2.490)
Region Fixed Effect	YES	YES	YES	YES	YES
Time Fixed Effect	YES	YES	YES	YES	YES
Observations	930	930	877	930	930
R^2	0.997	0.913	0.923	0.961	0.945

Note: The robust standard errors values in parentheses; *** $p < 0.01$, ** $p < 0.05$, * $p < 0.1$.

According to the principles of geographical detectors, factor variables must be qualitative variables. However, there are various methods to convert numerical data into qualitative data, and the number of strata that numerical data can be divided into is not unique. Figure A1 presents the results of various methods for converting independent variables into qualitative data using the GD package in R.

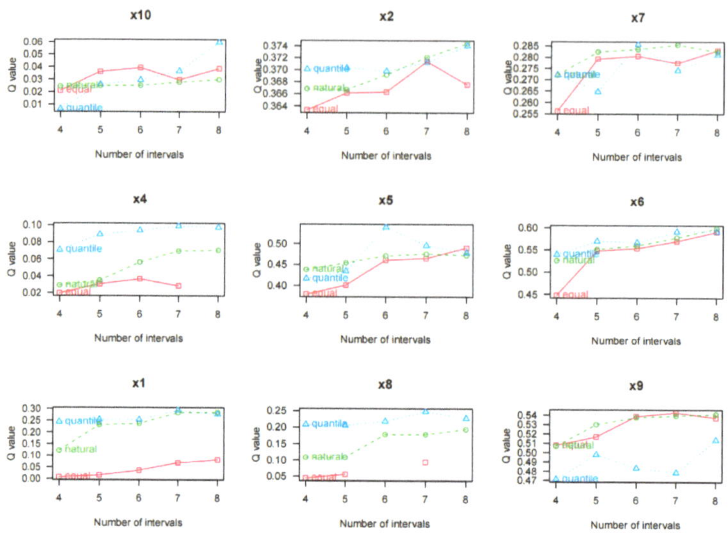

Figure A1. Optimal stratification geographical detector analysis results.

References

1. Renault, D.; Laparie, M.; McCauley, S.J.; Bonte, D. Environmental adaptations, ecological filtering, and dispersal central to insect in-vasions. *Annu. Rev. Entomol.* **2018**, *63*, 345–368. [CrossRef] [PubMed]
2. Perrings, C.; Williamson, M.; Barbier, E.B.; Delfino, D.; Dalmazzone, S.; Shogren, J.; Simmons, P.; Watkinson, A. Biological invasion risks and the public good: An economic perspective. *Conserv. Ecol.* **2002**, *6*, 1. [CrossRef]
3. Ehrenfeld, J.G. Ecosystem consequences of biological invasions. *Annu. Rev. Ecol. Evol. Syst.* **2010**, *41*, 59–80. [CrossRef]
4. Simberloff, D.; Martin, J.-L.; Genovesi, P.; Maris, V.; Wardle, D.A.; Aronson, J.; Courchamp, F.; Galil, B.; García-Berthou, E.; Pascal, M.; et al. Impacts of biological invasions: What's what and the way forward. *Trends Ecol. Evol.* **2013**, *28*, 58–66. [CrossRef]
5. Rai, P.K.; Singh, J.S. Invasive alien plant species: Their impact on environment, ecosystem services and human health. *Ecol. Indic.* **2020**, *111*, 106020. [CrossRef]
6. Lais, C.; Philip, E.H.; Ross, N.C.; Melina, K.; Alok, B.; Phillip, J.H.; Corey, J.A.B.; Paride, B.; Sven, B.; Guillaume, L.; et al. Benefits Do Not Balance Costs of Biological Invasions. *BioScience* **2024**, *74*, 340–344. [CrossRef]
7. Roy, H.E.; Pauchard, A.; Stoett, P.; Renard Truong, T.; Bacher, S.; Galil, B.; Hulme, P.; Ikeda, T.; Sankaran, K.V.; McGeoch, M. IPBES invasive alien species assessment: Full report. *Zenodo* **2024**, *11*, 2024. Available online: https://zenodo.org/records/11629357 (accessed on 15 February 2025).
8. Tang, X.; Yuan, Y.; Liu, X.; Zhang, J. Potential range expansion and niche shift of the invasive *Hyphantria cunea* between native and invasive countries. *Ecol. Èntomol.* **2021**, *46*, 910–925. [CrossRef]
9. Renault, D.; Angulo, E.; Cuthbert, R.N.; Haubrock, P.J.; Capinha, C.; Bang, A.; Kramer, A.M.; Courchamp, F. The magnitude, diversity, and distribution of the economic costs of invasive terrestrial invertebrates worldwide. *Sci. Total Environ.* **2022**, *835*, 155391. [CrossRef]
10. Cuthbert, R.N.; Pattison, Z.; Taylor, N.G.; Verbrugge, L.; Diagne, C.; Ahmed, D.A.; Leroy, B.; Angulo, E.; Briski, E.; Capinha, C.; et al. Global economic costs of aquatic invasive alien species. *Sci. Total Environ.* **2021**, *775*, 145238. [CrossRef]
11. Banha, F.; Diniz, A.M.; Del Amo, R.O.; Oliva-Paterna, F.J.; Anastácio, P.M. Perceptions and Risk Behaviors Regarding Biological Invasions in Inland Aquatic Ecosystems. *J. Environ. Manag.* **2022**, *308*, 114632. [CrossRef] [PubMed]
12. Melissa, J.W.; James, A.T.; Rebecca, S.E.; Juan, J.M.; Tarek, S.; Andrew, P.R.; John, M.K.; Craig, P.; Lloyd, D.S.; Jessica, V.; et al. Approaches for Estimating Benefits and Costs of Interventions in Plant Biosecurity Across Invasion Phases. *Ecol. Appl.* **2021**, *31*, e02319. [CrossRef]
13. Hassel, H.; Cedergren, A. A Framework for Evaluating Societal Safety Interventions. *Saf. Sci.* **2021**, *142*, 105393. [CrossRef]
14. Warren, L.O.; Tadic, M. Fall Webworm, *Hyphantria cunea* (Drury). *Ark. Agric. Exp. Stn. Bull.* **1970**, *759*, 1–106.
15. Umeya, K. Invasion and establishment of a new insect pest in Japan. *Adapt. Speciat. Fall Webworm* **1977**, 1–12.
16. Liu, F.; Su, H.; Liu, T.; Ding, N.; Fang, G. Assessment of Damage Caused by the American White Moth in China in 2021. *China For. Pest Dis.* **2023**, *42*, 15–21.

17. Ge, X.; He, S.; Zhu, C.; Wang, T.; Xu, Z.; Zong, S. Projecting the Current and Future Potential Global Distribution of *Hyphantria cunea* (lepidoptera: Arctiidae) Using CLIMEX. *Pest Manag. Sci.* **2018**, *75*, 160–169. [CrossRef]
18. Zhao, X.; Geng, Y.; Hu, T.; Xie, C.; Xu, W.; Zuo, Z.; Xue, M.; Hao, D. Ecological Strategies of *Hyphantria cunea* (lepidoptera: Arctiidae) Response to Different Larval Densities. *Front. Ecol. Evol.* **2023**, *11*, 1177029. [CrossRef]
19. Chen, Q.; Zhao, H.; Wen, M.; Li, J.; Zhou, H.; Wang, J.; Zhou, Y.; Liu, Y.; Du, L.; Kang, H.; et al. Genome of the webworm *Hyphantria cunea* unveils genetic adaptations supporting its rapid invasion and spread. *BMC Genom.* **2020**, *21*, 1–22. [CrossRef]
20. Cheng, A.T.; Sims, K.R.E.; Yi, Y. Economic development and conservation impacts of China's nature reserves. *J. Environ. Econ. Manag.* **2023**, *121*, 102848. [CrossRef]
21. Yu, C.; Zhang, Z.; Jeppesen, E.; Gao, Y.; Liu, Y.; Liu, Y.; Lu, Q.; Wang, C.; Sun, X. Assessment of the effectiveness of China's protected areas in enhancing ecosystem services. *Ecosyst. Serv.* **2023**, *65*, 101588. [CrossRef]
22. Meng, X.; Gao, X.; Lei, J.; Li, S. Development of a multiscale discretization method for the geographical detector model. *Int. J. Geogr. Inf. Sci.* **2021**, *35*, 1650–1675. [CrossRef]
23. Seebens, H.; Bacher, S.; Blackburn, T.M.; Capinha, C.; Dawson, W.; Dullinger, S.; Genovesi, P.; Hulme, P.E.; van Kleunen, M.; Kühn, I.; et al. Projecting the continental accumulation of alien species through to 2050. *Glob. Change Biol.* **2020**, *27*, 970–982. [CrossRef] [PubMed]
24. Yue, T.X.; Zhao, N.; Liu, Y.; Wang, Y.; Zhang, B.; Du, Z.; Fan, Z.; Shi, W.; Chen, C.; Zhao, M.; et al. A fundamental theorem for eco-environmental surface modelling and its applications. *Sci. China Earth Sci.* **2020**, *63*, 1092–1112. [CrossRef]
25. Cheng, F.; Zhao, G. Fine-scale simulation of population distribution based on zoning strategy and machine learning. *Sci. Surv. Mapp.* **2020**, *45*, 165–173.
26. Tan, J.; Xie, X.; Zuo, J.; Xing, X.; Liu, B.; Xia, Q.; Zhang, Y. Coupling random forest and inverse distance weighting to generate climate surfaces of precipitation and temperature with multiple-covariates. *J. Hydrol.* **2021**, *598*, 126270. [CrossRef]
27. Karger, D.N.; Wilson, A.M.; Mahony, C.; Zimmermann, N.E.; Jetz, W. Global daily 1 km land surface precipitation based on cloud cover-informed downscaling. *Sci. Data* **2021**, *8*, 307. [CrossRef]
28. Wen, X.; Ferreira, A.M.A.; Rae, L.M.; Saffari, H.; Adeel, Z.; Bakkensen, L.A.; Estrada, K.M.M.; Garfin, G.M.; McPherson, R.A.; Vargas, E.F. A comprehensive methodology for evaluating the economic impacts of floods: An application to Canada, Mexico, and the United States. *Sustainability* **2022**, *14*, 14139. [CrossRef]
29. Emmendorfer, L.R.; Dimuro, G.P. A point interpolation algorithm resulting from weighted linear regression. *J. Comput. Sci.* **2021**, *50*, 101304. [CrossRef]
30. Zhao, G.; Yang, M. Urban Population Distribution Mapping with Multisource Geospatial Data Based on Zonal Strategy. *ISPRS Int. J. Geo-Inf.* **2020**, *9*, 654. [CrossRef]
31. De Chaisemartin, C.; d'Haultfoeuille, X. Fuzzy differences-in-differences. *Rev. Econ. Stud.* **2018**, *85*, 999–1028. [CrossRef]
32. Nunn, N.; Qian, N. The potato's contribution to Population and urbanization: Evidence from a historical experiment. *Q. J. Econ.* **2011**, *126*, 593–650. [CrossRef] [PubMed]
33. Goodman-Bacon, A. Difference-in-differences with variation in treatment timing. *J. Econ.* **2021**, *225*, 254–277. [CrossRef]
34. Hortal, J.; Jiménez-Valverde, A.; Gómez, J.F.; Lobo, J.M.; Baselga, A. Historical bias in biodiversity inventories affects the observed environmental niche of the species. *Oikos* **2008**, *117*, 847–858. [CrossRef]
35. Gallardo, B.; Bacher, S.; Barbosa, A.M.; Gallien, L.; González-Moreno, P.; Martínez-Bolea, V.; Sorte, C.; Vimercati, G.; Vilà, M. Risks posed by invasive species to the provision of ecosystem services in Europe. *Nat. Commun.* **2024**, *15*, 2631. [CrossRef]
36. Diagne, C.; Leroy, B.; Gozlan, R.E.; Vaissière, A.-C.; Assailly, C.; Nuninger, L.; Roiz, D.; Jourdain, F.; Jarić, I.; Courchamp, F. InvaCost, a public database of the economic costs of biological invasions worldwide. *Sci. Data* **2020**, *7*, 277. [CrossRef]
37. Paini, D.R.; Sheppard, A.W.; Cook, D.C.; De Barro, P.J.; Worner, S.P.; Thomas, M.B. Global threat to agriculture from invasive species. *Proc. Natl. Acad. Sci. USA* **2016**, *113*, 7575–7579. [CrossRef]
38. Contreras-Díaz, R.G.; Nori, J.; Chiappa-Carrara, X.; Peterson, A.T.; Soberón, J.; Osorio-Olvera, L. Well-intentioned Initiatives Hinder Understanding Biodiversity Conservation: Cloaked iNaturalist Information for Threatened Species. *Biol. Conserv.* **2023**, *282*, 110042. [CrossRef]
39. Diagne, C.; Leroy, B.; Vaissière, A.-C.; Gozlan, R.E.; Roiz, D.; Jarić, I.; Salles, J.-M.; Bradshaw, C.J.A.; Courchamp, F. High and rising economic costs of biological invasions worldwide. *Nature* **2021**, *592*, 571–576. [CrossRef]
40. Mu, H.; Li, X.; Wen, Y.; Huang, J.; Du, P.; Su, W.; Miao, S.; Geng, M. A Global Record of Annual Terrestrial Human Footprint Dataset from 2000 to 2018. *Sci. Data* **2022**, *9*, 176. [CrossRef]

biology

MDPI

Article

Distribution and Ecological Risk of *Ludwigia peploides* in South Korea

Aram Jo [1,2,*], Soo In Lee [1], Donghui Choi [1], Youngha Kim [3], Yong Ho Lee [4] and Sun Hee Hong [4]

[1] Invasive Alien Species Team, National Institute of Ecology, Seocheon 33657, Republic of Korea
[2] Department of Biological Science, Kongju National University, Kongju 32588, Republic of Korea
[3] National Ecosystem Survey Team, National Institute of Ecology, Seocheon 33657, Republic of Korea
[4] School of Applied Science in Natural Resources & Environment, Hankyong National University, Anseong 17579, Republic of Korea
* Correspondence: joaram@nie.re.kr; Tel.: +82-41-950-5818

Simple Summary: The water primrose (*Ludwigia peploides*) is believed to have been planted as an ornamental aquatic plant in South Korea. It spread to natural ecosystems through rivers, and its distribution is gradually expanding in Suwon, Hwaseong, Busan, and Jeju. However, there has been no specific study on the ecological risk of *L. peploides* introduced into South Korea. This study, therefore, investigates the distribution status and ecological risks of *L. peploides* in South Korea through field surveys and allelopathic material analysis, as well as assessing abiotic risk factors. The distribution was confirmed at a total of 19 sites. The species was observed to primarily inhabit aquatic environments but were also observed to inhabit moist terrestrial areas such as riverbanks or wet grasslands. The results of this study are expected to aid in the identification of the current distribution and potential ecological risks of *L. peploides* in South Korea, providing vital information for the development of effective management strategies.

Abstract: The number of alien species introduced into South Korea continues to increase over the years. In particular, several plants have been introduced as ornamentals. *Ludwigia peploides*, which is native to the Americas and Australia, is believed to have been planted as an ornamental aquatic plant called "water primrose" and "primrose". It spread to natural ecosystems through rivers, and its distribution is gradually expanding in Suwon, Hwaseong, Busan, and Jeju. However, there has been no specific study on the ecological risk of *L. peploides* introduced into South Korea. This study, therefore, investigates the distribution status and ecological risks of *L. peploides* in South Korea through field surveys and allelopathic material analysis, as well as assessing abiotic risk factors. The distribution was confirmed at a total of 19 sites, with high-density mats of a single species forming along the water's edge and on the water surface. The maximum distribution area was 13,922 m^2 in Chilgok Reservoir in Anseong. Stems and plant fragments transported along waterways were continuously forming colonies through vegetative propagation. When evaluating the overall risk, it is determined that *L. peploides* has a high potential to cause significant damage to the ecosystem if not managed promptly. Therefore, continuous monitoring is necessary to effectively manage and prevent the habitat expansion of *L. peploides*. The results of this study are expected to aid in the identification of the current distribution and potential ecological risks of *L. peploides* in South Korea, providing essential data for ecological risk assessment and the development of effective management strategies.

Keywords: *Ludwigia peploides*; invasive alien plants; ecological risk; field survey

1. Introduction

The disappearance of the geographical boundaries for species due to globalization, along with habitat environmental changes caused by climate change and global warming, provides opportunities for some alien species to migrate and establish themselves in new habitats.

In some cases, these species, referred to as invasive alien species (IAS), negatively impact the richness and abundance of native species, potentially driving them to extinction and significantly threatening the biodiversity [1–4]. Additionally, the economic damage caused by IAS, affecting agriculture, livestock industries, infrastructure, and more, exceeds USD 1.28 trillion globally [5]. The international community has made efforts to prevent the invasion and spread of alien species. Several parties gathered and adopted the Kunming-Montreal Global Biodiversity Framework (GBF) in Montreal, Canada, in 2022. Among the 23 targets of the framework, the sixth is to "reduce the introduction of IAS by 50% and minimize their impact". This goal is to be achieved by 2030 in 19 countries, including South Korea [6].

The number of confirmed alien species has dramatically increased in South Korea from 1109 species in 2011 to 2653 species in 2021 [7]. Among the several alien species that have entered the country, the Ministry of Environment designates those considered highly harmful to the ecosystem as "ecosystem-disturbing species" based on the Act on the Conservation and Use of Biological Diversity (hereafter, the "Biodiversity Act"). These species are prioritized for management, and their import, cultivation, sale, transportation, and release are legally prohibited [8]. Each year, the Ministry of Environment conducts an ecological risk assessment of specific alien species reported to be harmful, as per Article 21-2 of the Biodiversity Act, to determine whether they should be designated as ecosystem-disturbing species. This assessment involves a literature review on the impact of the species in invaded regions and field surveys and analyses to determine if they pose similar risks to the domestic ecosystem.

Aquatic and riparian ecosystems are highly vulnerable to the invasion of invasive alien aquatic plants (IAAPs) [9,10]. River corridors and waterways are among the most heavily invaded ecosystems globally [11]. The primary reasons for this are the heterogeneous and diverse structures of river ecosystems, high connectivity, and the ease of propagule transportation [12,13]. *Ludwigia* genus are among the most aggressive, invasive plants in the world [14,15]. One of them, *Ludwigia peploides* ((Kunth) P.H. Raven, 1963) is a perennial water plant belonging to the Onagraceae family, native to North America, South America, and Australia. It has now spread worldwide to the United States, Australia, New Zealand, France, Belgium, and Greece in southern Europe due to the increased use of *Ludwigia* species as ornamental aquatic plants [16,17]. It exhibits a great deal of morphological variation depending on habitat conditions with stems tending to be erect in shallow waters and spreading laterally underwater [14,18]. Stems that spread at the surface of the water produce new roots at the nodes, and cut stems and leaves can also form new plants through vegetative propagation. Seeds and plant parts can spread long distances along the water. High-density mats of a single species formed on the water surface can replace the habitat of existing native plants, reduce species diversity, and cause several problems such as increasing sedimentation and restricting water flow when they proliferate at reservoir entrances or waterways [14,16,17,19].

In South Korea, there is a record of this plant being studied for horticultural purposes in 2003, and it has been distributed and planted as ornamentals, such as "water primrose" and "primrose". *L. peploides* have spread into natural ecosystems through rivers, and their distribution is gradually expanding in Suwon, Hwaseong, and Busan [20]. However, there has been no specific study on the ecological risk of *L. peploides* introduced into South Korea. This study, therefore, investigates the distribution status and ecological risks of the *L. peploides* in South Korea through field survey and allelopathic material analysis, as well as assessing abiotic risk factors such as water quality, temperature, depth, and flow rate. This is the first ecological risk analysis of *L. peploides* in South Korea, addressing biotic and abiotic invasion factors.

2. Materials and Methods

The investigation items for assessing the ecological risk *of L. peploides* followed the ecological risk assessment table outlined in regulation No. 239 (2023) of the National

Institute of Ecology (NIE). The risks not observed in domestic populations were compiled through a review of the literature.

2.1. Distribution Status and Biological Risk Factors

To identify the distribution of *L. peploides*, the areas where the distribution has earlier been reported were collected from the Guide for Nationwide Survey of Non-native Species in Korea (2015–2022) and the National Ecosystem Survey (2014–2022), along with literature reviews of papers and reports documenting its distribution in South Korea. Among the reported areas, four sites with high population densities—Boryeong, Osan, Busan, and Anseong—were selected for fixed plot survey (one point in the lentic zone and three points in the lotic zone). Quadrats (1 m × 1 m, n = 6) were established, and vegetation surveys were conducted using the Braun-Blanquet method. The relative frequency (RF) and relative coverage (RC) of species within the quadrats were used to calculate the species diversity index (Shannon–Wiener index) and species evenness index. Differences in these indices based on the presence of *L. peploides* were tested using one-way ANOVA (R, ver. 4.3.1, San Francisco, CA, USA). A flora survey was conducted to identify species co-occurring with *L. peploides*, and the importance value (IV) was calculated using the average of RF and RC to determine the importance of each species within the populations.

Fixed survey areas were set in Boryeong, Osan, Busan, and Anseong to investigate the spread of *L. peploides*. Within these survey areas, the distribution area was measured three times. The first measurement was from May to June, the second from July to August, and the third from September to October. The distribution area observed above the water surface was determined by visual inspection and drone shooting. The distribution area calculations were performed using Quantum Geographic Information System (QGIS ver. 3.16.8). The populations were continuously observed along the 1.5 km from Gonam 2 Bridge to Yeonji Bridge in Bongdangcheon Stream in Boryeong. However, for the quantitative analysis, a certain section (110 m × 30 m) was selected as the survey area where distribution monitoring was more manageable. The survey area (380 m × 5 m) was set along the stream flowing through the area around Malgeumteo Park in Osancheon Stream in Osan. In Busan, the survey area was set along approximately 640 m of the Suyeong River flowing from Sincheon Bridge towards Hoedong Reservoir. Chilgok Reservoir in Anseong was the only lentic zone among four fixed survey areas; the entire reservoir was, therefore, set as the scope of the survey. Additionally, to assess productivity, spots with 100% coverage of *L. peploides* were selected during its growing season in July and September at Chilgok Reservoir, which had the largest distribution area among the fixed survey areas. Quadrats (1 m × 1 m, n = 5) were used to measure the number of flowers, fruits, and seeds per unit area (1 m^2) at each spot.

L. peploides has been reported to exhibit high invasiveness, including reducing the germination rates of surrounding aquatic plants and exerting toxic effects by using allelopathic compounds, which are more abundant compared to other aquatic plants [21,22]. Allelopathic interactions may play a key role in the success of invasive plants because they can alter physiological processes and thereby influence community structure [23,24]. Secondary compounds previously researched in the *Ludwigia* genus were polyphenols, flavonoids, quercitrin, prunin, myricitrin, etc. [21,25]. Additionally, the research on selecting secondary compounds for evaluating the risk of invasive alien plants indicated that when compounds such as total phenolic compounds (TPCs), caffeic acid, and p-coumaric acid are present in high concentrations, there is a higher likelihood of negative effects on native plants [26]. In this study, therefore, six compounds were included for the analysis of allelopathic compounds: the total flavonoid content (TFC), total phenolic content (TPC), caffeic acid, p-coumaric acid, myricitrin, and prunin. Leaf and root samples for allelopathic analysis were collected in August from Boryeong, Osan, and Suwon, and the analysis was conducted at the National Instrumentation Center for Environmental Management (NICEM, Seoul, Republic of Korea). TFC and TPC analyses were performed using a spectrophotometer (Spectramax ABS Plus, Molecular Devices, San Jose, CA, USA), while the other compounds were analyzed using HPLC (high-performance liquid chromatography, Ultimate3000, Thermo Dionex, Walthamma, MA, USA). Differences in allelopathic compounds by plant part (leaves and roots) and by habitat were tested using one-way ANOVA (R, ver. 4.3.1) to understand the mechanism of allelopathy and variations in compound types and levels across different regions.

2.2. Abiotic Risk Factors

In the risk assessment of invasive alien species, the habitat characteristics are fundamental abiotic factors that should be investigated. For non-native aquatic plants, it is important to determine whether they grow under the same aquatic ecosystem conditions reported to be their native range or if they adapt and establish themselves in new domestic environments. Additionally, water quality analysis can provide valuable information for identifying suitable habitats for their spread and establishment. At the five sites in Osan, Boryeong, Busan, Daegu, and Anseong, 2 L water samples were collected for analysis of water quality using sterile sampling bottles in September and October after the rainy season. The samples were stored in the dark at 4 °C until analysis. Water quality analysis of samples was performed by Gyeonggi Research Institute of Environmental Sciences (GRIES, Seoul, Korea). The total organic carbon content (TOC) was measured using the high-temperature combustion oxidation method, and the total phosphorus (TP) was determined using ultraviolet–visible spectrophotometry. The chemical oxygen demand (COD) was measured using the acidified potassium permanganate method, and biochemical oxygen demand (BOD) was determined using the amount of dissolved oxygen consumed by aerobic microbial growth and respiration. Suspended solids (SSs) were measured using the gravimetric method with glass fiber filters. Additionally, a multi-parameter water quality meter (YSI-556MPS) was used to measure water temperature, pH, water depth, and flow rate at the habitats to determine how these parameters influence the distribution of *L. peploides*.

3. Results

3.1. Distribution Status

A total of 25 sites were identified with reported occurrences of *L. peploides* through literature reviews of papers and reports (Figure 1). During the initial field survey (April to early May 2023), few individuals were found at most sites except in Anseong, Osan, and Busan. However, in the second field survey (June to August), distribution was confirmed along the Seoho Stream in Suwon, from the Jangan-gu to the Paldal-gu and Gwonseon-gu, and also found in Daegu and Jeju. Only a very small number of individuals were found in Ilwang Reservoir in Suwon; a large population was previously reported, which appears to have resulted from the continuous removal of aquatic plants. Distribution was confirmed at 19 out of the 25 sites in the field survey.

3.2. Biological Risk Factors

A total of 25 *L. peploides* primarily inhabit aquatic environments but can also form populations on moist terrestrial areas such as riverbanks or wet grasslands. It is primarily found in wetlands and water–land boundaries in its native range [17]. It is relatively frost-resistant and has been reported to grow in water up to 3 m deep [27]. It has, however, been observed to inhabit waters up to 80 cm deep and is found at the water edge in rivers and reservoirs or in shallow land exposed to receding water levels in South Korea (Figure 2). As previously reported [28], domestic populations also demonstrated morphological variations, with erect stems, when growing on land and sprawling stems on the water surface.

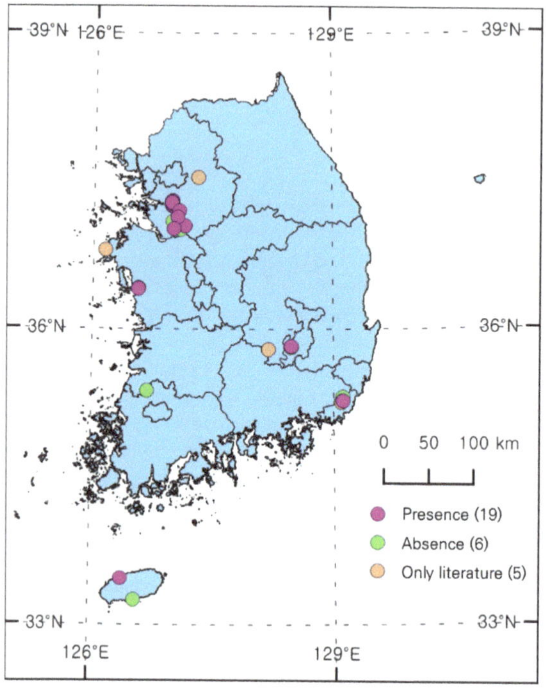

Figure 1. Distribution status of *L. peploides* in South Korea.

Figure 2. Habitats of *L. peploides* in South Korea. Boryeong: measuring the depth of the habitat of *L. peploides* in the water edge of the river. Osan: *L. peploides* inhabiting the boundary between the river and land. Suwon: large blooming populations of *L. peploides* in shallow land exposed to receding water levels under the bridge. Anseong: based on the road running horizontally in the middle, the reservoir below is entirely covered with the blooming *L. peploides* with yellow flowers.

The vegetation survey demonstrated that species evenness was significantly lower where *L. peploides* appeared ($p < 0.0954$). However, while species diversity tended to be slightly low, it did not show a significant difference ($p < 0.123$). The flora survey was conducted to identify species co-occurring with *L. peploides*. A total of 165 species of plants across 123 genera were observed, and no endangered plants were observed. However, the IV of emergent hydrophytes such as *Zizania latifolia*, *Phragmites australis*, *Typha orientalis*, and *Miscanthus sacchariflorus* presented a significant decrease.

The distribution area of *L. peploides* was calculated within the fixed plot survey. In the Bongdangcheon Stream in Boryeong, the initial survey presented the largest area of 155 m^2. However, the population was submerged below the water surface due to the higher water level caused by the summer rainy season. Further, the distribution tended to be pushed out by the competition with the co-occurring emergent plants compared to other habitats (Figure 3). However, after the rainy season, the distribution displayed dense growth on the open water surface. A similar pattern was observed in the Chilgok Reservoir of Anseong City, with an initially large area decreasing during the rainy season. In the case of Osancheon Stream in Osan and Suyeong River in Busan City, the plants presented maximum growth in August, with areas of 802 m^2 and 2585 m^2, respectively. However, in the Osancheon Stream, the distribution area decreased significantly due to river landscape management, and, in the Suyeong River, the distribution area decreased due to high water levels.

Small patches of *L. peploides* were continuously found outside the fixed plot along the water in the Bongdangcheon Stream and Suyeong River. There was a total of 38 patches in the Bongdangcheon Stream, most of which covered areas of less than 10 m^2. However, it was found that more than half of these patches had a cover greater than eight in the Braun-Blanquet method (Figure 4). The patches were continuously observed along approximately 1.8 km from Sincheon Bridge to the entrance of Hoedong Reservoir in Suyeong River. Seed production was measured in Chilgok Reservoir in Anseong to assess productivity. In July, the number of flowers per m^2 was 36 (±8), the number of fruits was 316 (±128), and the number of seeds was 15,561 (±7163). In September, the number of flowers per m^2 was 15 (±2), the number of fruits was 63 (±12), and the number of seeds was 753 (±358).

The analysis of allelopathic compounds in the populations of *L. peploides* from Boryeong, Osan, and Suwon revealed high levels of TFC and TPC, with significantly more detected in the leaves than in the roots ($p < 0.001$, Table 1). This trend was consistent across all regions, although there were significant differences in content levels between regions ($p < 0.001$). The TFC detected in the leaves had an average value of 8.40 mg/g, with Boryeong showing the highest value at 9.56 mg/g. The TPC had an average value of 39.45 mg/g, with Boryeong and Suwon showing very high values exceeding 40 mg/g. Similarly, Boryeong had the highest amount of caffeic acid at 219.72 mg/kg, with more than seven times the amount detected in the leaves compared to the roots. P-coumaric acid was detected only in the roots across all regions, with an average value of 16.17 mg/kg, while prunin was detected only in the leaves across all regions, with an average value of 514.44 mg/kg.

Table 1. The content level of allelopathic compounds between three regions (mean ± SE, $n = 3$). TFC: total flavonoid content; TPC: total phenolic content. The hyphen symbol indicates that the compound was not detected.

Region	Part	TFC (mg/g)	TPC (mg/g)	Caffeic Acid (mg/kg)	Myricitrin (mg/kg)	p-Coumaric Acid (mg/kg)	Prunin (mg/kg)
Boryeong	Leaf	9.56 ± 0.02	48.29 ± 0.68	219.72 ± 1.34	-	-	457.60 ± 3.80
	Root	2.18 ± 0.02	7.06 ± 0.11	27.65 ± 1.80	-	17.73 ± 0.24	-
Osan	Leaf	7.77 ± 0.10	27.96 ± 0.08	142.14 ± 0.42	-	-	499.81 ± 1.30
	Root	0.78 ± 0.04	0.43 ± 0.02	-	-	15.94 ± 0.06	-
Suwon	Leaf	7.88 ± 0.01	42.09 ± 0.33	147.94 ± 1.13	8681.84 ± 26.13	-	585.90 ± 2.57
	Root	1.27 ± 0.04	0.41 ± 0.01	19.58 ± 0.03	-	14.84 ± 0.25	-

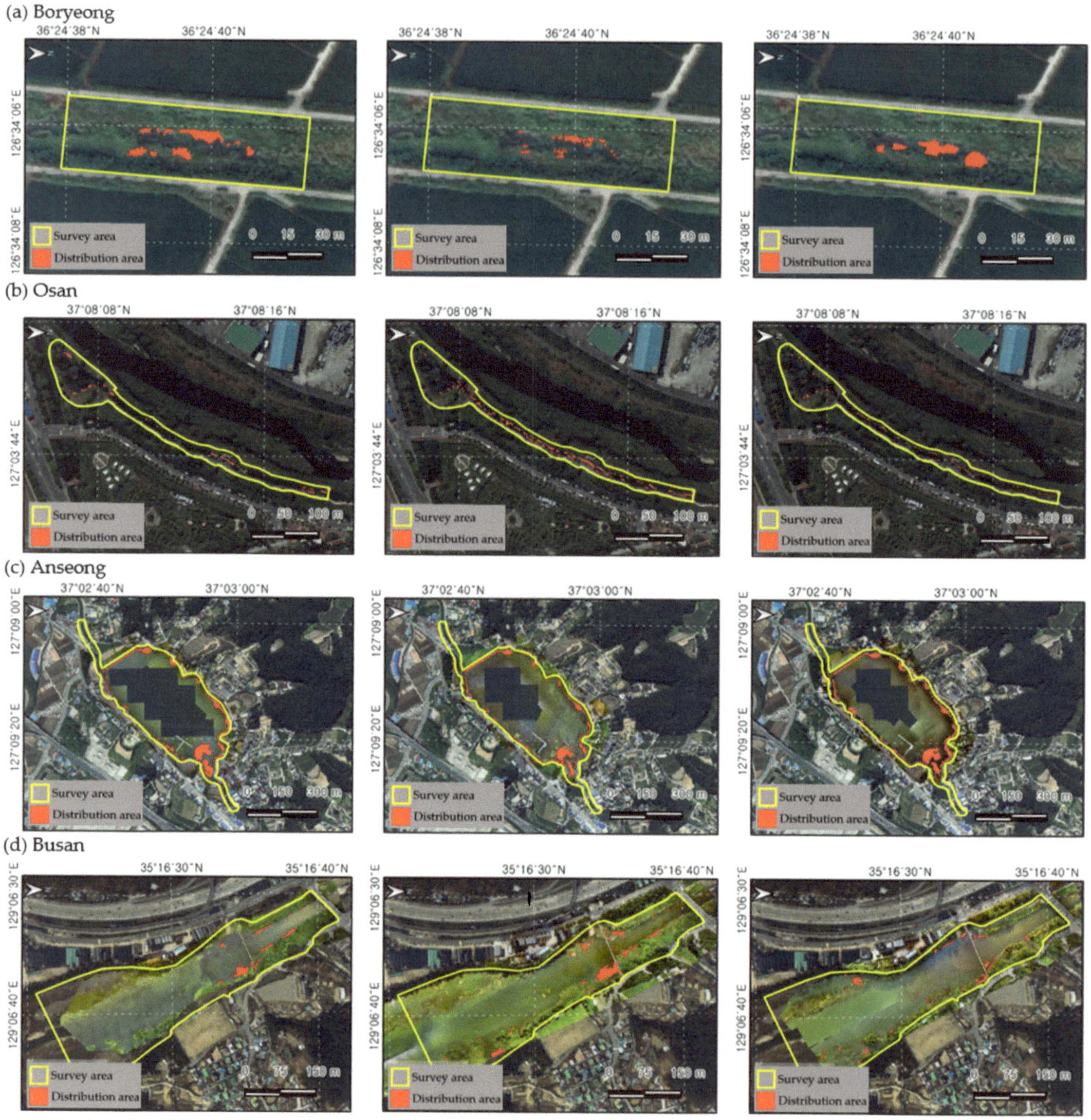

Figure 3. The results of the fixed plot distribution area survey to investigate the spread of *L. peploides* over time. (**a**) Bongdangcheon Stream, Byeong, (**b**) Osancheon Stream, Osan, (**c**) Chilgok Reservoir, Anseong, and (**d**) Sutyeong River, Busan. The yellow zone represents the fixed plot survey area, and the red zone shows the distribution area of *L. peploides*. From the left, the first column represents the first survey, measured from May to June; the second column represents the second survey, measured from July to August; and the third column represents the third survey, measured from September to October.

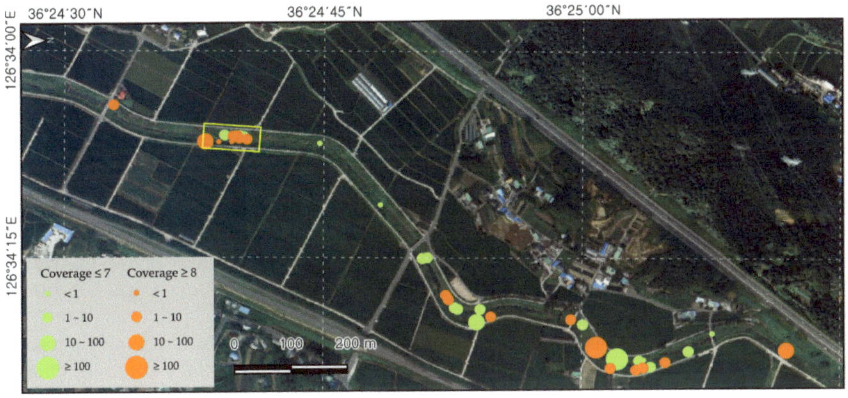

Figure 4. Distribution patches of *L. peploides* along the water in Bongdangcheon Stream, Boryeong. The yellow zone represents the fixed plot survey area, and the orange and circles shows the patches of *L. peploides*.

3.3. Abiotic Risk Factors

Analysis of water quality at five sites—Osan, Boryeong, Busan, Daegu, and Anseong—demonstrated that most parameters were in the slightly good range (Table 2). However, the levels of TOC and TP were found to be at very poor levels (exceeding the standards of 8 and 0.5, respectively) in the Osancheon Stream. Similarly, the COD of the Chilgok Reservoir was at a very poor level (exceeding the standard of 10). The pH was in the range of pH 8 to 9 at all sites, which was more alkaline than the common range of general aquatic plants (pH 6.5 to 7.5). Furthermore, when measuring water depth as per the presence or absence of *L. peploides*, the average water depth of areas where the plant was present was 39.3 cm ± 11.9 (n = 15); this was lower than the water depth of areas where the plant was absent (65.3 cm ± 34.5 (n = 15)). During the early stages of growth, *L. peploides* was primarily observed at the edges of slow-flowing rivers and reservoirs, and, during the middle and late growth stages, it appeared to spread across the water surface.

Table 2. The results of the water quality analysis at five sites. TOC: total organic carbon, TP: total phosphorus, COD: chemical oxygen demand, BOD: biochemical oxygen demand, SS: suspended solids.

Parameter (mg/L)	Osan	Boryeong	Busan	Dague	Anseong	Average
TOC	8.3	2.7	2.2	6.1	3.9	4.64
TP	0.879	0.059	0.27	0.653	0.077	0.3876
COD	10.5	4.4	5.8	9	13.3	8.6
BOD	3.3	1.5	2.1	3	2.6	2.5
SS	5	3.2	67	27.8	9.2	22.44

4. Discussion

The genus *Ludwigia* is one of the largest and most diverse groups within the Onagraceae family. These plants are known as one of the taxa that are very difficult to identify due to their morphological similarity and wide morphological variation depending on the habitat environment [20,28]. Additionally, almost all species in the Oligospermum section, a subunit of the genus *Ludwigia*, actively hybridize, and hybrid populations adapted to their habitats can increase their invasiveness [29,30]. *L. peploides*, first identified in Suwon, is distributed in temperate and tropical regions and grows in both terrestrial and aquatic environments. As dense mats of *Ludwigia* spp. spread, they physically restrict the habitats of fish and aquatic invertebrates, and if they occur along fish migration routes, they can limit fish movement, potentially reducing biodiversity in the invaded areas [31,32]. Additionally, due to their rapid reproduction and the ability to propagate vegetative propagation from cut stems, controlling them is very challenging. Due to its harmful impact on ecosystems and agriculture, *L. peploides* is included in the European list of invasive alien species of Union concern (the Union list), and its possession, import, sale, breeding, and release into natural ecosystems are prohibited [33]. It is also included in Portugal's national invasive species list, where its possession, breeding, and trade are prohibited [34]. In the UK, under Schedule 9 of the Wildlife and Countryside Act 1981, it is illegal to plant this species in the wild in England and Wales [35]. In New Zealand, it is classified as an "Unwanted Organism" under the Biosecurity Act 1993, and permission must be obtained for its movement, trade, and release [36]. In the wetlands of Portland, USA, labor costs for removing *L. peploides* exceeded KRW 57 million won, and physical removal costs from 2012 to 2014 amounted to approximately KRW 5 million won per acre; yet complete removal was not achieved [29,37].

It is understood that *L. peploides* has been introduced to South Korea as an ornamental plant and is currently observed nationwide including Suwon, Osan, Boryeong, Daegu, Anseong, and Jeju. The plant prefers slow-flowing rivers, wetlands, and stagnant reservoirs and is observed concentrated on the border between land and water surface of rivers or reservoirs. In South Korea, it is believed that a population initially planted in Ilwang Reservoir in Suwon has spread downstream to Seohocheon Strem [20]. Similarly, small patches were consistently found along the water in Boryeong and Busan. Therefore, although it is currently forming large populations in certain regions nationwide, it may continue to expand its distribution along the water. As a result of observing the change in distribution area in four fixed survey areas with high population density, it appears that July to September is the maximum growth period in South Korea. Although the rise in water levels during the rainy season influenced the growth and inflorescence of *L. peploides*, it did not significantly impact the subsequent distribution area expansion. However, the result of the habitat survey of *L. peploides* demonstrates that the population formation began with rooting on the soil at the border between land and water and spread to shallow water during the early stage of growth. It, therefore, indicates that the change in water level, affected by rain and reservoir water use, is the greatest factor in the formation of populations of *L. peploides* in South Korea.

It has been reported that *L. peploides* has tolerance to various water quality environments [21,38]. Similarly, the analysis of domestic habitats demonstrated that it actively grew even in poor water quality conditions, such as high levels of TOC and TP. It, therefore, appears that water pollution does not have a significant impact on the domestic habitat of *L. peploides*. Additionally, secondary compounds were analyzed to determine the allelopathic action of *L. peploides* on surrounding plants, and significant amounts of phenolic acids were detected, particularly in the leaves. These results are consistent with previous studies that found *L. peploides* has high invasiveness due to its higher TPC compared to other aquatic plants [22]. Phenolic compounds significantly contribute to the adaptation and growth of invasive alien plants, particularly aquatic plants, in new habitats and play a crucial role in protecting them from predation or pathogenic invasion. Invasive plants, therefore, with high phenolic compound content may have a competitive advantage over native plants [21,39]. However, further research is needed to understand the specific mechanisms by which the secondary compounds of *L. peploides* affect native plants coexisting in its habitats and the surrounding environment in South Korea.

The loss of biodiversity due to *L. peploides* was observed to be weak in this study. The results of the vegetation survey demonstrated that although the IV of co-occurring aquatic plants such as reed and bulrush decreased, there was no significant difference in species diversity. However, the high-density population mat of *L. peploides* can quickly cover the water surface, physically displacing native plant habitats and blocking light essential for submerged species. This can alter the habitat environment and potentially reduce the biodiversity of not only plants but also fish and aquatic invertebrates [14,16,21,30]. In addition, although no endangered plants were observed during the survey of flora, there have been instances in Belgium where *L. peploides* invaded protected areas with high biodiversity [19]. Since this study was conducted based on one year of observation, additional research is needed in the future to specifically, identify the risks and the impact of *L. peploides* on ecosystems in South Korea.

5. Conclusions

Once invaded, *L. peploides* can rapidly expand its distribution area through seed reproduction and vegetative propagation. It can survive over winter both under the water surface and below ground and repeat the generation the following year. It is, therefore, difficult to completely eradicate it. When evaluating the overall risk, it is determined that *L. peploides* has a high potential to cause significant damage to the ecosystem if not managed promptly. Therefore, continuous monitoring is necessary to effectively manage and prevent habitat expansion of *L. peploides*. Since this study was conducted based on one year of observation, additional research is needed in the future to specifically identify the risks and the impact of *L. peploides* on ecosystems in South Korea. The results of this study can help identify the current distribution and potential ecological risks of *L. peploides* in South Korea, providing essential data for ecological risk assessment. As the first nationwide survey of *L. peploides* distribution, it can be used in the future to prioritize regions for management to prevent further spread and serve as foundational data for developing effective management strategies. Additionally, this study will lay the groundwork for building risk assessment data for aquatic alien plants, aiding in the future development of more advanced and effective risk assessment methods.

Author Contributions: Conceptualization, A.J. and S.I.L.; formal analysis, A.J.; investigation, A.J., S.I.L., D.C., Y.K., S.H.H. and Y.H.L.; writing—original draft preparation, A.J.; writing—review and editing, A.J.; visualization, D.C.; administration, A.J. All authors have read and agreed to the published version of the manuscript.

Funding: This work was supported by a grant from the National Institute of Ecology (NIE), funded by the Ministry of Environment (MOE) of the Republic of Korea (NIE-A-2024-08 and NIE-C-2024-09).

Institutional Review Board Statement: Not applicable.

Informed Consent Statement: Not applicable.

Data Availability Statement: All data generated or analyzed during this study are included in this published article.

Conflicts of Interest: The authors declare no conflicts of interest. The funders had no role in the design of the study; in the collection, analyses, or interpretation of data; in the writing of the manuscript; or in the decision to publish the results.

References

1. Pyšek, P.; Jarošík, V.; Hulme, P.E.; Pergl, J.; Hejda, M.; Schaffner, U.; Vilà, M. A global assessment of invasive plant impacts on resident species, communities and ecosystems: The interaction of impact measures, invading species' traits and environment. *Glob. Chang. Biol.* **2012**, *18*, 1725–1737. [CrossRef]
2. Vilà, M.; Espinar, J.L.; Hejda, M.; Hulme, P.E.; Jarošík, V.; Maron, J.L.; Pergl, J.; Schaffner, U.; Sun, Y.; Pyšek, P. Ecological impacts of invasive alien plants: A meta-analysis of their effects on species, communities and ecosystems. *Ecol. Lett.* **2011**, *14*, 702–708. [CrossRef] [PubMed]
3. Mollot, G.; Pantel, J.H.; Romanuk, T.N. The effects of invasive species on the decline in species richness. *Adv. Ecol. Res.* **2017**, *56*, 61–83.
4. Vilà, M.; Basnou, C.; Pyšek, P.; Josefsson, M.; Genovesi, P.; Gollasch, S.; Nentwig, W.; Olenin, S.; Roques, A.; Roy, D. How well do we understand the impacts of alien species on ecosystem services? A pan-European, cross-taxa assessment. *Front. Ecol. Environ.* **2010**, *8*, 135–144. [CrossRef]
5. Diagne, C.; Leroy, B.; Vaissière, A.C.; Gozlan, R.E.; Roiz, D.; Jarić, I.; Salles, J.M.; Bradshaw, C.J.A.; Courchamp, F. High and rising economic costs of biological invasions worldwide. *Nature* **2021**, *592*, 571–576. [CrossRef]
6. Convention on Biological Diversity. Decision adopted by the Conference of the Parties to the Convention on Biological Diversity [CBD/COP/DEC/15/4]. Kunming-Montreal Global Biodiversity Framework. 2022. Available online: https://www.cbd.int/doc/decisions/cop-15/cop-15-dec-04-en.pdf (accessed on 28 June 2023).
7. National Institute of Ecology (NIE). *Guidebook for National Inhabitants Survey of Alien Species*; National Institute of Ecology: Chungcheongnam-do, Republic of Korea, 2021; 161p, ISBN 979-11-6698-022-0. (In Korean)

8. Kim, D.E. Management system of invasive alien species threatening biodiversity in Korea and suggestions for the improvement. *J. Environ. Impact Assess.* **2018**, *27*, 33–55.
9. Hussner, A.; Heidbüchel, P.; Coetzee, J.; Gross, E.M. From introduction to nuisance growth: A review of traits of alien aquatic plants which contribute to their invasiveness. *Hydrobiologia* **2021**, *848*, 2119–2151. [CrossRef]
10. Santos, M.J.; Anderson, L.W.; Ustin, S.L. Effects of invasive species on plant communities: An example using submersed aquatic plants at the regional scale. *Biol. Invasions* **2011**, *13*, 443–457. [CrossRef]
11. Hejda, M.; Chytrý, M.; Pergl, J.; Pyšek, P. Native-range habitats of invasive plants: Are they similar to invaded-range habitats and do they differ according to the geographical direction of invasion? *Divers. Distrib.* **2015**, *21*, 312–321. [CrossRef]
12. Bolpagni, R. Towards global dominance of invasive alien plants in freshwater ecosystems: The dawn of the Exocene? *Hydrobiologia* **2021**, *848*, 2259–2279. [CrossRef]
13. Gurnell, A.; Thompson, K.; Goodson, J.; Moggridge, H. Propagule deposition along river margins: Linking hydrology and ecology. *J. Ecol.* **2008**, *96*, 553–565. [CrossRef]
14. Thiébaut, G.; Dutartre, A. Management of invasive aquatic plants in France. In *Aquatic Ecosystem Research Trends*; Nairne, G.H., Ed.; Nova Science Publishers, Inc.: New York, NY, USA, 2009; pp. 25–46. ISBN 978-1-60692-772-4.
15. Thouvenot, L.; Haury, J.; Thiébaut, G. A success story: Water primroses, aquatic plant pests. *Aquat. Conserv. Mar. Freshw. Ecosyst.* **2013**, *23*, 790–803. [CrossRef]
16. Zotos, A.; Sarika, M.; Lucas, E.; Dimopoulos, P. *Ludwigia peploides* subsp. montevidensis: A New Alien. Taxon. for the Flora of Greece and the Balkans. *J. Biol. Res.* **2006**, *5*, 71–78.
17. Buzjak, S.; Sedlar, Z. *Ludwigia peploides* (Kunth.) P.H. Raven–floating water primrose, a new species in Croatian Flora from the list of invasive alien species of union concern. *Nat. Croat.* **2018**, *27*, 351–356. [CrossRef]
18. Ruaux, B.; Greulich, S.; Haury, J.; Berton, J.P. Sexual reproduction of two alien invasive *Ludwigia* (Onagraceae) on the middle Loire River, France. *Aquat. Bot.* **2009**, *90*, 143–148. [CrossRef]
19. Robert, H.; Lafontaine, R.-M.; Beudels-Jamar, R.C.; Delsinne, T. *Risk Analysis of the Water Primrose Ludwigia peploides (Kunth) P.H. Raven—Risk Analysis Report of Non-Native Organisms in Belgium*; Royal Belgian Institute of Natural Sciences for the Federal Public Service Health, Food Chain Safety and Environment: Brussels, Begium, 2013; 35p.
20. Kim, H.W.; Son, D.C.; Park, S.H.; Jang, C.S.; Sun, E.M.; Jo, H.Y.; Yun, S.M.; Chang, K.S. Unrecorded alien plant in South Korea: *Ludwigia peploides* subsp. *montevidensis* (Spreng.) P.H. Raven. *Korean J. Plant Res.* **2019**, *32*, 201–206.
21. Dandelot, S.; Robles, C.; Pech, N.; Cazaubon, A.; Verlaque, R. Allelopathic potential of two invasive Alien *Ludwigia* spp. *Aquat. Bot.* **2008**, *88*, 311–316. [CrossRef]
22. Grutters, B.M.C.; Saccomanno, B.; Gross, E.M.; Van de Waal, D.B.; van Donk, E.; Bakker, E.S. Growth strategy, phylogeny and stoichiometry determine the allelopathic potential of native and non-native plants. *Oikos* **2017**, *126*, 1770–1779. [CrossRef]
23. Inderjit; Duke, S.O. Ecophysiological aspects of allelopathy. *Planta* **2003**, *217*, 529–539.
24. Bais, H.P.; Vepachedu, R.; Gilroy, S.; Callaway, R.M.; Vivanco, J.M. Allelopathy and exotic plant invasion: From molecules and genes to species interactions. *Science* **2003**, *301*, 1377–1380. [CrossRef]
25. Thiébaut, G.; Thouvenot, L.; Rodríguez-Pérez, H. Allelopathic effect of the invasive *Ludwigia hexapetala* on growth of three macrophyte species. *Front. Plant Sci.* **2018**, *9*, 1835. [CrossRef] [PubMed]
26. Ministry of Environment. *Genetic and Ecological Risk Assessment of the Invasion of Alien Plants on Native Species*; Ministry of Environment: Tokyo, Japan, 2006.
27. Lambert, E.; Dutartre, A.; Coudreuse, J.; Haury, J. Relationships between the biomass production of invasive *Ludwigia* species and physical properties of habitats in France. *Hydrobiologia* **2010**, *656*, 173–186. [CrossRef]
28. Zardini, E.M.; Peng, C.I.; Hoch, P.C. Chromosome numbers in *Ludwigia* sect. *Oligospermum* and Sect. *Oocarpon* (Onagraceae). *Taxon* **1991**, *40*, 221–230. [CrossRef]
29. Staunch, A. The Blue Heron Wetland Restoration Project: Eradication of *Ludwigia peploides* ssp. *montevidensis* from the Blue Heron Wetlands of Portland, OR. Master's Thesis, Alexander Staunch, Portland State University, Portland, OR, USA, 2015; Volume 47.
30. Dandelot, S.; Verlaque, R.; Dutartre, A.; Cazaubon, A. Ecological, dynamic and taxonomic problems due to *Ludwigia* (Onagraceae) in France. *Hydrobiologia* **2005**, *551*, 131–136. [CrossRef]
31. Dandelot, S. Les *Ludwigia* spp. du Sud de la France: Historique, Biosyste'Matique et E'Cologie. Ph.D. Thesis, Université Paul Cézanne, Aix-Marseille III, Aix-Marseille, France, 2004; 218p. (In French).
32. Legrand, C. *Pour Contrôler la Prolifération des Jussies (Ludwigia spp.) dans les Zones Humides Méditerranéennes: Guide Technique*; Agence Méditerranéenne de L'environnement: Montpellier, France, 2002; 68p. (In French)
33. EUR-Lex. Regulation (EU) 2016/1141 of the European Parliament and of the Council of 13 July 2016. Available online: https://eur-lex.europa.eu/legal-content/EN/TXT/PDF/?uri=CELEX:02016R1141-20220802&from=EN (accessed on 29 November 2023).
34. Invasoras.pt. Nova Legislação Em Vigor Sobre Espécies Exóticas Invasoras. Available online: https://invasoras.pt/pt/nova-legisla%C3%A7%C3%A3o-em-vigor-sobre-esp%C3%A9cies-ex%C3%B3ticas-invasoras (accessed on 29 November 2023).
35. GB Non-Native Species Secretariat. Identification Guide: Ludwigia grandiflora (Water Primrose). Available online: https://www.nonnativespecies.org/assets/Uploads/ID_Ludwigia_grandiflora_Water_Primrose-2-v2.pdf (accessed on 29 November 2023).
36. Grewell, B.J.; Thomason, M.J.S.; Netherland, M.D. *Establishing Research and Management Priorities for Invasive Water Primroses (Ludwigia spp.)*; U.S. Army Corps of Engineers Engineer Research and Development Center: Vicksburg, MS, USA, 2016; 53p.

37. Ministry for Primary Industries Manatū Ahu Matua. Handling Unwanted Organisms. Available online: https://www.mpi.govt.nz/biosecurity/how-to-find-report-and-prevent-pests-and-diseases/handling-unwanted-organisms (accessed on 29 November 2023).
38. EPPO (European and Mediterranean Plant Protection Organization). *Ludwigia grandiflora* and *L. peploides* Onagraceae—Water Primroses. Available online: https://gd.eppo.int/taxon/LUDPE/documents (accessed on 28 June 2023).
39. Kim, Y.O.; Lee, E.J. Comparison of phenolic compounds and the effects of invasive and native species in East Asia: Support for the novel weapons hypothesis. *Ecol. Res.* **2011**, *26*, 87–94. [CrossRef]

biology

MDPI

Article

Distribution, Effect, and Control of Exotic Plants in Republic of Korea

Bong Soon Lim [1], Ji Eun Seok [1], Chi Hong Lim [1], Gyung Soon Kim [2], Hyun Chul Shin [2] and Chang Seok Lee [1,*]

1 Department of Bio & Environmental Technology, Seoul Women's University, Seoul 01797, Republic of Korea; bs6238@swu.ac.kr (B.S.L.); jus826@swu.ac.kr (J.E.S.)
2 National Institute of Ecology, Seocheon 33657, Republic of Korea; othello1@nie.re.kr (H.C.S.)
* Correspondence: leecs@swu.ac.kr; Tel.: +82-2-970-5666

Simple Summary: This study was carried out to clarify the spatial distribution of exotic plants at national, regional, and local levels, as well as their ecological impacts, and to prepare a strategy to reduce the impacts in Republic of Korea. A review of the biological characteristics of invasive plants showed that therophytes, annual plants, plants that disperse seeds by gravity (D4), erect form (E), and nonclonal growth form (R5) occupied the highest proportion. Exotic plants usually preferred disturbed areas such as lowlands, roadsides, and bare ground. At the national level, the distribution of exotic plants tended to be dominated by topographic conditions and increased around urbanized areas, agricultural fields, and coastal areas. At the regional level, they appeared in artificial plantations, vegetation due to disturbance, and vegetation established on lower slopes compared with upper slopes. At the local level, exotic plants appeared abundantly in the introduced vegetation, whereas they were rare in the native vegetation. Restorative treatments recovered species composition close to the reference vegetation and species diversity reduced by invasive species.

Abstract: This study was carried out to clarify the spatial distribution of exotic plants at national, regional, and local levels, as well as their ecological impacts, and to prepare a strategy to reduce the impacts in Republic of Korea. This study was attempted at the national, regional, and local levels throughout Republic of Korea. Compositae occupied the highest percentage among invading exotic plants in Republic of Korea. A review of the biological attributes of exotic plants based on the dormancy form, longevity, disseminule form, growth form, and radicoid form showed that therophytes, annual plants, plants that disperse seeds by gravity (D_4), erect form (E), and nonclonal growth form (R_5) occupied the highest proportion. At the national level, the spatial distribution of exotic plants tended to depend on topographic conditions such as elevation and slope degree, and to increase around urbanized areas, agricultural fields, and coastal areas. The habitat types that exotic plants established were similar in their native habitat and in Korea, where they invaded. They preferred disturbed land such as roadsides, bare ground, agricultural fields, and so on. The spatial distribution of vegetation types dominated by exotic plants was restricted in the lowland. The proportion of the exotic/native plants tended to proportionate reversely to the vegetation type richness (the number of vegetation types); that is, the ecological diversity. The proportion of the exotic plants was higher in artificial plantations, vegetation due to disturbance, and vegetation established on lower slopes compared with upper slopes. Even at the local level, the exotic plants appeared abundantly in the introduced vegetation, while they were rare in the native ones. In the vegetation infected by exotic species, not only the species composition changed significantly, but the species diversity also decreased. Restorative treatment by introducing mantle vegetation around the hiking trail inhibited the establishment of exotic plants. Further, the restoration practice recovered the similarity of the species composition compared to the reference vegetation and increased the species diversity.

Keywords: control; disturbance; ecological restoration; exotic plant; invasion

1. Introduction

The geographical distribution ranges of many species are restricted by the dispersal barriers of major environmental factors, including climate. As a result of geographical isolation, evolution proceeded in different patterns in each major area of the world [1–3]. Humans have greatly changed this pattern by transporting species throughout the world. Before industrialization, people moved cultivated plants and domestic animals from place to place when they established new agricultural fields and colonies. In modern times, a wide variety of species have been introduced, intentionally and accidentally, into places other than their native habitat [4–10].

Such exotic plants usually expand their distribution range beyond the place of initial establishment by leveraging their advantageous life-history strategies. In particular, the disturbed land provides microsites advantageous for exotic species equipped with opportunistic or ruderal life-history strategies [7,8,11–13]. Exotic plants with favorable life-history strategies in disturbed environments have reduced or replaced native species and changed ecosystem functions, raising concerns about the conservation of native ecosystems [7,8,13–20].

One of the major threats to biodiversity in the world is the direct destruction of their habitats by people through inadequate resource use or pollution. Another serious, but underestimated, problem is the threat to natural and seminatural habitats through the invasion of exotic species, which is a potentially lasting and pervasive threat [9,21–24]. When exploitation or pollution stops, the ecosystem often begins to recover; therefore, it is not a lasting threat. However, even if the introduction of alien organisms stops, the existing exotic species do not disappear; rather, they sometimes continue to spread and consolidate, and thus they are judged as a more serious threat [6,25–27].

Exotic species have the potential to invade other ecosystems and directly or indirectly affect native biota. They have, in fact, invaded all types of ecosystems on the planet, affecting their native biota. These species have been involved in many hundreds of extinctions, especially under islands conditions, whether it is on real islands or ecological islands. The environmental costs include the irreversible loss of native species and ecosystems [7,28–30].

Exotic plants can cause various problems in plant communities composed of native species: excluding native species; altering the habitat, hydrology, and nutrient cycle; significantly impacting biodiversity [8,15,31–33]. Exotic species can transform the structure and function of ecosystems by suppressing or excluding native species, either directly by competing for resources or indirectly by changing the way nutrients cycle through the system [28,33]. Increasing global domination by a few invasive species risks producing a homogenous world rather than a world characterized by great biological diversity and local distinctiveness [7,8,33].

The rate at which new invasions are detected is increasing over time across many ecosystems and regions. The observed rate is often increasing exponentially. Efforts to search for exotic species have increased in recent years, possibly contributing to the observed increase in invasions within some taxonomic groups. However, the overall pattern is consistent, arising from conspicuous and well-known taxa and in well-studied systems, indicating a dramatic increase in the invasion rates in the last half of the twentieth century [34–38].

Along with an apparent increase in the rate of invasions, the strong impacts of invasions have drawn widespread public and scientific attention, propelling policymaking and management actions. Therefore, many countries now have policies to prevent the risk of future invasions and to control the impact and spread of established non-native species [39–41].

The prevention of new invasions has a clear priority in new policies. Management measures against established invasions can also have the advantage of providing post-settlement options, but such efforts are often specific to the particular species and are potentially costly and long-term proposals [42–44]. The successful eradication of existing invasions is an impossible outcome in many instances, despite the significant improvements in the ability to detect new invasions early. Furthermore, a separate effort may be required for each invasion case, whether an eradication program or an effort to control the spread and abundance, even if it is the same species. In contrast, strategies for preventing new invasions target major transfer mechanisms, or vectors. These vector managements can be used to interfere with the transfer of a particular target species; at the same time, they are designed to prevent the wholesale transfer of diverse assemblages, including both target and non-target species, providing a robust and efficient management approach [7,28,36,45–47].

Invasive plants are typically controlled by applying physical or chemical methods [48,49], but they require repeated application [48]; despite that, the effect is not significant [50]. Despite the considerable time and resources required to develop and implement secure and effective eradication or control agents, these programs are often unsuccessful. Furthermore, without careful selection and execution, many chemical, mechanical, and biological control methods can be detrimental to non-target species and ecosystem health, which can facilitate the process of invasion. However, implementing restoration after the invasive species are eradicated or restoring ecosystems to their full potential are useful as a strategy to control invasive species [28,29,42,47,50–55].

Once the exotic species are established, the cost of management is dramatically increased and it is almost impossible to eradicate species that have begun to spread. Moreover, eradication can also create a bare ground, and thus promote reinvasion [56,57]. Therefore, management strategies should be prioritized over eradication measures. Restoration strategies reflecting ecological theory are based on limiting the assembly and invasion of exotic species [58–61].

The objective of this paper is to answer the following big questions of invasion ecology based on the results obtained from Republic of Korea: (1) Which taxa invade? (2) What types of ecosystems are vulnerable to invasive species and their impacts? (3) What is the ecological impact of their invasion? (4) How can we manage invasions?

To arrive at the goal, first of all we analyzed the family composition and the spatial distribution of the exotic plants at the national level. Secondly, we carried out a vegetation survey in eight major mountains of Seoul, and we analyzed the relationship between the occupancy ratio of the exotic plants and the landscape structure at the regional level based on the data. Thirdly, we analyzed the relationship between the performance of an exotic plant and the landscape structure on the mountain to clarify the environmental factors and the influences on the establishment and expansion of the exotic plant at the local level. Fourthly, we investigated the effects that the exotic plant caused based on species composition and diversity in a forest ecosystem and a riparian ecosystem. Finally, we evaluated the effect of restoration as a control measure of the exotic plant in a forest ecosystem and a riparian ecosystem.

2. Materials and Methods

2.1. Study Area

Study areas were selected at the national, regional, and local levels. The whole national territory was regarded as the study area at the national level. Seoul and Mt. Mido, which is located in the urban center of Seoul, were selected as the study areas at the regional and local levels, respectively. Korea is located in the middle latitudes of the Northern Hemisphere, Temperate Zone, and thus has four distinct seasons (between 38°21′ N, 126°11′ E and 33°12′ N, 129°52′ E; Figure 1). Annual mean temperatures in Korea range between 10 °C and 16 °C, except in the high mountain areas. The monthly mean temperatures range from 20 °C to 26 °C in the month of August, and from −5 °C to 5 °C in January. Annual precipitation is recorded as about 1500 mm and 1300 mm in the southern and the central parts of Korea, respectively. Winter precipitation is less than 10% of the total precipitation. Humidity reaches 70–80% nationwide in July, which is the highest. In contrast, monthly mean humidity remains at its lowest level in January and April, at 30–40% levels [62].

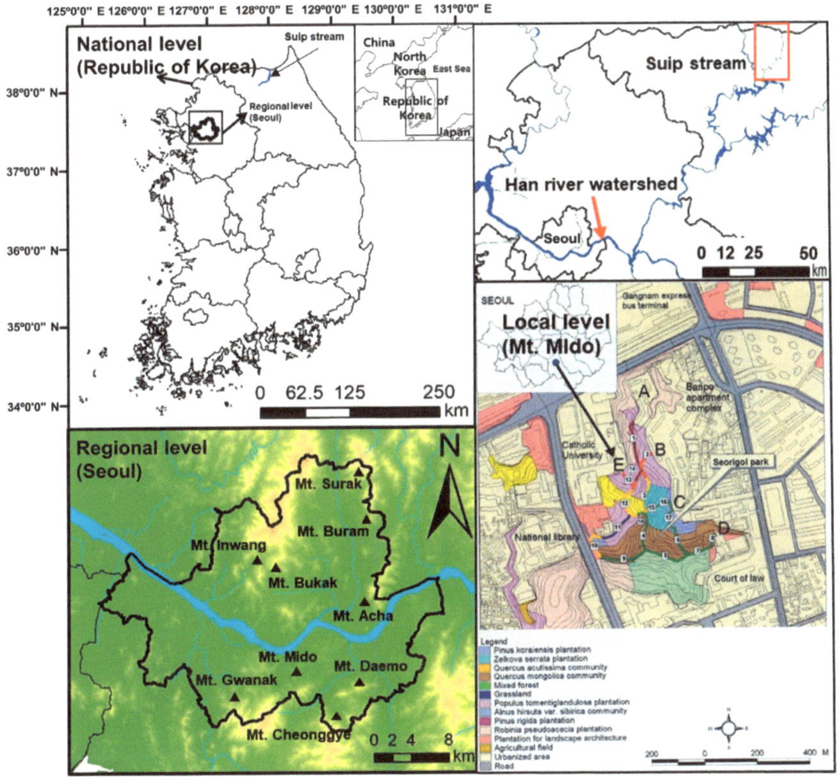

Figure 1. Maps showing the study areas at the national (**upper left**), regional (**lower left** and **upper right**), and local (**lower right**) levels. A map showing a study area at the local level represents the spatial distribution of vegetation and the cover of *Ageratina altissima* in the study area, Mt. Mido located in southern Seoul. A, B, C, D, and E indicate the entrances of the urban park and numbers in the quadrangle show the section number of the hiking trail.

The Korean peninsula has experienced frequent human interferences as a corridor, which links between continent and sea. Moreover, Korea experienced the most rapid economic growth in the world and depends highly on trade. Those societal factors urge human disturbance as well. In addition, Korea also experienced severe artificial disturbance through wars, including the Second World War and the Korean War. All of those factors have contributed to the invasion and expansion of exotic species.

Seoul, the capital of Republic of Korea, is located in the central part of the Korean Peninsula and covers 605 km^2 of land (126°46′15″ to 127°11′15″ E longitude, 37°25′50″ to 37°41′45″ N latitude; Figure 1). The mountainous vegetation of Seoul consists of five major plant communities distributed along an elevational gradient, with the *Pinus densiflora* community in the mountain peaks, the *Quercus mongolica* community in the upper slopes, the *Q. aliena* community in the lower slopes, and the *Zelkova serrata* and *Carpinus laxiflora* communities in the mountain valleys. *Alnus japonica* stands remain in the plains and valleys of the lowland that have escaped development, such as the cultural heritage areas [63–65]. Much of the natural landscape in the Seoul metropolitan area has disappeared due to extensive deforestation for fuel, building material, and other purposes during the 20th century [66]. The population of Seoul has increased from 2.4 million in 1960 to 9.8 million as of 2010 [67]. During this period, the proportion of green space has decreased from 70% in 1960 to 28% in 2017, mostly for housing [63,66,68]. The central government of Republic of Korea has designated most of the forested mountains located in the suburbs as green belts to prevent further loss of green space. Under the Seoul Metropolitan Government's current Green Belt Ordinance, commercial, industrial, and urban development is not allowed in these forests [68].

Mt. Mido is located in Southern Seoul (Figure 1). Mt. Mido is an ecological island enclosed by a residential area (Banpo apartment complex), educational (Catholic University) and public facilities (National Library), a public facility (Court of Law), and a transportation facility (Gangnam Express Bus Terminal) in the east, west, south, and north, respectively. The vegetation of this site is composed of a plantation and secondary forest (Figure 1; [64]). The plantation is composed of *Robinia pseudoacacia, Populus tomentoglandulosa, Z. serrata,* and *Pinus koraiensis* plantations. The secondary forest is composed of *Q. acutissima,* and *Q. mongolica* communities. The vegetation of this site, especially the site facing the hiking trail, is severely affected by frequent visits by nearby residents, the indiscriminate introduction of sports and recreational facilities, and the indiscreet introduction of non-native plants for gardening and landscaping.

The effects of the exotic plants on the forest vegetation and the restorative treatment as a control plan of the exotic plants were investigated in Mt. Mido (Figure 1). The effect of the exotic plants on the riparian vegetation was investigated in the Han River watershed (Figure 1).

The effect of the restorative treatment as a control measure of the exotic plants was investigated in the Dongmun stream, which is located in Munsan-eub, Paju-si, Gyunggi province in central–western Korea. "Eub" and "Si" are the administrative units, which correspond to "town" and "city", respectively. Although the riparian zone of the Dongmun stream was covered with natural vegetation, which were dominated by *Salix pierotii* (Korean willow), *S. gracilistyla,* and *Phragmites communis,* exotic plants, including *Ambrosia trifida* and *Sicyos angulatus,* are emerging as the new dominant species, as they have expanded their range rapidly in recent years [69,70]. The Suip stream, which is located in Bangsan-myun, Yanggu-gun, Gangwon province in eastern Korea (where "Myun" and "gun" are the administrative units, which correspond to "town" and "county", respectively), was selected as the reference stream, as it was left to its natural process for more than 70 years after the Korean War.

2.2. *Survey on Forest Vegetation*

Distribution of exotic plants was investigated at three spatial scales, at the national (Republic of Korea), regional (Seoul, the capital of Republic of Korea), and local (Mt. Mido) levels. Ecological information for analysis on the nationwide distributional characteristics of the exotic plant species was obtained from the Ecobank of the National Institute of Ecology of Korea (http://www.nie-ecobank.kr, accessed on 30 May 2023). A field survey on distribution of exotic plant species was carried out based on the guideline for monitoring wild plants that disturb ecosystems [71]. Survey sites were designated at regular intervals throughout the whole national territory in Republic of Korea (refer to Figure 1).

The habitat types of each exotic plant were classified into the mountainous area, riparian zone, agricultural field, roadside, bare ground, and so on. Species nomenclature followed the Korean Plant Names Index [72]. The percentage of exotic plant species to native plant species was obtained based on flora survey data in each site. The field survey was conducted by recording the distribution area and coverage of the plants appearing in those quadrats after installing three quadrats (5×5 m^2) at the locations selected in each grid. The ratio of exotic plants was calculated as the ratio of the number of exotic plants to the total number of plant species appearing in the target area.

We divided life-forms based on dormancy form, longevity, disseminule form, growth form, and radicoid form [73]. The dormancy form was divided into therophyte, wintering therophyte, hemicryptophyte, geophyte, chamaephyte, hydrophyte, and phanerophyte. The disseminule form was divided into disseminated widely by wind or water (D1), disseminated by attaching to or eaten by animals and man (D2), disseminated by mechanical propulsion of the dehiscence of fruits (D3), and having no special modification for dissemination (D4). The life-form based on longevity was divided into annual, wintering annual, biennial, perennial, and woody plants. The growth form was divided into the tussock form (T), branched form (B), procumbent form (P), erect form (E), partial-rosette form (Pr), and pseudo-rosette form (Ps). The radicoid form was divided into moderate extent of rhizomatous growth (R2), narrowest extent of rhizomatous growth (R3), clonal growth by stolons and struck roots (R4), and nonclonal growth (R5). An ecological map obtained from Seoul City (www.seoul.go.kr, accessed on 30 May 2023) was used to identify vegetation types and landscape boundaries in Seoul. Landscape ecological analyses of the maps were determined with the ArcView GIS software [74].

Vegetation data for study at the regional level were collected in Mts. Acha, Bukak, Buram, Cheonggye, Daemo, Gwanak, Inwang, and Surak, with five plots at each site in Seoul (Figure 1). A vegetation survey was conducted in forty plots, with five plots in each site [75].

The field survey on Mt. Mido selected for study at the local level was carried out from May to September 2002 and a resurvey was done from May to September 2017. A vegetation map (scale of 1:5000) of the study site was constructed with the GIS (Geographic Information System) program supported by ArcView [74] based on an urban ecological map [64] and field surveys. The distribution map of *Ageratina altissima* was prepared by representing the cover class of Braun–Blanquet [76], evaluated by the field survey on the maps at 1:5000 scales. Sections of the hiking trail were divided by expressing the homogeneous range of the cover class of *A. altissima*. The vegetation survey was conducted at 6, 21, and 10 plots in the *A. altissima*, *Q. mongolica*, and *Q. acutissima* stands, respectively, for a total of 37 plots.

The plot size was 20 m × 20 m in the studies of both regional and local levels. The vegetation survey was conducted by applying the phytosociological procedure of Braun–Blanquet [76]. Dominance of each species in each plot was evaluated on an ordinal scale, and each ordinal scale was converted to the median value of the percent of the cover range in each cover class. Relative coverage was considered as the importance value of each species. Relative percentage in percent was calculated by dividing the cover fraction of each species by the summed cover of all species in each plot, and then multiplying the value by 100. A matrix of importance values for all species in all plots was prepared and used as data for ordination using detrended correspondence analysis (DCA) [77]. To describe and compare species diversity and dominance among sites, rank–abundance curves [65,78,79] were prepared.

Restorative treatment was practiced by introducing *Rhododendron yedoense* var. *poukhanense* with a one m breadth beside the hiking trail to prevent the expansion of artificial disturbance on the trail.

2.3. Survey on the Riparian Vegetation

The cover class of all plant species in all plots installed randomly was recorded [76,80]. Plot sizes were 1 m × 1 m in the riparian zones dominated by herbaceous vegetation immediately adjacent to stream channels, 5 m × 5 m in the shrub lands, and 10 m × 10 m in the forests distant from the stream channels. Nomenclature followed the KNA (Korea National Arboretum) [72]. The cover class was estimated with the ordinal scale (from 1 for <1% to 5 for >75%) of Braun–Blanquet [76].

Experimental restoration was practiced by introducing *S. pierotii* and *S. gracilistyla* on the bank covered with *A. trifida* and *S. angulatus* with a 20 cm interval.

2.4. Statistics Analysis

We obtained Pearson's correlation coefficient to deduce the relationship between the percentage of the exotic plants and environmental factors ($\alpha = 0.05$). One-way analysis of variance (ANOVA) was used to compare the difference in the percentage of the exotic plants among the plant communities different in backgrounds and topographic locations ($\alpha = 0.05$). The difference in values among the sites was tested by Scheffe's test.

Data were analyzed using SPSS (version 24, SPSS Inc., Chicago, IL, USA) and R software version 4.2.3 (R Project for Statistical Computing).

3. Results

3.1. A Comparison among Taxa of Exotic Plants

Compositae occupied the highest percentage among exotic plants at 20.4% and followed in order by Gramineae (19.9%), Cruciferae (9.0%), Leguminosae (7.0%), Solanaceae (4.2%), Caryophyllaceae (3.4%), Amaranthaceae (3.1%), Polygonaceae (2.8%), Scrophulariaceae (2.8%), Malvaceae (2.5%), Convolvulaceae (2.5%), Chenopodiaceae (2.0%), Umbelliferae (2.0%), and so on (Figure 2, Appendix A).

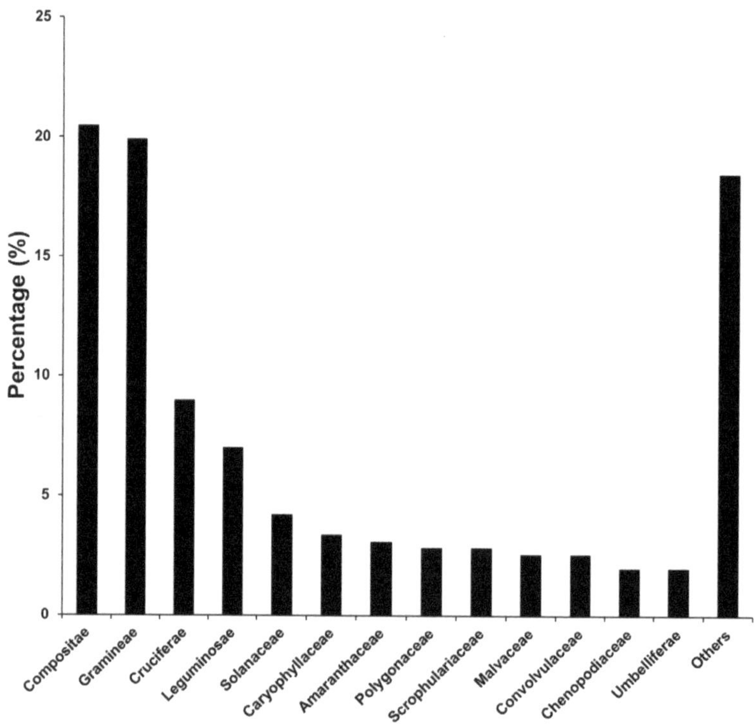

Figure 2. Proportion of exotic plants by family found in Republic of Korea.

3.2. Life-Form Composition of Exotic Plants

In the dormancy form composition of the invading exotic plants in Korea, therophytes occupied the highest percentage, followed by wintering therophytes, hemicryptophytes, geophytes, chamaephytes, hydrophytes, and phanerophytes (Figure 3). Based on the longevity, annual plants occupied more than half, followed by perennial, biennial, wintering annual, and woody plants (Figure 3). The percentage of the disseminule form of the invading exotic plants in Korea was higher in the order of dispersal by gravity (D4), dispersal by wind and water (D1), dispersal by animal (D2), and dispersal by elasticity (D3) (Figure 3). The percentage of the growth form was higher in the order of the erect form (E), procumbent form (P), branched form (B), tussock form (T), partial-rosette form (Pr), and pseudo-rosette form (Ps). The radicoid form of the invading exotic plants in Korea was higher in the order of the R5, plants isolated without any connecting organ, R3 with a narrow rhizome, R4 with a stolon, and R2 with a relatively wide rhizome (Figure 3).

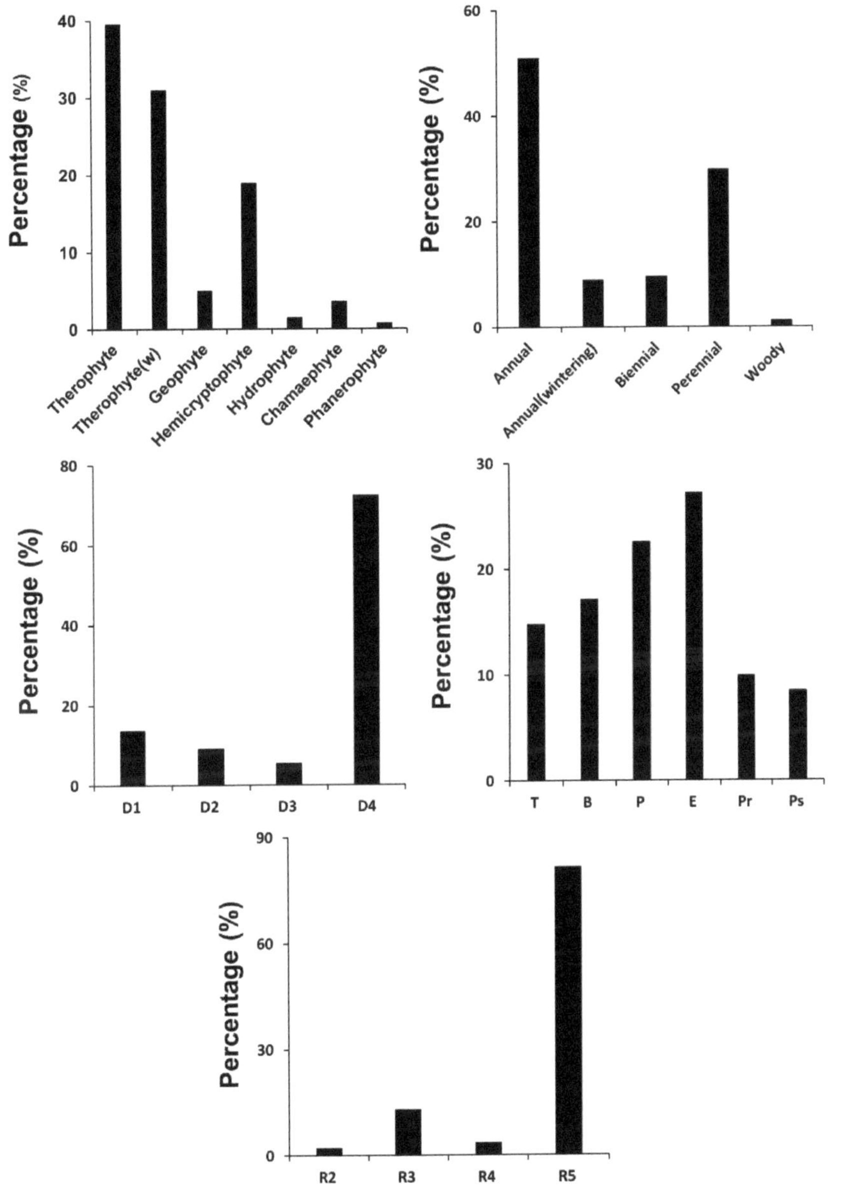

Figure 3. Dormancy form composition of the invading exotic plants in Korea. D1: disseminated widely by wind or water; D2: disseminated attaching to or eaten by animals and man; D3: disseminated by the mechanical propulsion of the dehiscence of fruits; D4: having no special modification for dissemination. T: tussock form; B: branched form; P: procumbent form; E: erect form; Pr; partial-rosette form; Ps: pseudo-rosette form; R2: moderate extent of rhizomatous growth; R3: narrowest extent of rhizomatous growth; R4: clonal growth by stolons and struck roots; R5: nonclonal growth.

3.3. Habitat Types of Exotic Plants

Habitat types in the original habitats of exotic plants that established in Korea showed a higher frequency in roadsides, bare ground, agricultural fields, pastures, riversides, polluted soils, wetlands, and landfill areas (Figure 4). Habitat types invaded in Korea showed a trend similar to that in their original habitats, as the percentage followed the order of roadsides, bare ground, seashores, landfill areas, pastures, wetlands, agricultural fields, forest edges, and riversides (Figure 4).

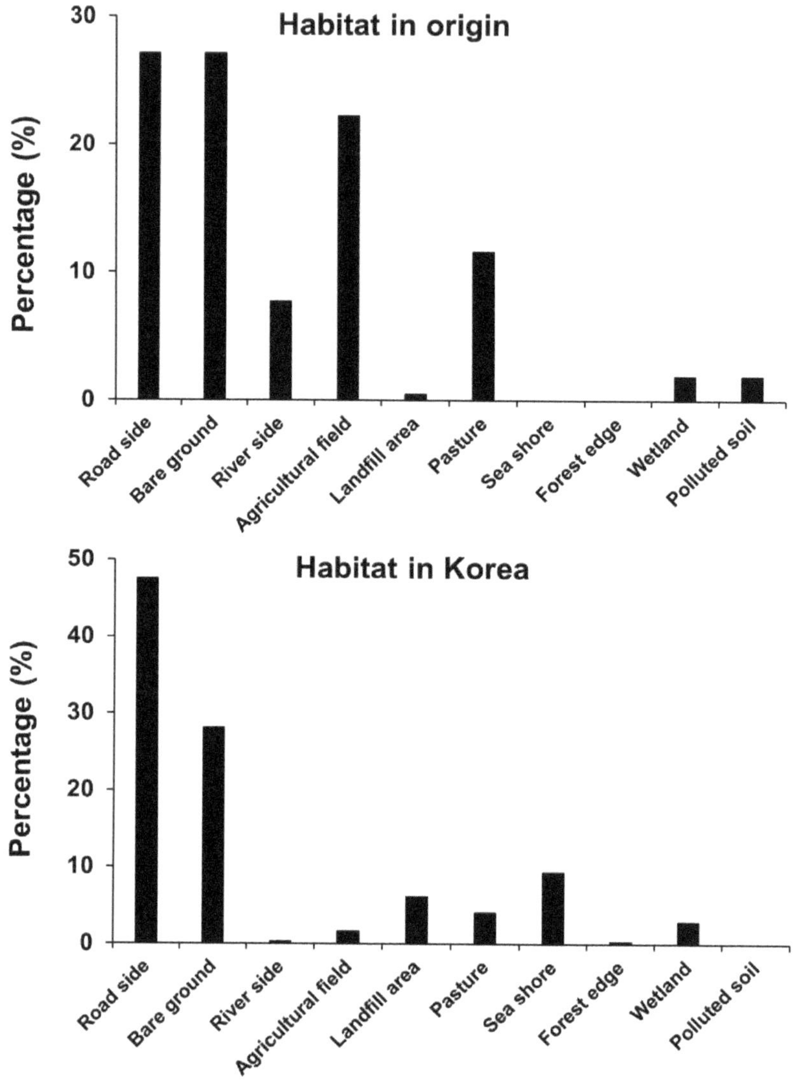

Figure 4. Habitat type composition of the invading exotic plants in Korea in their original place (**upper**) and Korea (**lower**).

3.4. *Spatial Distribution of Exotic Plants at the National Level*

Spatial distribution of exotic plants, elevation, slope degree, percentage of urbanized land, and percentage of agricultural fields throughout the whole national territory of Republic of Korea is depicted in the grid maps (Figure 5). The occupied rate of the exotic plants tended to depend on the elevation and slope degree, and to increase around urbanized areas, agricultural fields, and coastal areas (Figure 5).

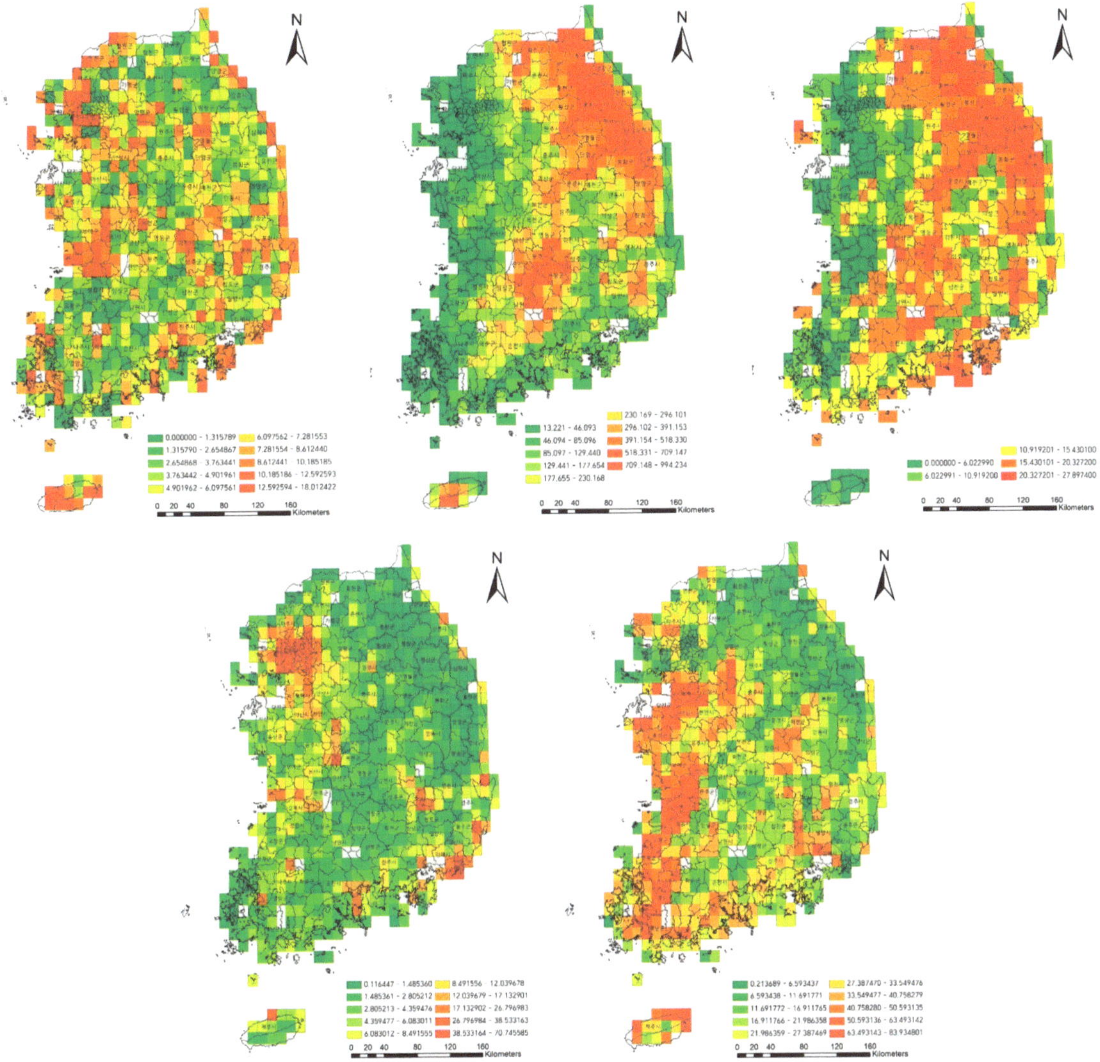

Figure 5. Spatial distribution of percentage of the exotic plant species (**upper left**), elevation (**upper center**), slope degree (**upper right**), urbanized rate (**lower left**), and percentage of agricultural fields (**lower right**) throughout the whole national territory of Republic of Korea. Korean letters ("-군", "-시") on maps are the administrative units, which correspond to "county" and "city", respectively.

As a result of the correlation analysis, the percentage of exotic plants showed significant correlation with the elevation (negative logarithmic), slope degree (negative linear), the percentage of urbanized land (second function), and the percentage of agricultural fields (linear) (Figure 6).

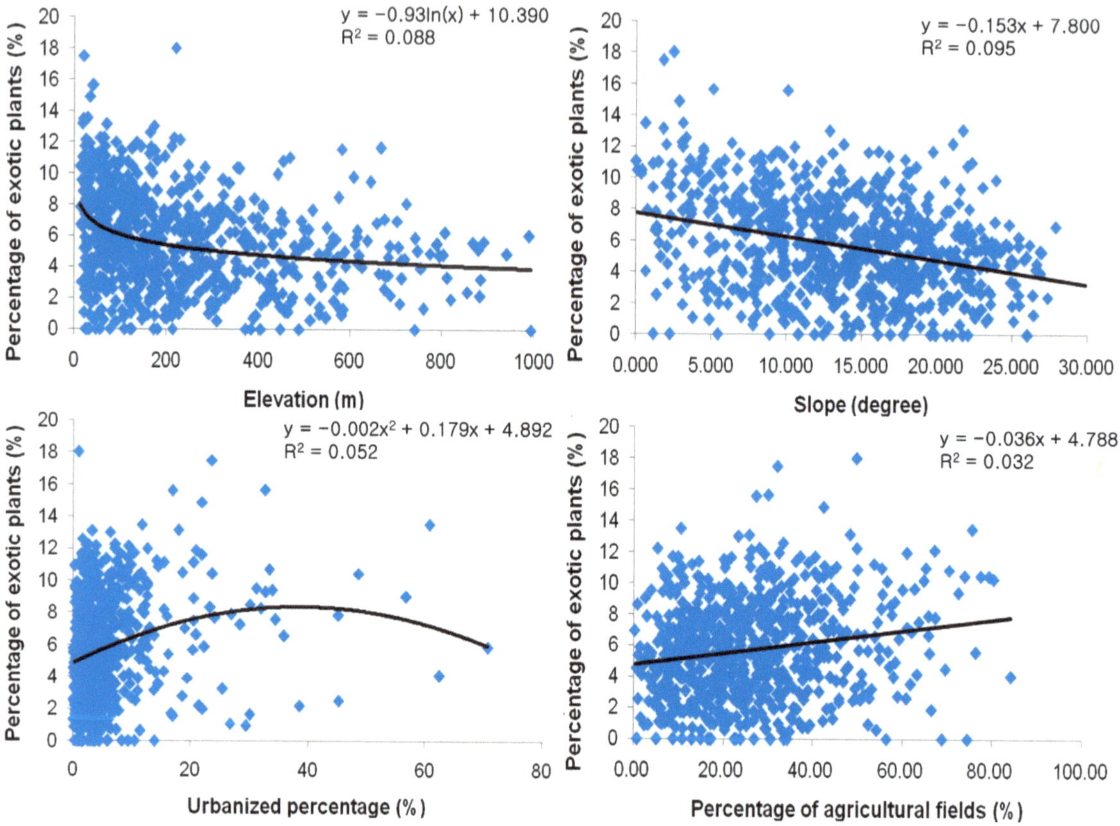

Figure 6. Relationships between elevation, slope, percentage of urbanized land, and percentage of agricultural fields and percentage of exotic plants ($\alpha = 0.05$).

3.5. Spatial Distribution of Exotic Plants at the Regional Level

The spatial distribution of plant communities dominated by exotic plants at the regional level showed that their distribution was restricted to lowlands compared to uplands (Figure 7).

The percentage of exotic to native plants investigated in eight major mountains tended to proportionate reversely to the vegetation type richness (Figure 8).

Comparing the percentages of the exotic plants among plant communities with different traits, the ratio tended to be higher following the order of artificial plantations (constructed by introducing *Alnus hirsuta*, *Alnus hirsuta* var. *sibirica*, *R. pseudoacacia*, *P. tomentiglandulosa*, *Pinus rigida*, *P. koraiensis*, *Larix leptolepis*, and *Betula platyphylla* var. *japonica*, respectively), vegetation due to disturbance (dominated by *Sorbus alnifolia*, *Arudinella hirsuta*, and *Lespedeza cyrtobotrya*, respectively), vegetation established in lower slopes (dominated by *A. japonica*, *Prunus sargenntii*, *Q. acutissima*, *Fraxinus mandshurica*, and *P. densiflora*, respectively), and natural vegetation established in upper slopes (dominated by *Q. mongolica*, *Betula davurica*, *Q. variabilis*, *P. densiflora*, *Q. serrata*, *C. laxiflora*, *Q. aliena*, and *Z. serrata*, respectively) (Figure 9).

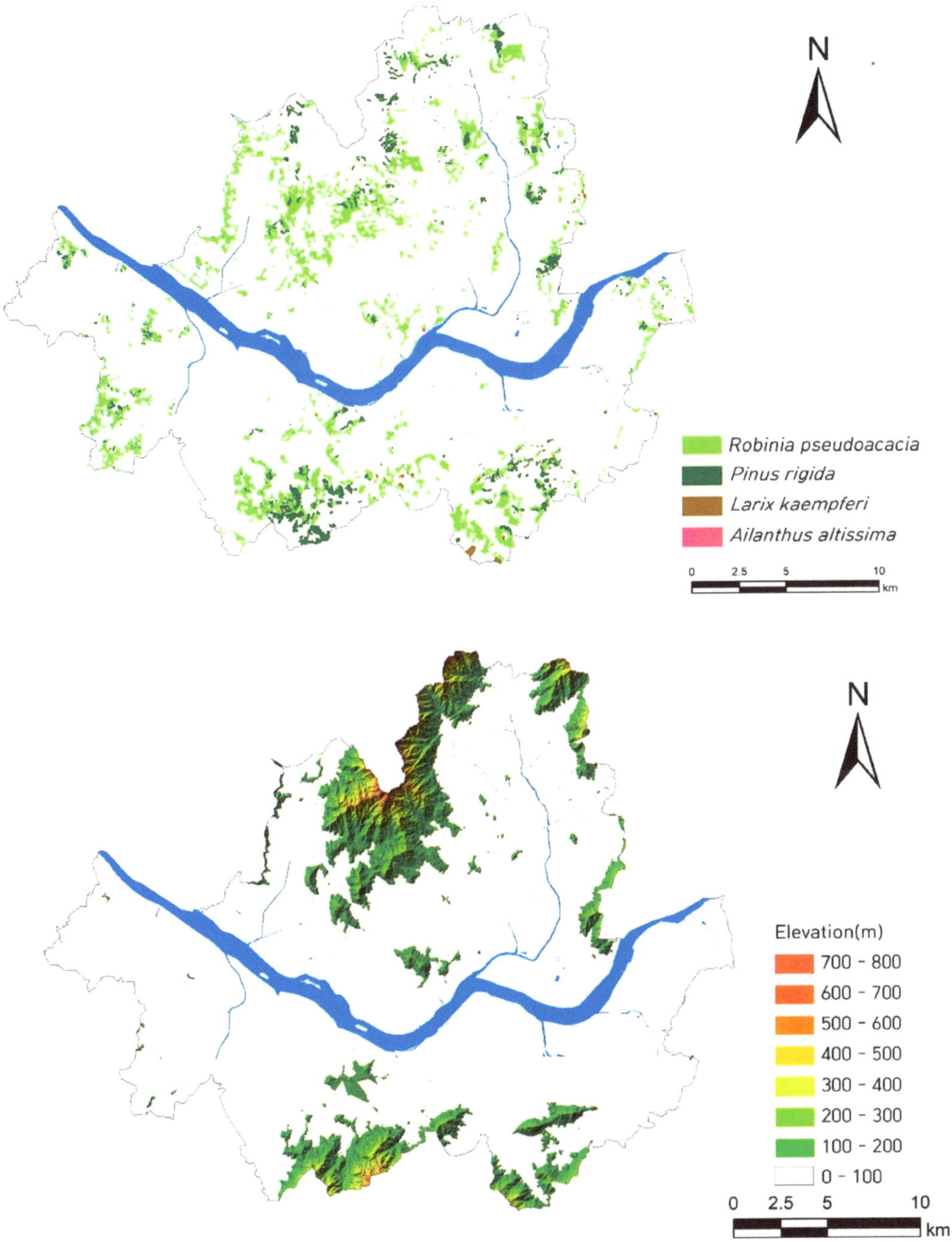

Figure 7. Spatial distribution of plant communities dominated by exotic plants (**upper**) and elevation (**lower**) in the Seoul Metropolitan Area, central–western Korea. Their distribution is restricted to the lowlands.

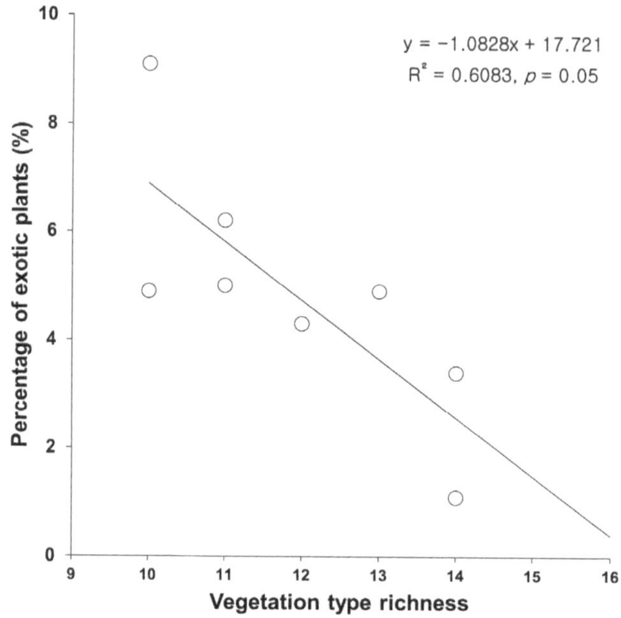

Figure 8. Relationship between percentage of exotic plants and the number of vegetation types in eight major mountains of Seoul.

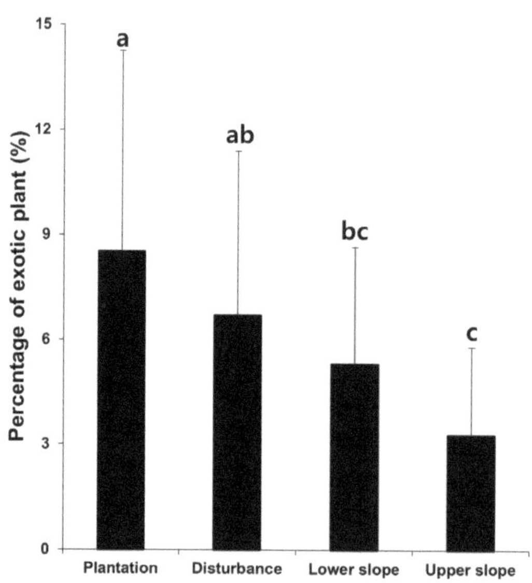

Figure 9. A comparison of the percentage of exotic plant species among plant communities with different traits in eight major mountains of Seoul. Each bar expresses the mean and standard deviation of the mean. An ANOVA test was conducted on the percentages of the exotic plants in the plant communities different in establishment backgrounds and topographic locations at α = 0.05; the means with the same alphabetical character (in superscript) for each parameter were not different from each other. Error bars indicate the standard deviation of the mean percentage of exotic plants for each vegetation type. Disturbance: plant communities due to disturbance; Lower slope: plant communities established in lower slopes; Upper slope: plant communities established in upper slopes.

3.6. Spatial Distribution of A. altissima as an Exotic Plant

In 18 sections (Figure 1), divided by a homogeneous range of cover classes, cover classes of *A. altissima* ranged from none to IV. The cover class of *A. altissima* was the highest in sections 1 and 14 as cover class IV, and sections 2 and 13 (cover class III), section 11 (cover class II), followed by sections 3, 10, 12, and 15 (cover class I). On the other hand, *A. altissima* did not appear in sections 5, 6, 7, 8, and 9, which are located around the summit, and sections 16 and 17, which are located amidst young Korean pine plantations afforested recently (Figure 10).

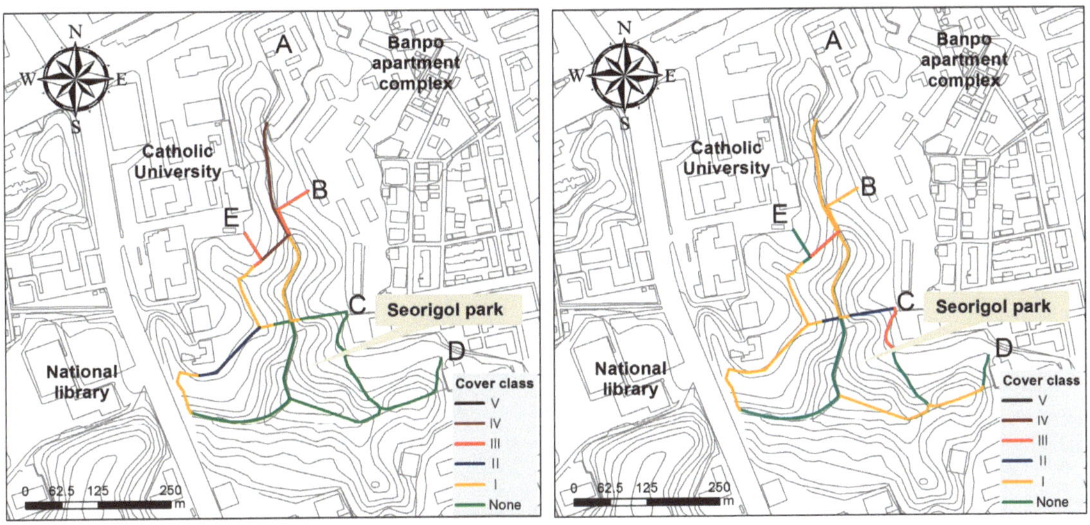

Figure 10. Maps showing the cover class of an exotic plant, *A. altissima*, before (**left**) and after (**right**) a restorative treatment, which was implemented by introducing the Korean azalea beside the hiking trail. The section numbers of the hiking trail were shown in Figure 1. A, B, C, D, and E indicate the entrances of the urban park.

3.7. The Ecological Effects of Exotic Plant Infection

The effects of the exotic plant infection were investigated in terms of changes of species composition and species diversity. The effect in the forest ecosystem was investigated by comparing the species composition and diversity of a plant community dominated by *A. altissima* with three native plant communities dominated by *Q. mongolica* and *Q. acutissima*.

As a result of the DCA ordination based on the vegetation data, the species composition of stands infected with a tree-of-heaven showed a large difference from that of Mongolian oak stands, which was established on the upper slope. Although it was not a large difference, it also showed a difference in species composition from stands of *Q. acutissima*, which were established on the mid-slopes (Figure 11).

The effects of the exotic plant infection were also investigated in terms of changes in species composition and diversity in the riparian ecosystem. The effect on the riparian ecosystem was investigated by comparing the species composition and diversity of two plant communities dominated by *A. trifida* and *S. angulata* with those of three native plant communities dominated by *S. pierotii*, *S. gracilistyla*, and *Phragmites japonica*, which represent the vegetation zones dominated by herbaceous plant, shrub, and tree in the riparian zone.

As the result of the DCA ordination based on the vegetation data, the species composition of the stands infected with two exotic plants was remarkably different from those of *Phragmites japonica* (herb), *S. gracilistyla* (shrub), and *S. pierotii* (tree), which dominate the native riparian vegetation in Republic of Korea (Figure 12).

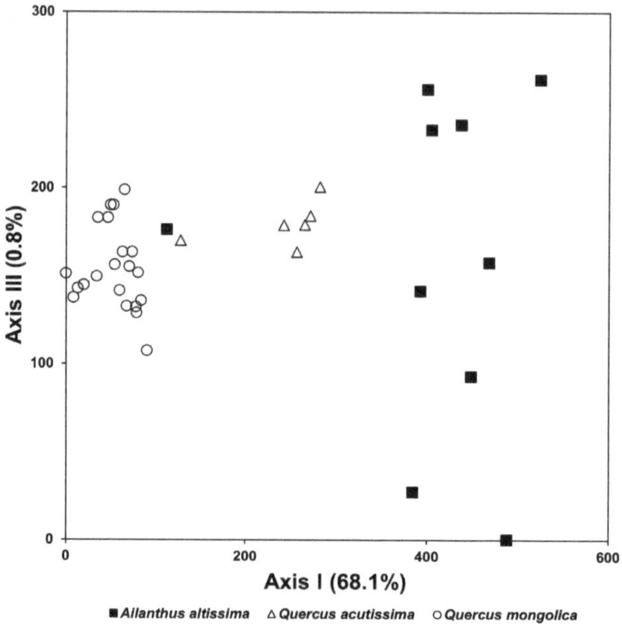

Figure 11. DCA ordination of stands based on vegetation data collected from native oak (*Q. acutissima* and *Q. mongolica*) stands and stands infected with an exotic plant, tree-of-heaven, based on vegetation data.

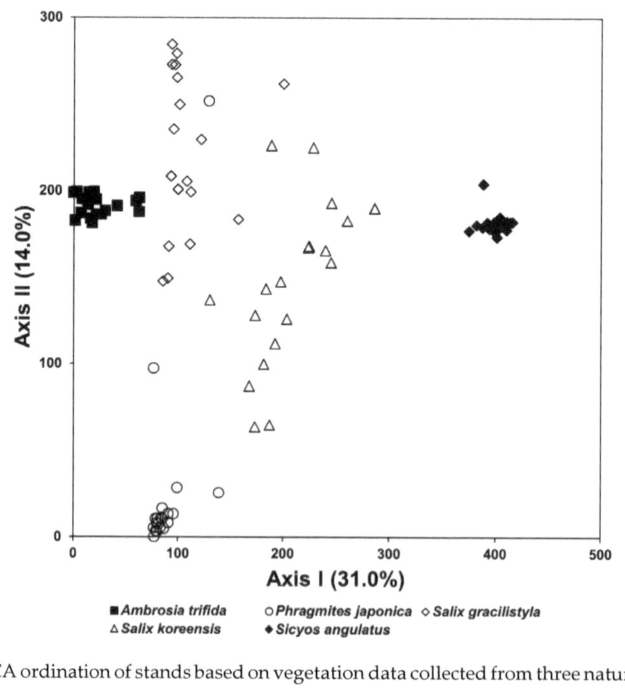

Figure 12. DCA ordination of stands based on vegetation data collected from three natural vegetation types dominated by *S. pierotii*, *S. gracilistyla*, and *P. japonica* and two vegetation types infected with two exotic plants, *A. trifida* and *S. angulata*.

Species diversity was compared by the species rank–abundance curve, and the richness of the stands infected with a tree-of-heaven was lower than those of the three oak communities, and the slope of the curve that the exotic plant community formed was steeper than those done by the native plant communities (Figure 13).

The richness of the stands infected with two exotic plants established in the riparian zone was lower than those of three native riparian vegetation stands, and the slope of the curves that the two exotic plants communities formed were steeper than those done by the native riparian plant communities (Figure 14).

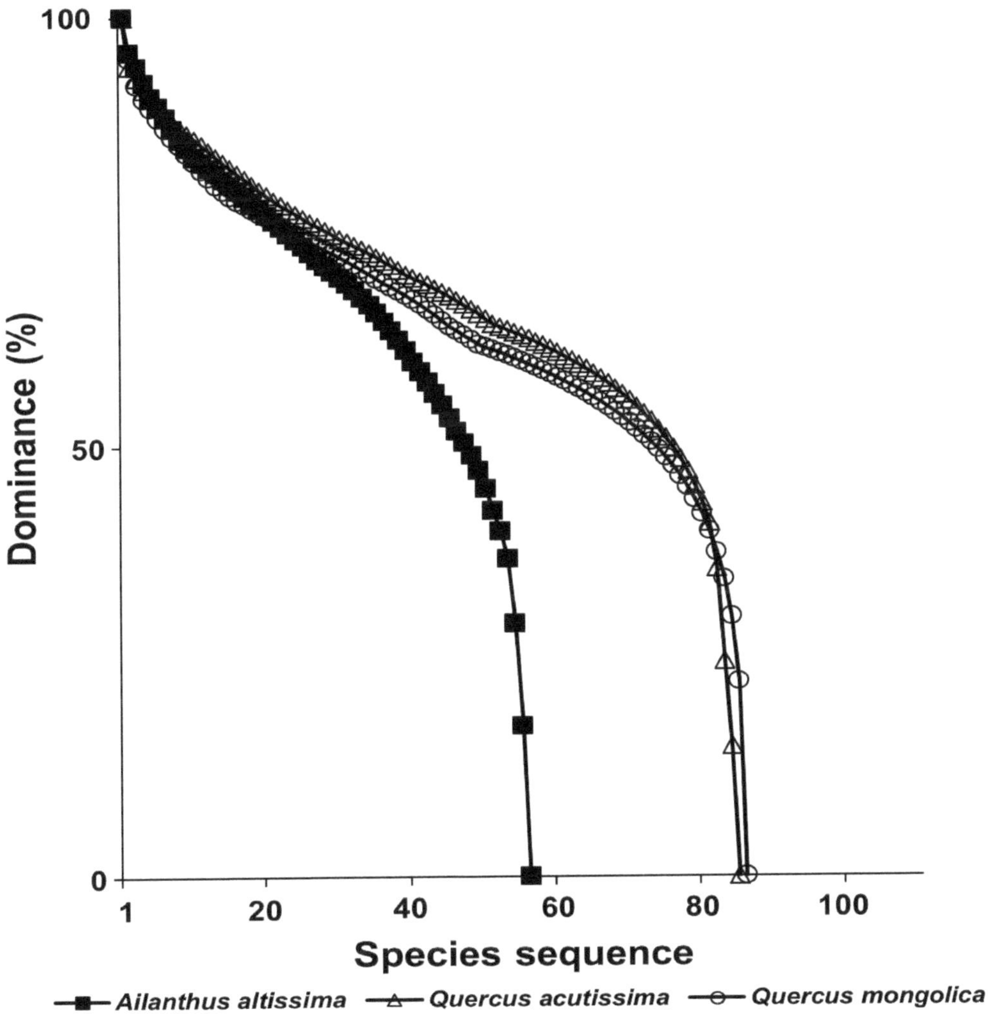

Figure 13. A comparison of species rank–abundance curves of *A. altissima*, *Q. acutissima*, and *Q. mongolica* stands.

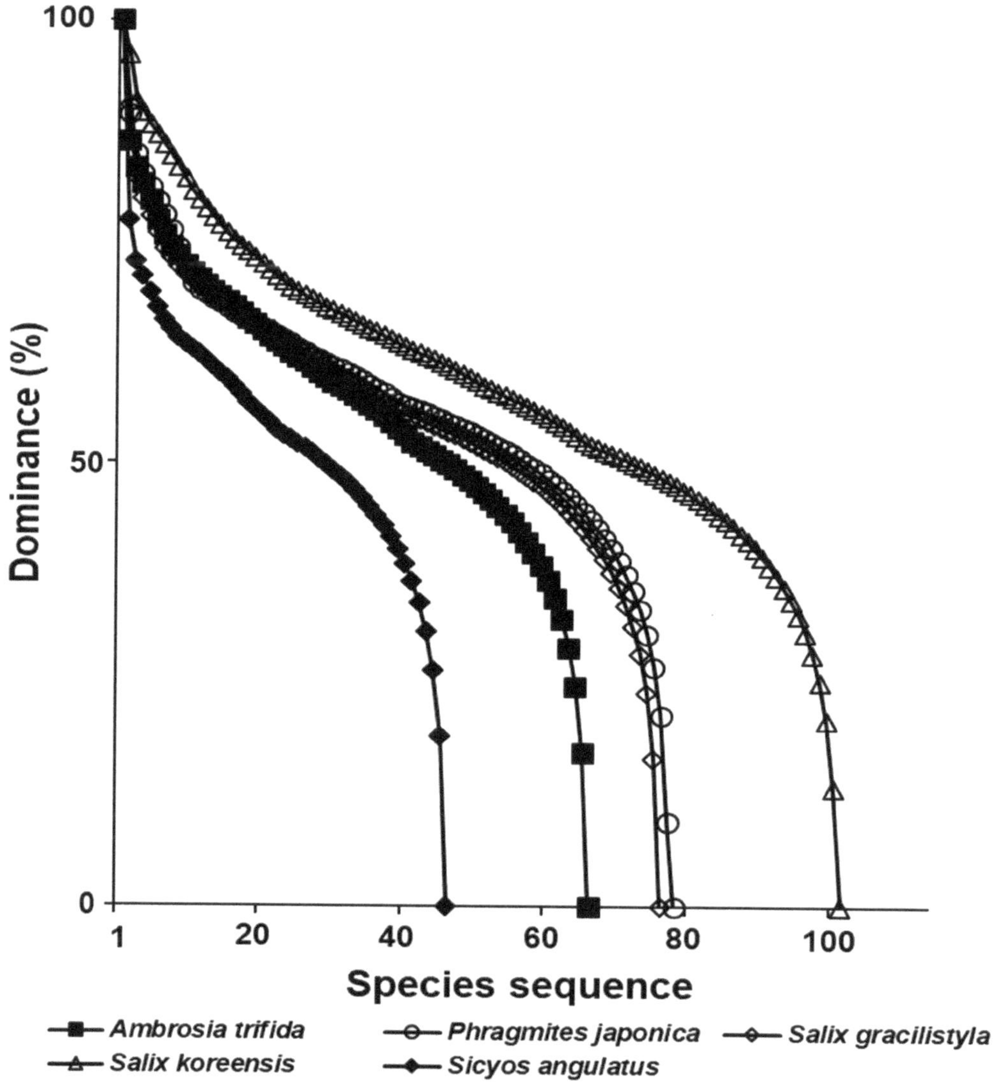

Figure 14. A comparison of species rank–abundance curves of two exotic species stands established in the riparian zone (*A. trifida* and *S. angulata*) and three native riparian vegetation stands (*S. pierotii*, *S. gracilistyla*, and *P. japonica*).

3.8. The Effects of Ecological Restoration for Control of Exotic Plant Species

Restorative treatment by introducing mantle vegetation around the hiking trail reduced the cover of *A. altissima* in sections 1, 2, 11, 13, and 14. However, the cover increased in sections 5, 7, 16, 17, and 18 despite the treatment. On the other hand, sections 3, 4, 6, 8, 9, 10, 12, and 15 did not show any change (Figure 10).

As the result of the DCA ordination based on the vegetation data obtained from the nonrestored stand, which were dominated by *A. trifida* and *S. angulate*, the sites were restored by imitating the natural river dominated by *S. pierotii*; in the reference river, the nonrestored stands showed a species composition remarkably different from that of the reference stands, while the restored sites showed a species composition similar to that of the reference stands (Figure 15).

The richness of the restored sites was higher than that of the reference stands. The richness of the nonrestored stands was the same as that of the reference, but the slope of the curves that the former stand formed was steeper than that of the reference stand (Figure 16).

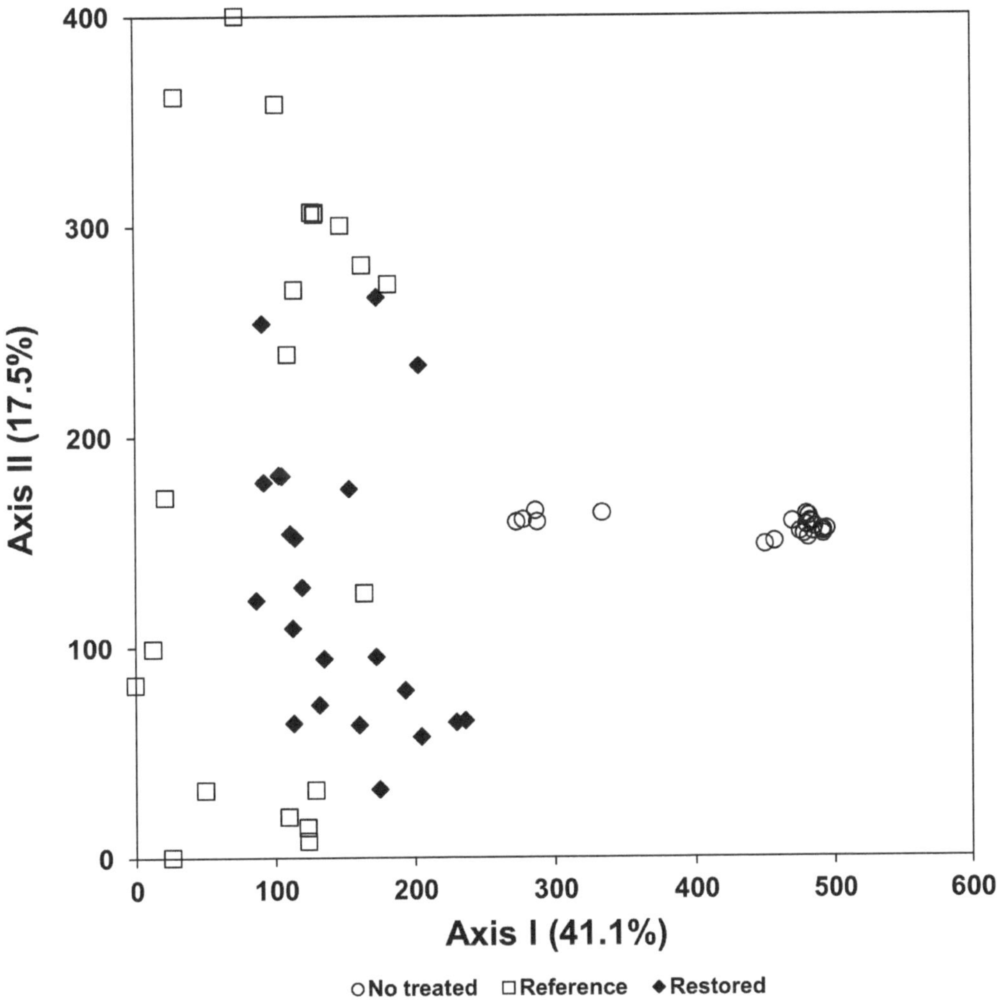

Figure 15. Stand ordination of the restored, non-restored, and reference stands. Non-restored stands were dominated by *A. trifida* or *S. angulate*, whereas the reference stands were dominated by *S. pierotii* and the restored sites were treated based on the reference information obtained from the natural river. Data of the reference stands were collected in the Suip stream, a natural stream which was left in its natural process for more than 70 years since the Korean War.

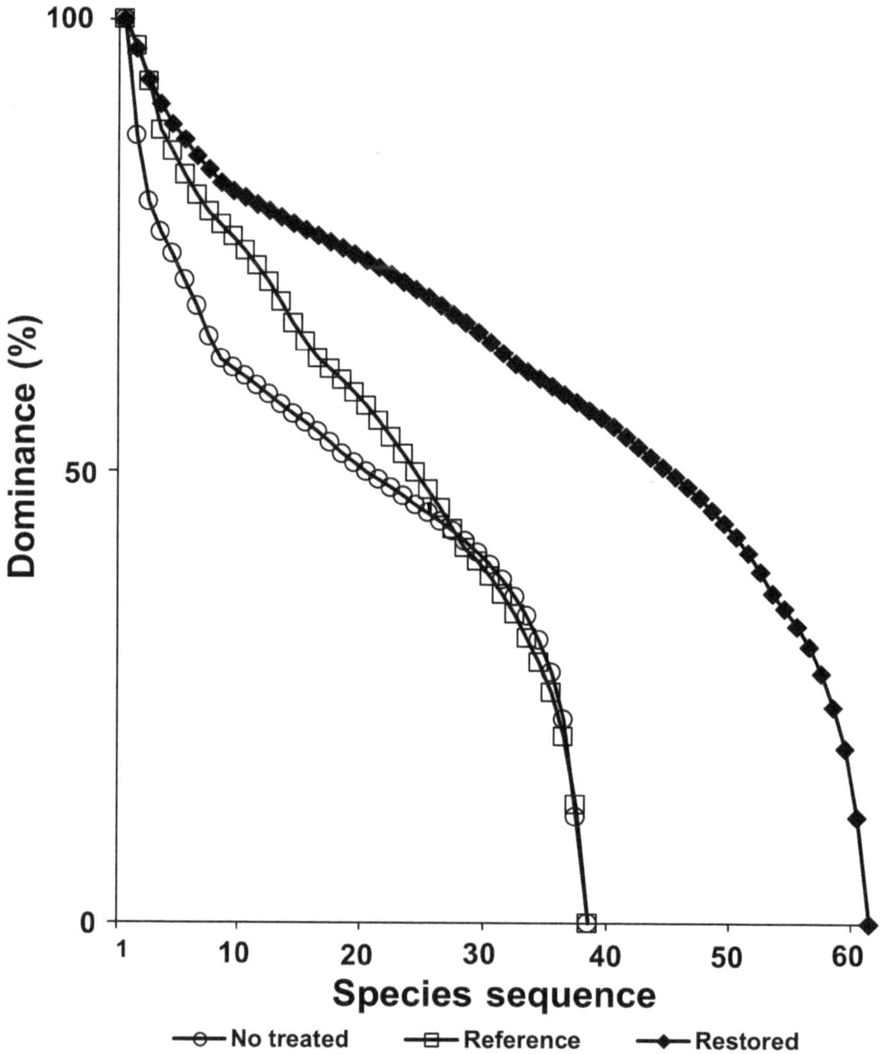

Figure 16. A comparison of species rank–abundance curves among the sites restored by introducing *S. pierotii* (Sp) and *S. gracilistyla* (Sg), nonrestored sites dominated by *A. trifida* (At) and *S. angulatus* (Sa), and reference sites.

4. Discussion

Ecologists studying exotic species try to address the following basic questions: Which taxa often invade? How fast do they invade? What kinds of ecosystems are vulnerable to exotic taxa and their impacts? What are the ecological effects of their invasion? How can we contain, control, or eradicate harmful invaders? Our discussion was focused on those five questions.

4.1. Which Taxa Invade?

Among 357 exotic plants investigated in Korea, compositae occupied the highest percentage among exotic plants at 20.4%, followed in order by Gramineae (19.9%), Cruciferae (9.0%), Leguminosae (7.0%), Solanaceae (4.2%), Caryophyllaceae (3.4%), Amaranthaceae

(3.1%), and so on (Figure 2, [81,82]). On the other hand, reviewing the biological attributes of exotic plants based on the dormancy form, longevity, disseminule form, growth form, and radicoid form showed that therophytes, annual plants, plants that disperse seeds by gravity (D4), erect forms (E), and R_5 forms occupied the highest proportion (Figure 3).

Taxa, which account for a higher percentage, usually have better dispersal agents, higher reproductive capacity, and shorter life cycles [83]. McNeeley [28] suggested the following five predictions regarding invasive species: Firstly, the probability of a species becoming invasive increases as the initial population size increases, so species that are introduced intentionally and kept under cultivated or maintained under animal husbandry over a long period are more likely to establish. Secondly, species with larger native geographic ranges are more likely to become invasive than those with smaller native ranges. Thirdly, species invading a country or region should be considered as a high risk of becoming invasive in an ecologically or climatically similar country or region. Fourthly, species with specialized pollinators are less likely to become invasive unless they are introduced together. Finally, successful invasions usually require that the new habitat conditions are similar to those at the origin, especially in terms of climate conditions. Another group of exotic species are those that have expanded their ranges within the continental areas because they fit the ways in which humans have altered the environment [84]. The other special class of exotic species includes those that have close relatives in the native biota. When exotic species hybridize with indigenous species and varieties, unique genotypes may be removed from local populations and taxonomic boundaries may become obscured [85].

The results of this study, showing that compositae account for the highest proportion of invading taxa in Korea (Figure 2), with the life-form composition of therophytes, annual plants, plants that disperse seeds by gravity (D4), erect forms (E), and R_5 forms occupying the highest proportion (Figure 3), are well in line with the general characteristics of the exotic plants mentioned before.

4.2. *How Fast Do They Invade?*

The rate of spread is a function of both reproduction and dispersal; there are species that reproduce quickly and spread much faster [86]. In order to determine the rate of spread of plants, information on rare dispersal events that can send plants over an unusually long distance is required. While the rate of dispersal is critical, other factors, such as reproductive maturity age, disturbance frequency, intensity of habitat disturbance, and fecundity, are also important. Seeds can often be transported over long distances by agents such as water, wind, vehicles, or livestock at very high speeds [28].

Among the invading exotic plant species in Korea, therophytes and annular plants accounted for a high proportion, and the disseminule form of seeds was high in plants that disperse seeds by gravity (D4) (Figure 3). These biological attributes correspond to favorable conditions to reproduce quickly and spread more rapidly [86].

On the other hand, as the type of habitat where the exotic plant species established included roadsides, bare ground, and coastal areas, accounting for a high proportion, the result was similar to the origin of the habitat types of those plant species (Figure 4). These results mean that such exotic plants are properly finding their preferred places, which is interpreted as a result of active human and material exchanges around the world. In reality, the increased mobility of people and their goods increases the likelihood of moving species around the world [27,28].

4.3. *What Types of Ecosystems Are Susceptible to Exotic Plant Species and Their Impacts?*

One reason that exotic species can so easily invade and dominate new habitats and replace native species is that there are no natural predators, pests, and parasites in the new habitat. Human activity may create unusual environmental conditions, such as nutrient pulses, increased disturbances including fire, or increased light intensity, to which exotic species can adapt more easily than native species. The higher occurrence of exotics is often found in habitats that have been further altered by human interference. Fragmented forests, suburban developments, and easy access to landfills have allowed for an increase in the numbers and ranges of so-called wandering species [12,33,44,87].

All ecosystems, including those in well-protected national parks, could potentially be invaded, but some seem more vulnerable than others. Evolutionarily and geographically isolated ecosystems, especially oceanic islands, are particularly vulnerable. Urban–industrial areas, habitats suffering from periodic disturbance, ports, lagoons, estuaries, and water fronts, where the effects of natural and artificial disturbances often coexist, are also particularly vulnerable to invasions [88,89]. Virtually all ecological communities are vulnerable to invasion to some extent, and artificial disturbance increases the vulnerability of most ecosystems. Therefore, the continued expansion of human activity is likely to increase the vulnerability of ecological communities to invasion [28,44,90].

In the result of this study, exotic species tended to be distributed in places with low elevations and gentle slopes, around cities and agricultural fields, and along the coastal area at the nation level (Figure 5). Therefore, the percentage of exotic species tended to be proportionate to the elevation and slope degree (negatively), and the urbanization ratio and percentage of agricultural fields (positively) (Figure 6). The vegetation type dominated by exotic plant species investigated in Seoul was restricted to the lowlands, with frequent disturbances (Figure 7). The proportion of exotic plant species surveyed in eight mountains within Seoul city showed the opposite trend to the richness of vegetation types that those mountains possessed (Figure 8). The low richness of vegetation types means that the lowlands of the mountain have been transformed into urbanized areas, including residential areas. As a result of comparing the ratio of exotic plant species by vegetation type, exotic species showed a high rate in the artificial afforestation and the vegetation type caused by disturbance, whereas they appeared at a low rate in natural vegetation (Figure 9). In an isolated mountain surrounded by urbanized areas, the coverage of *A. altissima* surveyed along the hiking trail was high along the low-lying hiking trail close to the residential area, and it was low or did not appear around the mountain peak dominated by natural vegetation (Figure 10). These results are evidence to demonstrate that the invasion of exotic plants begins at sites disturbed by human influence, and that the spread is also due to the effects of such disturbances.

The forest edge, including the hiking trail, is vulnerable to the invasion of exotic plants due to physical disturbance and nutrient input, as a place where organisms, matter, and energy are exchanged between the two habitats [91–94]. Thus, exotic plants are abundant while species diversity is lower in the forest edge compared with the undisturbed forest interior [95–98]. In reality, experimental manipulation, which removed the forest canopy and undergrowth, facilitated the invasion of exotic plants [99,100]. This phenomenon suggests that the invasion and expansion of exotic plants are closely related to artificial disturbances. The results of this study resemble the facts commonly known regarding the invasion and expansion of the exotic plants.

A. altissima, as an exotic plant species, appears more abundantly at the lower elevations, where frequent artificial interferences are prevailing, than at the higher elevations, where such impacts are lessened (Figure 10). On the other hand, they appeared abundantly in the introduced vegetation, such as the black locust plantation, but they did not appear or were rare in the natural one, such as the oak forest (Figure 10). However, an exceptional phenomenon was found at an entrance C. The lack of *A. altissima* in the entrance was due to the dense coverage of the recently afforested Korean pine stand (Figure 10). In this respect, the light condition is also important [65].

4.4. What Is the Ecological Impact of Exotic Species Invasion?

Every exotic species alters the species composition of native biological communities in some way. Whether it becomes invasive, and thus harmful, depends on the specific nature of the exotic species, the vulnerability of the host ecosystem, and opportunity [101–104]. Changes in the ecosystems may be initiated by natural disturbances, such as storms, earthquakes, volcanic eruptions, fires, climate, or management regimes, but are reinforced or accelerated by the invasion of exotic species. Land transformation and invasions are interlinked, bringing more opportunities for invasion [27,33,105,106].

The species composition of an ecosystem at any given location and time is determined depending on the current environmental conditions, the levels and types of disturbance, the balance of loss and recruitment, and the composition of the regional species pool. Increasing levels of human-induced ecosystem transformation may accelerate environmental change, and the dramatic increase in the intentional and accidental biota transport across the world inevitably will increase the regional species pool, whereas it will perhaps also decrease native species and ultimately decrease the global species pool. The combination of these factors lays the base for a radical alteration of an ecosystem [12,28,107–110].

In a result of this study, the species composition of vegetation types infected with a tree-of-heaven showed a difference from that in the reference area, where the exotic species did not invade (Figure 11). The species composition of riparian vegetation infected with giant ragweed and *S. angulatus* also changed significantly compared with that in the reference area, where they did not invade (Figure 12). In both forest ecosystems and riverside ecosystems, species diversity of vegetation types in which exotic species invaded decreased (Figures 13 and 14).

4.5. Restoration as a Tool to Inhibit Invasion and Expansion of Invasive Species

Invasive species provide a serious challenge to environmental managers because of their explosive growth [111,112]. It is known that disturbed ecosystems are usually more vulnerable to the infestation of exotic plants than undisturbed ecosystems [44,87,113]. The results of this study showed that the invasion and spread of exotic species were closely related to disturbance at national, regional, and local levels. In this respect, it is imperative to protect disturbed ecosystems from human interference, such as excessive use and management, to prevent the invasion and expansion of exotic species. It is imperative to foster closed undisturbed conditions to discourage disturbance-adapted exotic plants [44,114]. In fact, a comprehensive restoration of entire ecosystems may be necessary [115,116]. Artificial interference for forests is currently declining in rural areas in Korea, but interference due to forest management still remains in urban areas, which may cause further invasion of exotic plants. Therefore, we propose a management plan with ecological restoration principles reflected to address ecosystems infected with exotic species [30,117,118].

How do we return a disturbed ecosystem to a stable ecosystem? Can we contain, control, or eradicate harmful invaders through such restoration? The goal of restoration ecology is to aid the recovery of an ecosystem that has been disturbed or damaged by external influences, such as fire, logging, mining, agriculture, or urban development. A major goal of restoration practitioners is to return a habitat to a more desirable condition, involving a particular species composition, community structure, and/or set of ecosystem functions [119–121].

Restoration is used to maintain the overall health and sustainability of an ecosystem, and thus can be the most effective way to increase the resilience and resistance of an ecosystem to invasion by exotic species. Although restoration is used for various purposes, it is particularly being used as a way to combat exotic species, especially under the stresses of climate change and the expanding influence of human activities. Optimal, balanced, and stable ecological conditions arising as a result of restoration can more effectively limit the spread and settlement of exotic species [50,122–124].

In a result of this study, the restorative treatment that introduced mantle vegetation around the hiking trail where *A. altissima* invaded reduced the coverage of the exotic plant (Figure 10). In addition, the riparian vegetation, where giant ragweed invaded, showed significantly different species composition from the reference vegetation and showed low species diversity (Figures 12 and 14). However, the restoration implemented by introducing Korean willow restored species composition similar to the reference vegetation and increased species diversity (Figures 15 and 16). This result suggests that plantings of willow could help restore riparian zones that have been overtaken by giant ragweed. In reality, willows can act as a nurse crop for other understory plants by ameliorating high light, temperature, and soil moisture [125,126]. In particular, the decline of giant ragweed dominance by willow shading might increase plant diversity in this restored riparian ecosystem. Based on the results, we suggest a comprehensive restoration plan which recovers the integrity of the ecosystems disturbed due to various human activities to inhibit the invasion and expansion of exotic species [29,95,127,128].

5. Conclusions

Compositae occupied the highest percentage among exotic plants, followed by Gramineae, Cruciferae, Leguminosae, and so on. At the national level, the spatial distribution of exotic plants tended to depend on topographic conditions, such as the elevation and slope degree, and to increase around urbanized areas, agricultural fields, and coastal areas. The habitat types that exotic plants established were similar in their native habitat and in Korea, where they invaded. They preferred disturbed lands such as roadsides, bare ground, agricultural fields, and so on. The spatial distribution of vegetation types dominated by exotic plants was restricted in the lowlands. The proportion of the exotic/native plants tended to proportionate reversely to the vegetation type richness; that is, ecological diversity. The proportion of the exotic plants was higher in the artificial plantations, vegetation due to disturbance, and vegetation established in lower slopes compared with the upper slopes. Even at the local level, exotic plants appeared abundantly in the introduced vegetation, whereas they were rare in the native ones. In the vegetation infected by exotic species, not only the species composition changed significantly, but also the species diversity decreased. Synthesizing these results, the invasion of exotic plants begins at sites disturbed by human influence, and the spread is also due to the effects of such disturbances. Restorative treatment by introducing mantle vegetation around the hiking trail inhibited the establishment of exotic plants. Further, the restoration practice recovered the similarity of the species composition compared to the reference vegetation and increased the species diversity. Based on the results, we suggest a comprehensive restoration plan which recovers the integrity of the ecosystems disturbed due to various human activities to inhibit the invasion and expansion of exotic species.

Author Contributions: Conceptualization, B.S.L. and C.S.L.; methodology, B.S.L. and J.E.S.; validation, J.E.S. and B.S.L.; formal analysis, B.S.L. and C.H.L.; investigation, G.S.K., H.C.S., C.S.L., B.S.L. and C.H.L.; data curation, B.S.L.; writing—original draft preparation, C.S.L. and B.S.L.; writing—review and editing, G.S.K., H.C.S. and C.S.L.; visualization, J.E.S. and B.S.L.; supervision, C.S.L.; project administration, C.S.L. All authors have read and agreed to the published version of the manuscript.

Funding: This work was supported by the Korea Environment Industry & Technology Institute (KEITI) through the Wetland Ecosystem Value Evaluation and Carbon Absorption Value Promotion Technology Development Project, funded by the Korea Ministry of Environment (2022003630002).

Institutional Review Board Statement: Not applicable.

Informed Consent Statement: Not applicable.

Data Availability Statement: Raw exotic plant data is available on IKAS (https://kias.nie.re.kr/home/main/main.do accessed on 3 June 2023).

Conflicts of Interest: The authors declare no conflict of interest.

Appendix A. A List of Exotic Plants Investigated in Republic of Korea

Af: Africa, Am: America, As: Asia, Au: Australia, Eu: Europe, Med: Mediterranean

Family Name	Scientific Name	Origin
Amaranthaceae	*Amaranthus albus*	Am
Amaranthaceae	*Amaranthus arenicola*	Am
Amaranthaceae	*Amaranthus blitum*	Eu
Amaranthaceae	*Amaranthus hybridus*	Am
Amaranthaceae	*Amaranthus palmeri*	Am
Amaranthaceae	*Amaranthus patulus*	Am
Amaranthaceae	*Amaranthus powellii*	Am
Amaranthaceae	*Amaranthus retroflexus*	Am
Amaranthaceae	*Amaranthus spinosus*	Am
Amaranthaceae	*Amaranthus viridis*	Am
Amaranthaceae	*Celosia argentea*	Am
Amaryllidaceae	*Zephyranthes candida*	Am
Boraginaceae	*Amsinckia lycopsoides*	Am
Boraginaceae	*Asperugo procumbens*	Eu–As–Af
Boraginaceae	*Symphytum officinale*	Eu
Campanulaceae	*Triodanis perfoliata*	Am
Cannabaceae	*Cannabis sativa*	As
Caryophyllaceae	*Cerastium glomeratum*	Eu
Caryophyllaceae	*Dianthus armeria*	Eu
Caryophyllaceae	*Holosteum umbellatum*	Eu–As
Caryophyllaceae	*Saponaria officinalis*	Eu
Caryophyllaceae	*Scleranthus annuus*	Eu
Caryophyllaceae	*Silene alba*	Eu
Caryophyllaceae	*Silene antirrhina*	Am
Caryophyllaceae	*Silene armeria*	Eu
Caryophyllaceae	*Silene gallica*	Eu–As
Caryophyllaceae	*Spergula arvensis*	Eu
Caryophyllaceae	*Spergularia rubra*	Eu–As
Caryophyllaceae	*Vaccaria vulgaris*	Eu
Chenopodiaceae	*Atriplex hastata*	Eu
Chenopodiaceae	*Chenopodium album*	Eu–As
Chenopodiaceae	*Chenopodium ambrosioides*	Am
Chenopodiaceae	*Chenopodium ficifolium*	Eu
Chenopodiaceae	*Chenopodium glaucum*	Eu
Chenopodiaceae	*Chenopodium hybridum*	Eu–As
Chenopodiaceae	*Chenopodium pumilio*	Au
Commelinaceae	*Commelina benghalensis*	As–Af
Commelinaceae	*Commelina diffusa*	etc.
Commelinaceae	*Tradescantia reflexa*	Am
Compositae	*Achillea millefolium*	Eu
Compositae	*Ageratum conyzoides*	Am
Compositae	*Ambrosia artemisiifolia*	Am
Compositae	*Ambrosia trifida* f. *integrifolia*	Am
Compositae	*Ambrosia trifida*	Am
Compositae	*Anthemis arvensis*	Eu
Compositae	*Anthemis cotula*	Eu
Compositae	*Aster novi-belgii*	Am
Compositae	*Aster pilosus*	Am
Compositae	*Aster subulatus*	Am
Compositae	*Aster subulatus* var. *sandwicensis*	Am
Compositae	*Bidens frondosa*	Am
Compositae	*Bidens pilosa*	Am
Compositae	*Bidens pilosa var. minor*	Am
Compositae	*Bidens polylepis*	Am
Compositae	*Bidens subalternans*	Am
Compositae	*Carduus crispus* f. *albus*	Eu–As
Compositae	*Carduus crispus*	Eu–As
Compositae	*Carduus nutans*	Eu
Compositae	*Centaurea cyanus*	Eu

Family Name	Scientific Name	Origin
Compositae	*Chrysanthemum leucanthemum*	Eu
Compositae	*Cirsium arvense*	Eu
Compositae	*Cirsium vulgare*	Eu
Compositae	*Conyza bonariensis*	Am
Compositae	*Conyza canadensis*	Am
Compositae	*Conyza parva*	Am
Compositae	*Conyza sumatrensis*	Am
Compositae	*Coreopsis lanceolata*	Am
Compositae	*Coreopsis tinctoria*	Am
Compositae	*Cosmos bipinnatus*	Am
Compositae	*Cosmos sulphureus*	Am
Compositae	*Crassocephalum crepidioides*	Af
Compositae	*Crepis tectorum*	Eu
Compositae	*Dracopis amplexicaulis*	Am
Compositae	*Eclipta alba* var. *erecta*	Am
Compositae	*Erechtites hieracifolia*	Am
Compositae	*Erigeron annuus*	Am
Compositae	*Erigeron philadelphicus*	Am
Compositae	*Erigeron strigosus*	Eu
Compositae	*Eupatorium rugosum*	Am
Compositae	*Euthamia graminifolia*	Am
Compositae	*Galinsoga ciliata*	Am
Compositae	*Galinsoga parviflora*	Am
Compositae	*Gamochaeta pensylvanica*	Am
Compositae	*Gnaphalium calviceps*	Am
Compositae	*Gnaphalium purpureum*	Am
Compositae	*Helianthus debilis*	Am
Compositae	*Helianthus tuberosus*	Am
Compositae	*Hieracium caespitosum*	Eu
Compositae	*Hypochaeris radicata*	Eu
Compositae	*Lactuca scariola*	Eu
Compositae	*Lapsana communis*	Eu
Compositae	*Matricaria inodora*	Eu
Compositae	*Matricaria matricariodes*	As
Compositae	*Parthenium hysterophorus*	Am
Compositae	*Rudbeckia bicolor*	Am
Compositae	*Rudbeckia hirta*	Am
Compositae	*Rudbeckia laciniata* var. *hortensis*	Am
Compositae	*Senecio inaequidens*	Af
Compositae	*Senecio scandens*	As
Compositae	*Senecio vulgaris*	Eu
Compositae	*Solidago altissima*	Am
Compositae	*Solidago serotina*	Am
Compositae	*Sonchus asper*	Eu
Compositae	*Sonchus oleraceus*	Eu
Compositae	*Tagetes minuta*	Am
Compositae	*Taraxacum laevigatum*	Eu
Compositae	*Taraxacum officinale*	Eu
Compositae	*Tragopogon dubius*	Eu
Compositae	*Verbesina alternifolia*	Am
Compositae	*Xanthium canadense*	Am
Compositae	*Xanthium italicum*	Am
Compositae	*Xanthium strumarium*	As
Convolvulaceae	*Convolvulus arvensis*	Eu
Convolvulaceae	*Cuscuta pentagona*	Am
Convolvulaceae	*Ipomoea hederacea*	Am
Convolvulaceae	*Ipomoea hederacea* var. *integriuscula*	Am
Convolvulaceae	*Ipomoea lacunosa*	Am
Convolvulaceae	*Ipomoea purpurea*	Am
Convolvulaceae	*Ipomoea triloba*	Am

Family Name	Scientific Name	Origin
Convolvulaceae	*Jacquemontia tamnifolia*	Am
Convolvulaceae	*Quamoclit coccinea*	Am
Crassulaceae	*Sedum mexicanum*	Am
Cruciferae	*Alliaria petiolata*	Eu–As
Cruciferae	*Barbarea verna*	Eu
Cruciferae	*Barbarea vulgaris*	Eu
Cruciferae	*Brassica juncea*	As
Cruciferae	*Cakile edentula*	Am
Cruciferae	*Camelina microcarpa*	Eu
Cruciferae	*Cardaria draba*	Eu
Cruciferae	*Chorispora tenella*	Eu–As
Cruciferae	*Coronopus didymus*	Eu
Cruciferae	*Descurainia pinnata*	Am
Cruciferae	*Diplotaxis muralis*	Eu
Cruciferae	*Erucastrum gallicum*	Eu
Cruciferae	*Lepidium apetalum*	Am
Cruciferae	*Lepidium bonariense*	Am
Cruciferae	*Lepidium campestre*	Eu
Cruciferae	*Lepidium latifolium*	Eu–As
Cruciferae	*Lepidium perfoliatum*	Eu–As
Cruciferae	*Lepidium ruderale*	Eu
Cruciferae	*Lepidium virginicum*	Am
Cruciferae	*Myagrum perfoliatum*	Eu
Cruciferae	*Nasturtium officinale*	Eu
Cruciferae	*Neslia paniculata*	Eu
Cruciferae	*Raphanus raphanistrum*	Eu
Cruciferae	*Rapistrum rugosum*	Eu
Cruciferae	*Rorippa sylvestris*	Eu
Cruciferae	*Sinapis arvensis*	Eu
Cruciferae	*Sinapis arvensis* var. *orientalis*	Eu
Cruciferae	*Sisymbrium altissimum*	Eu
Cruciferae	*Sisymbrium officinale*	Eu
Cruciferae	*Sisymbrium officinale* var. *leiocarpum*	Eu
Cruciferae	*Sisymbrium orientale*	Eu
Cruciferae	*Thlaspi arvense*	Eu
Cucurbitaceae	*Sicyos angulatus*	Am
Cyperaceae	*Carex scoparia*	Am
Euphorbiaceae	*Euphorbia dentata*	Am
Euphorbiaceae	*Euphorbia heterophylla*	Am
Euphorbiaceae	*Euphorbia hirta*	Am
Euphorbiaceae	*Euphorbia maculata*	Am
Euphorbiaceae	*Euphorbia prostrata*	Am
Euphorbiaceae	*Euphorbia supina*	Am
Fumariaceae	*Fumaria officinalis*	Eu
Geraniaceae	*Erodium cicutarium*	Eu
Geraniaceae	*Geranium carolinianum*	Am
Geraniaceae	*Geranium dissectum*	Eu
Gramineae	*Aegilops cylindrica*	Eu
Gramineae	*Agropyron repens*	Eu
Gramineae	*Agropyron repens* f. *aristatum*	Eu
Gramineae	*Aira caryophyllea*	Eu
Gramineae	*Alopecurus japonicus*	As
Gramineae	*Alopecurus myosuroides*	Eu–As
Gramineae	*Alopecurus pratensis*	Eu–As
Gramineae	*Andropogon virginicus*	Am
Gramineae	*Anthoxanthum odoratum*	Eu–As
Gramineae	*Arrhenatherum elatius*	Eu
Gramineae	*Arrhenatherum elatius* var. *bulbosum*	Eu
Gramineae	*Avena fatua*	Eu–As
Gramineae	*Avena sativa*	Eu–As

Family Name	Scientific Name	Origin
Gramineae	*Briza minor*	Eu
Gramineae	*Bromus carinatus*	Eu–Am
Gramineae	*Bromus inermis*	Eu
Gramineae	*Bromus mollis*	Eu
Gramineae	*Bromus racemosus*	Eu
Gramineae	*Bromus rigidus*	Eu
Gramineae	*Bromus secalinus*	Eu
Gramineae	*Bromus sterilis*	Eu
Gramineae	*Bromus tectorum*	Eu
Gramineae	*Bromus tectorum* var. *glabratus*	Eu–As
Gramineae	*Bromus unioloides*	Am
Gramineae	*Catapodium rigidum*	Eu
Gramineae	*Cenchrus longispinus*	Am
Gramineae	*Chloris virgata*	Am
Gramineae	*Coix lacryma-jobi*	As
Gramineae	*Dactylis glomerata*	Eu–As
Gramineae	*Dactyloctenium aegyptium*	As
Gramineae	*Dichanthelium acuminatum*	Am
Gramineae	*Eragrostis curvula*	Af
Gramineae	*Eremochloa ophiuroides*	As
Gramineae	*Festuca arundinacea*	Eu
Gramineae	*Festuca megalura*	Am
Gramineae	*Festuca myuros*	Eu
Gramineae	*Festuca pratensis*	Eu
Gramineae	*Glyceria declinata*	Eu
Gramineae	*Holcus lanatus*	Eu
Gramineae	*Hordeum jubatum*	Eu
Gramineae	*Hordeum murinum*	Eu–As
Gramineae	*Hordeum pusillum*	Am
Gramineae	*Leptochloa fusca*	As
Gramineae	*Lolium multiflorum*	Eu
Gramineae	*Lolium multiflorum* var. *ramosum*	Eu
Gramineae	*Lolium perenne*	Eu
Gramineae	*Lolium rigidum*	Med
Gramineae	*Lolium temulentum*	Eu
Gramineae	*Panicum dichotomiflorum*	Am
Gramineae	*Panicum miliaceum*	As
Gramineae	*Panicum virgatum*	Am
Gramineae	*Parapholis incurva*	Eu
Gramineae	*Paspalum dilatatum*	Am
Gramineae	*Paspalum distichum*	As
Gramineae	*Paspalum distichum* var. *indutum*	Am
Gramineae	*Paspalum notatum*	Am
Gramineae	*Paspalum urvillei*	Am
Gramineae	*Pennisetum flaccidum*	As
Gramineae	*Phalaris canariensis*	Eu–As
Gramineae	*Phalaris minor*	Eu–As
Gramineae	*Phleum paniculatum*	Eu
Gramineae	*Phleum pratense*	Eu
Gramineae	*Poa bulbosa* var. *vivipara*	Eu
Gramineae	*Poa compressa*	Eu
Gramineae	*Poa pratensis*	Eu
Gramineae	*Puccinellia distans*	Eu
Gramineae	*Rottboellia cochinchinensis*	etc.
Gramineae	*Saccharum arundinaceum*	As
Gramineae	*Sorghum halepense*	Eu
Gramineae	*Sorghum halepense* f. *muticum*	Eu
Gramineae	*Spartina anglica*	Eu
Guttiferae	*Hypericum perforatum*	Eu
Iridaceae	*Sisyrinchium angustifolium*	Am

Family Name	Scientific Name	Origin
Iridaceae	*Tritonia crocosmaeflora*	Eu
Juglandaceae	*Pterocarya stenoptera*	As
Labiatae	*Lamium purpureum*	Eu–As
Labiatae	*Lamium purpureum* var. *hybridum*	Eu
Labiatae	*Scutellaria baicalensis*	As
Leguminosae	*Amorpha fruticosa*	Am
Leguminosae	*Astragalus sinicus*	As
Leguminosae	*Lespedeza davidii*	As
Leguminosae	*Lespedeza floribunda*	As
Leguminosae	*Lespedeza lichiyuniae*	As
Leguminosae	*Lotus corniculatus*	Eu
Leguminosae	*Lotus uliginosus*	Eu–Af
Leguminosae	*Lupinus angustifolius*	Eu
Leguminosae	*Medicago lupulina*	Eu
Leguminosae	*Medicago minima*	Eu
Leguminosae	*Medicago polymorpha*	Eu
Leguminosae	*Medicago sativa*	Med
Leguminosae	*Melilotus alba*	As
Leguminosae	*Melilotus suaveolens*	As
Leguminosae	*Robinia pseudoacacia*	Am
Leguminosae	*Securigera varia*	Eu–As
Leguminosae	*Trifolium campestre*	Eu
Leguminosae	*Trifolium dubium*	Eu–As
Leguminosae	*Trifolium hybridum*	Eu–As
Leguminosae	*Trifolium incarnatum*	Eu
Leguminosae	*Trifolium pratense*	Eu
Leguminosae	*Trifolium repens*	Eu–Af
Leguminosae	*Trifolium resupinatum*	Eu–As
Leguminosae	*Vicia dasycarpa*	Eu
Leguminosae	*Vicia villosa*	Eu
Lythraceae	*Ammannia coccinea*	Am
Magnoliaceae	*Magnolia obovata*	As
Malvaceae	*Abutilon theophrasti*	As
Malvaceae	*Hibiscus trionum*	Eu
Malvaceae	*Malva neglecta*	Eu–As
Malvaceae	*Malva parviflora*	Eu
Malvaceae	*Malva pusilla*	Eu
Malvaceae	*Malva sylvestris* var. *mauritiana*	Eu
Malvaceae	*Modiola caroliniana*	Am
Malvaceae	*Sida rhombifolia*	Am
Malvaceae	*Sida spinosa*	Am
Molluginaceae	*Mollugo verticillata*	Am
Onagraceae	*Oenothera biennis*	Am
Onagraceae	*Oenothera erythrosepala*	Am
Onagraceae	*Oenothera laciniata*	Am
Onagraceae	*Oenothera rosea*	Am
Onagraceae	*Oenothera striata*	Am
Oxalidaceae	*Oxalis articulata*	Am
Oxalidaceae	*Oxalis corymbosa*	Am
Papaveraceae	*Papaver dubium*	Eu
Papaveraceae	*Papaver hybridum*	Eu
Papaveraceae	*Papaver rhoeas*	Eu
Phytolaccaceae	*Phytolacca americana*	Am
Phytolaccaceae	*Phytolacca esculenta*	As
Plantaginaceae	*Nuttallanthus canadensis*	Am
Plantaginaceae	*Plantago aristata*	Am
Plantaginaceae	*Plantago lanceolata*	Eu
Plantaginaceae	*Plantago virginica*	Am
Polygonaceae	*Fallopia convolvulus*	Eu–As
Polygonaceae	*Fallopia dentatoalata*	Eu

Family Name	Scientific Name	Origin
Polygonaceae	*Fallopia dumetorum*	Eu
Polygonaceae	*Persicaria capitata*	As
Polygonaceae	*Persicaria orientalis*	As
Polygonaceae	*Persicaria wallichii*	As
Polygonaceae	*Rumex acetosella*	Eu
Polygonaceae	*Rumex crispus*	Eu
Polygonaceae	*Rumex nipponicus*	As
Polygonaceae	*Rumex obtusifolius*	Eu–As
Ranunculaceae	*Ranunculus arvensis*	Eu
Ranunculaceae	*Ranunculus muricatus*	Eu
Rosaceae	*Potentilla amurensis*	Eu
Rosaceae	*Potentilla supina*	Eu
Rosaceae	*Rubus fruticosus*	Eu
Rosaceae	*Sanguisorba minor*	Eu
Rubiaceae	*Diodia teres* var. *hirsutior*	Am
Rubiaceae	*Diodia teres*	Am
Rubiaceae	*Diodia virginiana*	Am
Rubiaceae	*Oldenlandia corymbosa*	As–Af
Rubiaceae	*Sherardia arvensis*	Eu
Saururaceae	*Houttuynia cordata*	As
Scrophulariaceae	*Cymbalaria muralis*	Eu
Scrophulariaceae	*Gratiola officinalis*	Eu
Scrophulariaceae	*Lindernia anagallidea*	Am
Scrophulariaceae	*Lindernia dubia*	Am
Scrophulariaceae	*Verbascum thapsus*	Eu
Scrophulariaceae	*Veronica americana*	Am
Scrophulariaceae	*Veronica arvensis*	Eu–As
Scrophulariaceae	*Veronica hederaefolia*	Eu
Scrophulariaceae	*Veronica persica*	Eu–As
Scrophulariaceae	*Veronica serpyllifolia*	Eu
Simaroubaceae	*Ailanthus altissima*	As
Solanaceae	*Datura meteloides*	Am
Solanaceae	*Datura stramonium*	As
Solanaceae	*Datura stramonium* var. *chalybea*	Am
Solanaceae	*Nicandra physalodes*	Am
Solanaceae	*Physalis angulata*	Am
Solanaceae	*Physalis wrightii*	Am
Solanaceae	*Solanum americanum*	Am
Solanaceae	*Solanum carolinense*	Am
Solanaceae	*Solanum elaeagnifolium*	Am
Solanaceae	*Solanum nigrum* var. *humile*	Eu
Solanaceae	*Solanum photeinocarpum*	Am
Solanaceae	*Solanum rostratum*	Am
Solanaceae	*Solanum sarachoides*	Am
Solanaceae	*Solanum sisymbriifolium*	Am
Solanaceae	*Solanum viarum*	Am
Umbelliferae	*Anthriscus caucalis*	Eu
Umbelliferae	*Apium leptophyllum*	Am
Umbelliferae	*Bifora radians*	Med
Umbelliferae	*Chaerophyllum tainturieri*	Am
Umbelliferae	*Conium maculatum*	Eu
Umbelliferae	*Foeniculum vulgare*	Eu
Umbelliferae	*Lisaea heterocarpa*	As
Valerianaceae	*Valerianella olitoria*	Eu
Verbenaceae	*Verbena bonariensis*	Am
Verbenaceae	*Verbena brasiliensis*	Am
Violaceae	*Viola arvensis*	Eu
Violaceae	*Viola papilionacea*	Am

References

1. Tóth, E.G.; Tremblay, F.; Housset, J.M.; Bergeron, Y.; Carcaillet, C. Geographic isolation and climatic variability contribute to genetic differentiation in fragmented populations of the long-lived subalpine conifer Pinus cembra L. in the western Alps. *BMC Evol. Biol.* **2019**, *19*, 190. [CrossRef]
2. Dambros, C.; Zuquim, G.; Moulatlet, G.M.; Costa, F.R.C.; Tuomisto, H.; Ribas, C.C.; Azevedo, R.; Baccaro, F.; Bobrowiec, P.E.D.; Dias, M.S.; et al. The role of environmental filtering, geographic distance and dispersal barriers in shaping the turnover of plant and animal species in Amazonia. *Biodivers. Conserv.* **2020**, *29*, 3609–3634. [CrossRef]
3. Rincón-Barrado, M.; Olsson, S.; Villaverde, T.; Moncalvillo, B.; Pokorny, L.; Forrest, A.; Riina, R.; Sanmartín, I. Ecological and geological processes impacting speciation modes drive the formation of wide-range disjunctions within tribe Putorieae (Rubiaceae). *J. Syst. Evol.* **2021**, *59*, 915–934. [CrossRef]
4. Grove, R.H.; Burdon, J.J. *Ecology of Biological Invasions*; Cambridge University Press: Cambridge, UK, 1986.
5. Hedgpeth, J.W. Foreign invaders. *Science* **1993**, *261*, 34–35. [CrossRef]
6. Cronk, Q.C.; Fuller, J.L. *Plant Invaders. People and Plants Conservation Manual*; Earthscan: London, UK, 1995.
7. Mooney, H.A. Invasive alien species: The nature of the problem. In *Invasive Alien Species*; Island Press: Washington, DC, USA, 2005; Volume 63, pp. 1–15.
8. Rawlins, K.; Griffin, J.; Moorhead, D.; Bargeron, C.; Evans, C. *EDDMapS: Invasive Plant Mapping Handbook*; Center for Invasive Species and Ecosystem Health, The University of Georgia: Athens, GA, USA, 2011.
9. World Wide Fund (WWF). *Living Planet Report—2020: Bending the Curve of Biodiversity Loss*; World Wide Fund (WWF): Gland, Switzerland, 2020.
10. Ellis, E.C. Land Use and Ecological Change: A 12,000-Year History. *Annu. Rev. Environ. Resour.* **2021**, *46*, 1–33. [CrossRef]
11. Johnstone, I.M. Plant invasion windows: A time-based classification of invasion potential. *Biol. Rev.* **1986**, *61*, 369–394. [CrossRef]
12. Hobbs, R.J.; Huenneke, L.F. Disturbance, Diversity, and Invasion: Implications for Conservation. *Conserv. Biol.* **1992**, *6*, 324–337. [CrossRef]
13. Mayfield, A.E.; Seybold, S.J.; Haag, W.R.; Johnson, M.T.; Kerns, B.K.; Kilgo, J.C.; Larkin, D.J.; Lucardi, R.D.; Moltzan, B.D.; Pearson, D.E.; et al. Impacts of Invasive Species in Terrestrial and Aquatic Systems in the United States. In *Invasive Species in Forests and Rangelands of the United States: A Comprehensive Science Synthesis for the United States Forest Sector*; Poland, T.M., Patel-Weynand, T., Finch, D.M., Miniat, C.F., Hayes, D.C., Lopez, V.M., Eds.; Springer International Publishing: Cham, Switzerland, 2021; pp. 5–39.
14. Mooney, H.A.; Drake, J.A. *Ecology of Biological Invasions of North America and Hawaii*; Springer: New York, NY, USA, 1986.
15. Vitousek, P.M. Biological Invasions and Ecosystem Properties: Can Species Make a Difference? In *Ecology of Biological Invasions of North America and Hawaii*; Mooney, H.A., Drake, J.A., Eds.; Springer: New York, NY, USA, 1986; pp. 163–176.
16. Schofield, E.K. Effects of Introduced Plants and Animals on Island Vegetation: Examples from Galápagos Archipelago. *Conserv. Biol.* **1989**, *3*, 227–239. [CrossRef]
17. Simberloff, D.; Schmitz, D.C.; Brown, T.C. *Strangers in Paradise: Impact and Management of Nonindigenous Species in Florida*; Island Press: Washington, DC, USA, 1997.
18. Catford, J.A.; Bode, M.; Tilman, D. Introduced species that overcome life history tradeoffs can cause native extinctions. *Nat. Commun.* **2018**, *9*, 2131. [CrossRef]
19. Emery-Butcher, H.E.; Beatty, S.J.; Robson, B.J. The impacts of invasive ecosystem engineers in freshwaters: A review. *Freshw. Biol.* **2020**, *65*, 999–1015. [CrossRef]
20. Ruas, R.d.B.; Costa, L.M.S.; Bered, F. Urbanization driving changes in plant species and communities—A global view. *Glob. Ecol. Conserv.* **2022**, *38*, e02243. [CrossRef]
21. Coblentz, B.E. Exotic Organisms: A Dilemma for Conservation Biology. *Conserv. Biol.* **1990**, *4*, 261–265. [CrossRef]
22. Werren, G.L. *Environmental Weeds of the Wet Tropics Bioregion: Risk Assessment & Priority Ranking*; Citeseer: Cairns, Australia, 2001.
23. Convention on Biological Diversity (CBD). *Review of the Status and Trends of, and Major Threats to, Forest Biological Diversity*; Convention on Biological Diversity (CBD): Cairns, Australia, 2002; p. 164.
24. Pereira, H.M.; Navarro, L.M.; Martins, I.S. Global Biodiversity Change: The Bad, the Good, and the Unknown. *Annu. Rev. Environ. Resour.* **2012**, *37*, 25–50. [CrossRef]
25. Morris, R.J. Anthropogenic impacts on tropical forest biodiversity: A network structure and ecosystem functioning perspective. *Philos. Trans. R. Soc. B Biol. Sci.* **2010**, *365*, 3709–3718. [CrossRef]
26. Zietsman, L. *Observations on Environmental Change in South Africa*; SUN Press: Cape Town, South Africa, 2011.
27. Chu, E.W.; Karr, J.R. Environmental Impact: Concept, Consequences, Measurement. In *Reference Module in Life Sciences*; PMC COVID-19 Collection; Elsevier: Amsterdam, The Netherlands, 2016; pp. 1–22. [CrossRef]
28. McNeely, J.; Mooney, H.; Neville, L.; Schei, P.; Waage, J. *Global Strategy on Invasive Alien Species*; IUCN, Gland (Suiza) Global Invasive Species Programme: Cambridge, UK, 2001.
29. Weidlich, E.W.A.; Flórido, F.G.; Sorrini, T.B.; Brancalion, P.H.S. Controlling invasive plant species in ecological restoration: A global review. *J. Appl. Ecol.* **2020**, *57*, 1806–1817. [CrossRef]
30. Rai, P.K. Environmental Degradation by Invasive Alien Plants in the Anthropocene: Challenges and Prospects for Sustainable Restoration. *Anthr. Sci.* **2022**, *1*, 5–28. [CrossRef]

31. Huston, M.A. *Biological Diversity: The Coexistence of Species*; Cambridge University Press: Cambridge, UK, 1994.
32. Hull, R.B.; Gobster, P.H. Restoring Forest Ecosystems: The Human Dimension. *J. For.* **2000**, *98*, 32–36. [CrossRef]
33. Mooney, H.A.; Hobbs, R.J. *Invasive Species in a Changing World*; Island Press: Washington, DC, USA, 2000.
34. Mills, L.S.; Soule, E.M.; Doak, D.F. The Keystone-Species Concept in Ecology and Conservation. *BioScience* **1993**, *43*, 219–224. [CrossRef]
35. Rejmánek, M.; Randall, J.M. Invasive alien plants in California: 1993 summary and comparison with other areas in North America. *Madrono* **1994**, *41*, 161–177.
36. Williamson, M. Invasive species: Vectors management strategies. In *Invasion Vectors: A Conceptual Framework for Management*; Ruiz, G.M., Carlton, J.T., Eds.; Island Press: Washington, DC, USA, 2003; pp. 459–504.
37. Mormul, R.P.; Vieira, D.S.; Bailly, D.; Fidanza, K.; da Silva, V.F.B.; da Graça, W.J.; Pontara, V.; Bueno, M.L.; Thomaz, S.M.; Mendes, R.S. Invasive alien species records are exponentially rising across the Earth. *Biol. Invasions* **2022**, *24*, 3249–3261. [CrossRef]
38. Haubrock, P.J.; Ahmed, D.A.; Cuthbert, R.N.; Stubbington, R.; Domisch, S.; Marquez, J.R.G.; Beidas, A.; Amatulli, G.; Kiesel, J.; Shen, L.Q.; et al. Invasion impacts and dynamics of a European-wide introduced species. *Glob. Chang. Biol.* **2022**, *28*, 4620–4632. [CrossRef]
39. Shine, C.; Williams, N.; Gündling, L. *A Guide to Designing Legal and Institutional Frameworks on Alien Invasive Species*; IUCN: Gland, Switzerland, 2000.
40. National Invasive Species Council. *Meeting the Invasive Species Challenge: National Invasive Species Management Plan*; National Invasive Species Council: Washington DC, USA, 2001.
41. Early, R.; Bradley, B.A.; Dukes, J.S.; Lawler, J.J.; Olden, J.D.; Blumenthal, D.M.; Gonzalez, P.; Grosholz, E.D.; Ibañez, I.; Miller, L.P.; et al. Global threats from invasive alien species in the twenty-first century and national response capacities. *Nat. Commun.* **2016**, *7*, 12485. [CrossRef] [PubMed]
42. Mack, R.N.; Simberloff, D.; Mark Lonsdale, W.; Evans, H.; Clout, M.; Bazzaz, F.A. Biotic invasions: Causes, epidemiology, global consequences, and control. *Ecol. Appl.* **2000**, *10*, 689–710. [CrossRef]
43. Wittenberg, R.; JW Cock, M. *Invasive Alien Species: A Toolkit of Best Prevention and Management Practices*; CAB International: Oxon, UK, 2001.
44. Meyer, S.E.; Callaham, M.A.; Stewart, J.E.; Warren, S.D. Invasive Species Response to Natural and Anthropogenic Disturbance. In *Invasive Species in Forests and Rangelands of the United States: A Comprehensive Science Synthesis for the United States Forest Sector*; Poland, T.M., Patel-Weynand, T., Finch, D.M., Miniat, C.F., Hayes, D.C., Lopez, V.M., Eds.; Springer International Publishing: Cham, Switzerland, 2021; pp. 85–110.
45. Dybas, C.L. Harmful algal blooms: Biosensors provide new ways of detecting and monitoring growing threat in coastal waters. *BioScience* **2003**, *53*, 918–923. [CrossRef]
46. Chown, S.L.; Hodgins, K.A.; Griffin, P.C.; Oakeshott, J.G.; Byrne, M.; Hoffmann, A.A. Biological invasions, climate change and genomics. *Evol. Appl.* **2015**, *8*, 23–46. [CrossRef] [PubMed]
47. Dix, M.; Bass, B.; Covell, S. *Forest Service National Strategic Framework for Invasive Species Management*; USDA: Washington, DC, USA, 2013.
48. Derr, J.F. Common Reed (Phragmites australis) Response to Mowing and Herbicide Application. *Invasive Plant Sci. Manag.* **2008**, *1*, 12–16. [CrossRef]
49. Kettenring, K.M.; Adams, C.R. Lessons learned from invasive plant control experiments: A systematic review and meta-analysis. *J. Appl. Ecol.* **2011**, *48*, 970–979. [CrossRef]
50. Guo, Q.; Brockway, D.G.; Larson, D.L.; Wang, D.; Ren, H. Improving Ecological Restoration to Curb Biotic Invasion-A Practical Guide. *Invasive Plant Sci. Manag.* **2018**, *11*, 163–174. [CrossRef]
51. Dahlsten, D.L. Control of Invaders. In *Ecology of Biological Invasions of North America and Hawaii*; Mooney, H.A., Drake, J.A., Eds.; Springer: New York, NY, USA, 1986; pp. 275–302.
52. Genovesi, P.; Shine, C. *European Strategy on Invasive Alien Species: Convention on the Conservation of European Wildlife and Habitats (Bern Convention)*; Council of Europe: Strasbourg, France, 2004.
53. Simmons, M.T. Bullying the Bullies: The Selective Control of an Exotic, Invasive Annual (Rapistrum rugosum) by Oversowing with a Competitive Native Species (*Gaillardia pulchella*). *Restor. Ecol.* **2005**, *13*, 609–615. [CrossRef]
54. Cuda, J.; Charudattan, R.; Grodowitz, M.; Newman, R.; Shearer, J.; Tamayo, M.; Villegas, B. Recent advances in biological control of submersed aquatic weeds. *Aquat. Plant Manag.* **2008**, *46*, 15.
55. Greipsson, S. Phytoremediation. *Nat. Educ. Knowl.* **2011**, *3*, 7.
56. Buckley, Y.M.; Bolker, B.M.; Rees, M. Disturbance, invasion and re-invasion: Managing the weed-shaped hole in disturbed ecosystems. *Ecol. Lett.* **2007**, *10*, 809–817. [CrossRef]
57. Iannone Iii, B.V.; Galatowitsch, S.M. Altering Light and Soil N to Limit Phalaris arundinacea Reinvasion in Sedge Meadow Restorations. *Restor. Ecol.* **2008**, *16*, 689–701. [CrossRef]
58. Funk, J.L.; Cleland, E.E.; Suding, K.N.; Zavaleta, E.S. Restoration through reassembly: Plant traits and invasion resistance. *Trends Ecol. Evol.* **2008**, *23*, 695–703. [CrossRef] [PubMed]
59. Zedler, J. Ecological restoration: Guidance from theory. *San Fr. Estuary Watershed Sci.* **2005**, *3*, 1–31. [CrossRef]

60. Shea, K.; Chesson, P. Community ecology theory as a framework for biological invasions. *Trends Ecol. Evol.* **2002**, *17*, 170–176. [CrossRef]
61. Laughlin, D.C. Applying trait-based models to achieve functional targets for theory-driven ecological restoration. *Ecol. Lett.* **2014**, *17*, 771–784. [CrossRef]
62. Korea Meteorological Administration (KMA). Korea Meteorological Data Service. Available online: https://www.weather.go.kr/w/index.do (accessed on 4 May 2023).
63. Seoul City. *A Survey on Changes of Biological Species Distribution in Seoul*; Seoul City: Seoul, Republic of Korea, 1999.
64. Seoul City. *A Survey on Biotope and Establishment of Guideline to Construct Eco-Polis in Seoul*; Seoul City: Seoul, Republic of Korea, 2000.
65. Lee, C.; You, Y.; Robinson, G.R. Secondary Succession and Natural Habitat Restoration in Abandoned Rice Fields of Central Korea. *Restor. Ecol.* **2002**, *10*, 306–314. [CrossRef]
66. Yang, G.Y. *A Study on the Forest and the Distribution of Naturalized Plants in Seoul*; Chungang University: Seoul, Republic of Korea, 1989.
67. Seoul City. Statistics of Seoul. Available online: http://data.si.re.kr/2015br10-trendof-population-growth (accessed on 4 May 2023).
68. Kim, J.; Choe, S. *Seoul: The Making of a Metropolis*; Wiley: Hoboken, NJ, USA, 1997; Volume 43.
69. Lee, C.S.; Cho, Y.C.; Shin, H.C.; Moon, J.S.; Lee, B.C.; Bae, Y.S.; Byun, H.G.; Yi, H.B. Ecological response of streams in Korea under different management regimes. *Water Eng. Res.* **2005**, *6*, 131–147.
70. Lee, C.S.; Cho, Y.C.; Lee, A.N. Restoration Planning for the Seoul Metropolitan Area, Korea. In *Ecology, Planning, and Management of Urban Forests: International Perspectives*; Carreiro, M.M., Song, Y.C., Wu, J., Eds.; Springer: New York, NY, USA, 2008; pp. 393–419.
71. Ministry of Environment. *Monitoring of Invasive Alien Species Designated by the Wild Life Protection Act*; Ministry of Environment: Seoul, Republic of Korea, 2006.
72. Korea National Arboretum (KNA). Korean Plant Names Index. Available online: http://www.nature.go.kr/kbi/plant/pilbk/selectPlantPilbkGnrlList.do (accessed on 4 May 2023).
73. Raunkiaer, C. *The Life Forms of Plants and Statistical Plant Geography*; Charendon Press: Oxford, UK, 1934.
74. Environmental System Research Institute. *Arcview GIS*; Environmental System Research Institute: Redlands, CA, USA, 2005.
75. Seoul City. *Seoul Forest Ecosystem Survey Research Report*; Seoul City: Seoul, Republic of Korea, 1998.
76. Braun-Blanquet, J. Pflanzengesellschaft und Biozönose. *Pflanzensoziologie* **1964**, *3*, 1–6. [CrossRef]
77. Hill, M.O. Decorana. A Fortran program for detrended correspondence analysis and reciprocal averaging. *Vegetatio* **1979**, *42*, 47–58. [CrossRef]
78. Kent, M.; Coker, P. *Vegetation Description and Analysis: A Practical Approach*; WILEY-BLACKWELL: Chichester, UK, 1992.
79. Magurran, A.E. *Measuring Biological Diversity*; Blackwell: New York, NY, USA, 2004.
80. Ellenberg, D.; Mueller-Dombois, D. *Aims and Methods of Vegetation Ecology*; Wiley: New York, NY, USA, 1974.
81. National Institute of Environmental Research (NIER). *Survey for Ecological Impact by Naturalized Organisms (I)*; National Institute of Environmental: Seoul, Republic of Korea, 1995.
82. National Institute of Environmental Research (NIER). *Survey for Ecological Impact by Naturalized Organisms (II)*; National Institute of Environmental: Seoul, Republic of Korea, 1996.
83. Kim, J.M.; Yim, Y.J.; Jeon, E.S. *Invasive Alien Plants in Korea*; Science Books: Seoul, Republic of Korea, 2000.
84. SoulÉ, M.E. The Onslaught of Alien Species, and Other Challenges in the Coming Decades. *Conserv. Biol.* **1990**, *4*, 233–240. [CrossRef]
85. Cox, G.W. *Alien Species and Evolution: The Evolutionary Ecology of Exotic Plants, Animals, Microbes, and Interacting Native Species*; Island Press: Washington, DC, USA, 2004.
86. Richardson, D.M.; Rejmánek, M. Conifers as invasive aliens: A global survey and predictive framework. *Divers. Distrib.* **2004**, *10*, 321–331. [CrossRef]
87. Havel, J.E.; Kovalenko, K.E.; Thomaz, S.M.; Amalfitano, S.; Kats, L.B. Aquatic invasive species: Challenges for the future. *Hydrobiologia* **2015**, *750*, 147–170. [CrossRef] [PubMed]
88. Kowarik, I. Neophytes in Germany: Quantitative overview, introduction and dispersal pathways, ecological consequences and open questions. *Texte des Umweltbundesamtes Berlin* **1999**, *1*, 12–36.
89. Shiferaw, W.; Demissew, S.; Bekele, T. Invasive alien plant species in Ethiopia: Ecological impacts on biodiversity a review paper. *Int. J. Mol. Biol.* **2018**, *3*, 171–178. [CrossRef]
90. Daly, E.Z.; Chabrerie, O.; Massol, F.; Facon, B.; Hess, M.C.M.; Tasiemski, A.; Grandjean, F.; Chauvat, M.; Viard, F.; Forey, E.; et al. A synthesis of biological invasion hypotheses associated with the introduction–naturalisation–invasion continuum. *Oikos* **2023**, *5*, e09645. [CrossRef]
91. Wiens, J.A.; Nils Chr, S.; Van Horne, B.; Ims, R.A. Ecological Mechanisms and Landscape Ecology. *Oikos* **1993**, *66*, 369–380. [CrossRef]
92. Forman, R.T. *Land Mosaics: The Ecology of Landscapes and Regions*; Cambridge University Press: Cambridge, UK, 1995.
93. Pickett, S.T.A.; Cadenasso, M.L. Landscape Ecology: Spatial Heterogeneity in Ecological Systems. *Science* **1995**, *269*, 331–334. [CrossRef]

94. Miller, J.H.; Lemke, D.; Coulston, J. The invasion of southern forests by nonnative plants: Current and future occupation, with impacts, management strategies, and mitigation approaches. In *The Southern Forest Futures Project: Technical Report*; Wear, D.N., Greis, J.G., Eds.; Forest Service: Raleigh, NC, USA, 2013; pp. 397–456.
95. Hobbs, R.J. Disturbance of a precursor to weed invasion in native vegetation. *Plant Prot. Q.* **1991**, *6*, 99–104.
96. Abensperg-Traun, M.; Atkins, L.; Hobbs, R.; Steven, D. Exotic plant invasion and understorey species richness: A comparison of two types of eucalypt woodland in agricultural Western Australia. *Pac. Conserv. Biol.* **1998**, *4*, 21–32. [CrossRef]
97. Morgan, J.W. Patterns of invasion of an urban remnant of a species-rich grassland in southeastern Australia by non-native plant species. *J. Veg. Sci.* **1998**, *9*, 181–190. [CrossRef]
98. Grzędzicka, E. Assessment of Habitat Selection by Invasive Plants and Conditions with the Best Performance of Invasiveness Traits. *Diversity* **2023**, *15*, 333. [CrossRef]
99. Duggin, J.A.; Gentle, C.B. Experimental evidence on the importance of disturbance intensity for invasion of Lantana camara L. in dry rainforest–open forest ecotones in north-eastern NSW, Australia. *For. Ecol. Manag.* **1998**, *109*, 279–292. [CrossRef]
100. Cadenasso, M.L.; Pickett, S.T.A. Linking forest edge structure to edge function: Mediation of herbivore damage. *J. Ecol.* **2000**, *88*, 31–44. [CrossRef]
101. Lonsdale, W.M. Global patterns of plant invasions and the concept of invasibility. *Ecology* **1999**, *80*, 1522–1536. [CrossRef]
102. Davis, M.A.; Grime, J.P.; Thompson, K. Fluctuating resources in plant communities: A general theory of invasibility. *J. Ecol.* **2000**, *88*, 528–534. [CrossRef]
103. Katsanevakis, S.; Tempera, F.; Teixeira, H. Mapping the impact of alien species on marine ecosystems: The Mediterranean Sea case study. *Divers. Distrib.* **2016**, *22*, 694–707. [CrossRef]
104. Rojas-Sandoval, J.; Ackerman, J.D.; Marcano-Vega, H.; Willig, M.R. Alien species affect the abundance and richness of native species in tropical forests: The role of adaptive strategies. *Ecosphere* **2022**, *13*, e4291. [CrossRef]
105. Pyšek, P.; Richardson, D.M. Invasive Species, Environmental Change and Management, and Health. *Annu. Rev. Environ. Resour.* **2010**, *35*, 25–55. [CrossRef]
106. Ricciardi, A.; Iacarella, J.C.; Aldridge, D.C.; Blackburn, T.M.; Carlton, J.T.; Catford, J.A.; Dick, J.T.A.; Hulme, P.E.; Jeschke, J.M.; Liebhold, A.M.; et al. Four priority areas to advance invasion science in the face of rapid environmental change. *Environ. Rev.* **2020**, *29*, 119–141. [CrossRef]
107. Parendes, L.A.; Jones, J.A. Role of Light Availability and Dispersal in Exotic Plant Invasion along Roads and Streams in the H. J. Andrews Experimental Forest, Oregon. *Conserv. Biol.* **2000**, *14*, 64–75. [CrossRef]
108. Zhu, G.; Gutierrez Illan, J.; Looney, C.; Crowder, D.W. Assessing the ecological niche and invasion potential of the Asian giant hornet. *Proc. Natl. Acad. Sci. USA* **2020**, *117*, 24646–24648. [CrossRef]
109. Jungblut, S.; Beermann, J.; Boos, K.; Saborowski, R.; Hagen, W. Population development of the invasive crab Hemigrapsus sanguineus (De Haan, 1853) and its potential native competitor Carcinus maenas (Linnaeus, 1758) at Helgoland (North Sea) between 2009 and 2014. *Aquat. Invasions* **2017**, *12*. [CrossRef]
110. DeRoy, E.M.; Crookes, S.; Matheson, K.; Scott, R.; McKenzie, C.H.; Alexander, M.E.; Dick, J.T.A.; MacIsaac, H.J. Predatory ability and abundance forecast the ecological impacts of two aquatic invasive species. *NeoBiota* **2022**, *71*, 91–112. [CrossRef]
111. Kangas, P. *Ecological Engineering: Principles and Practice*; CRC Press: Boca Raton, FL, USA, 2003.
112. Correll, D.L. Principles of planning and establishment of buffer zones. *Ecol. Eng.* **2005**, *24*, 433–439. [CrossRef]
113. Baker, H.G. Patterns of plant invasion in North America. In *Ecology of biological invasions of North America and Hawaii*; Mooney, H.A., Drake, J.A., Eds.; Springer Science & Business Media: New York, NY, USA, 1986; pp. 44–57.
114. Berger, J.J. Ecological Restoration and NonIndigenous Plant Species: A Review. *Restor. Ecol.* **1993**, *1*, 74–82. [CrossRef]
115. Aronson, J.; Floret, C.; Le Floc'h, E.; Ovalle, C.; Pontanier, R. Restoration and Rehabilitation of Degraded Ecosystems in Arid and Semi-Arid Lands. I. A View from the South. *Restor. Ecol.* **1993**, *1*, 8–17. [CrossRef]
116. National Research Council. *Restoration of Aquatic Ecosystems: Science, Technology, and Public Policy*; The National Academies Press: Washington, DC, USA, 1992.
117. Bradshaw, A.D. Ecological principles and land reclamation practice. *Landsc. Plan.* **1984**, *11*, 35–48. [CrossRef]
118. Pabst, R.; Dias, F.S.; Borda-de-Água, L.; Rodríguez-González, P.M.; Capinha, C. Assessing and predicting the distribution of riparian invasive plants in continental Portugal. *Front. Ecol. Evol. Appl.* **2022**, *10*, 875578. [CrossRef]
119. McDonald, T.; Gann, G.; Jonson, J.; Dixon, K. *International Standards for the Practice of Ecological Restoration–Including Principles and Key Concepts*; Society for Ecological Restoration: Washington, DC, USA, 2016.
120. Gann, G.D.; McDonald, T.; Walder, B.; Aronson, J.; Nelson, C.R.; Jonson, J.; Hallett, J.G.; Eisenberg, C.; Guariguata, M.R.; Liu, J.; et al. International principles and standards for the practice of ecological restoration. Second edition. *Restor. Ecol.* **2019**, *27*, S1–S46. [CrossRef]
121. Kim, A.R.; Lim, B.S.; Seol, J.; Lee, C.S. Principle of restoration ecology reflected in the process creating the National Institute of Ecology. *J. Ecol. Environ.* **2021**, *45*, 12. [CrossRef]
122. Esler, K.J.; van Wilgen, B.W.; te Roller, K.S.; Wood, A.R.; van der Merwe, J.H. A landscape-scale assessment of the long-term integrated control of an invasive shrub in South Africa. *Biol. Invasions* **2010**, *12*, 211–218. [CrossRef]
123. Gaertner, M.; Biggs, R.; Te Beest, M.; Hui, C.; Molofsky, J.; Richardson, D.M. Invasive plants as drivers of regime shifts: Identifying high-priority invaders that alter feedback relationships. *Divers. Distrib.* **2014**, *20*, 733–744. [CrossRef]

124. Hobbs, R.J.; Richardson, D.M. Invasion ecology and restoreation ecology: Parallel evolution in two fields of endeavour. In *Fifty Years of Invasion Ecology: The Legacy of Charles Elton*; Richardson, D.M., Ed.; John Wiley & Sons: Hoboken, NJ, USA, 2011; pp. 61–70.
125. Dulohery, C.J.; Kolka, R.K.; McKevlin, M.R. Effects of a willow overstory on planted seedlings in a bottomland restoration. *Ecol. Eng.* **2000**, *15*, S57–S66. [CrossRef]
126. McLeod, R.W.; Kirkegaard, J.A.; Steel, C.C. Invasion, development, growth and egg laying by Meloidogyne javanica in Brassicaceae crops. *Nematology* **2001**, *3*, 463–472. [CrossRef]
127. Panetta, F.; Hopkins, A. Weeds in corridors: Invasion and management. *Nat. Conserv.* **1991**, *2*, 341–351.
128. Holmes, P.M.; Esler, K.J.; Gaertner, M.; Geerts, S.; Hall, S.A.; Nsikani, M.M.; Richardson, D.M.; Ruwanza, S. Biological Invasions and Ecological Restoration in South Africa. In *Biological Invasions in South Africa*; van Wilgen, B.W., Measey, J., Richardson, D.M., Wilson, J.R., Zengeya, T.A., Eds.; Springer International Publishing: Cham, Switzerland, 2020; pp. 665–700.

biology

MDPI

Article

An Annotated Checklist of Invasive Species of the Phyla Arthropods and Chordates in Panama

Digna Rodríguez-Gavilanes [1], Humberto A. Garcés Botacio [2], Rogemif Fuentes [3], Louise Rodriguez-Scott [4], Yostin Añino [5], Oscar G. López-Chong [6] and Enrique Medianero [7,8,*]

1 Escuela de Biología, Universidad de Panamá, Campus Octavio Méndez Pereira, Panama 0819-07289, Panama; dignamr09@gmail.com
2 Departamento de Biología Marina y Limnología, Universidad de Panamá, Campus Octavio Méndez Pereira, Avenida Simón Bolívar, Panama 0819-07289, Panama; humberto.garces@up.ac.pa
3 Fundación Los Naturalistas, David, Chiriquí 0426-01459, Panama; rogemifdaniel@gmail.com
4 Programa de Maestría en Ciencias Biológicas, Facultad de Ciencias, Naturales, Exactas y Tecnología, Universidad de Panamá, Panama 0819-07289, Panama; louise.rodriguez001@outlook.com
5 Museo de Invertebrado G.B.Fairchild, Universidad de Panamá, Panama 0819-07289, Panama; yostin0660@gmail.com
6 Colección de Aves, Instituto Smithsonian de Investigaciones Tropicales, Panama 0843-03092, Panama; lopezog@si.edu
7 Departamento de Ciencias Ambientales y Programa de Maestría en Entomología, Universidad de Panamá, Campus Octavio Méndez Pereira, Avenida Simón Bolívar, Panama 0819-07289, Panama
8 Miembro del Sistema Nacional de Investigación (SNI-SENACYT), Panama 0816-02852, Panama
* Correspondence: enrique.medianero@up.ac.pa

Simple Summary: Invasive species are considered a threat to the conservation of different environments. Annotating the numbers and species of these invasive organisms is critical to developing conservation strategies. This research gives background information on the types and possible origins of invasive species from the arthropod and chordate groups in Panama. The results indicated that approximately 141 exotic arthropod and chordate species have been reported as invasive species in Panama. Most of these species are believed to have been introduced via the Panama Canal Zone or accidentally. With the information compiled, this study will serve as preliminary data on the sources of introduction and will provide information for future research and plans to prevent the impact of those species.

Abstract: Invasive species are one of the five main causes of biodiversity loss, along with habitat destruction, overexploitation, pollution, and climate change. Numbers and species of invasive organisms represent one of the first barriers to overcome in ecological conservation programs since they are difficult to control and eradicate. Due to the lack of records of invasive exotic species in Panama, this study was necessary for identifying and registering the documented groups of invasive species of the Chordates and Arthropod groups in Panama. This exhaustive search for invasive species was carried out in different bibliographic databases, electronic portals, and scientific journals which addressed the topic at a global level. The results show that approximately 141 invasive exotic species of the Arthropoda and Chordata phyla have been reported in Panama. Of the 141 species, 50 species belonged to the Arthropoda phylum and 91 species belonged to the Chordate phylum. Panamanian economic activity could facilitate the introduction of alien species into the country. This study provides the first list of invasive exotic chordate and arthropod species reported for the Republic of Panama.

Keywords: invasive species; introduced species; biodiversity; Central America; tropical forest; pest

1. Introduction

Invasive species are one of the five main causes of biodiversity loss, along with habitat destruction, overexploitation, pollution, and climate change [1,2]. An invasive species is understood to be an exotic species that establishes itself in a natural or semi-natural ecosystem or habitat and is an agent inducing changes that affect native biodiversity [1]. The process of invasion is a progressive phenomenon that does not have to be unidirectional [3]. This means that not all introduced species (exotic) will become naturalized, nor will all naturalized ones become invasive. Additionally, there is not a consistent number of species that goes from one phase to another [3].

The problems associated with invasive species are summarized in three main aspects [2]: (1) big economic losses and ecological impacts caused by many invasive species; (2) the increase in the number of introduced species that become invasive and the problems associated with these species; and (3) the need to include invasive species in any ecological study, keeping in mind that these organisms are able to alter an ecosystem in which they are introduced [2].

Invasive species are one of the first challenges to overcome in ecological conservation programs since they are difficult to control and eradicate, mainly due to their reproductive strategies combined with their dispersal, establishment, and persistence [2]. Invasive alien species can be found in all groups of organisms [4]. Globally, 90% of vertebrate and plant introductions are intentional, and the remaining 10% are accidental [5]. After plants, arthropods and chordates include the largest number of invasive exotic species reported for Latin America and the Caribbean [4]. The animal groups which pertain to the Phylum Arthropoda contain the largest number of species known today [6]. Some species transmit diseases or are vectors of them, and some can affect crops, and many serve as natural indicators and pollinators. In addition, a few of these species play the role of primary, secondary consumers or decomposers [6]. Organisms found in the Chordate phylum have high diversities of ecological niches, with notable terrestrial and aquatic environment adaptations [7]. In terms of ecosystemic level, the introduction of mammals can affect food chain functioning, generating a cascade effect on the composition and abundance of invasive predator, herbivore, and plant species as well as nutrient cycles [7].

Panama has the highest number of known vertebrate animals of any country in Central America or the Caribbean and more bird species than the United States and Canada together [8]. In addition, it has 3.5% of flowering plants and 7.3% of fern and related species in the world [9]. Panama is the twenty-eighth country in the world with the greatest biological diversity [8]. However, in proportion to its size, it ranks tenth [8].

Panama's location and geography have facilitated the introduction of alien species [9]. Its geographical position, high land elevation, and maritime connectivity, as well as the existence of the Panama Canal, has helped biological invasions in Panama [9]. According to [10], the new locks used in the expansion of the Panama Canal allow the passage of ships that are 366 m longer and 49 m wider, facilitating higher transportation of exotic species through either the ship ballast or attachment to the ship's hull [9].

According to the IV Panama National Biodiversity Report, an estimate of approximately 324 exotic species have been introduced into Panama, the majority of which are plants. In addition, the introduction of exotic pet species has increased. Among the exotic species used as pets are birds, reptiles, and mammals [9]; however, it is not known how many of these exotic species have become invasive species. Due to the lack of records of invasive exotic species in Panama, this study is necessary to help answer how many invasive species of the group of chordates and arthropods have been registered in Panama.

This study provides the first list of invasive species in the Panamanian territory of two important groups of organisms: arthropods and chordates. This study also serves as preliminary data on the sources of introduction and provides information for future research and plans to prevent the negative impact of those species.

2. Materials and Methods

2.1. Study Area

Panama is geographically located at 7°12′07″ and 9°38′46″ north latitude and 77°09′24″ and 83°03′07″ west Neotropical Region longitude. Panama has a land area of 75,416.6 km^2 and is divided into ten provinces. Panama is in the central part of the American continent, in the most eastern and south part of Central America. It is bordered by the Caribbean Sea in the north, the Pacific Ocean in the south, Colombia in the east, and Costa Rica in the west [8].

The economy of Panama is based on four activities, one of them being the logistics industry, which is fundamentally based on the movement of cargo from all over the world [11]. This movement that takes place through ports, airports, railways, and the Panama Canal is one of the main activities that facilitates the introduction of invasive species [12].

2.2. Study Selection

The study was carried out through an exhaustive information review of the following sources:—Global Invasive Species Database: http://www.iucngisd.org/gisd/species.php?sc=965, accessed on 20 January 2023; Invasive Species Specialist Group IUCN/SSC; Invasive Species Specialist Group (ISSG); and Invasive Species Compendium (CABI): https://www.cabi.org/isc/datasheet/108530#tolistOfSpecies (accessed on 20 January 2023)—from bibliographic databases such as Web of Science (WOS), Science Direct, Scielo, PubMed, Redalyc, Dimensions, and Google Scholar. Using the Publish or Perish software version 8, the URLs resulting from the searches were downloaded in csv format, which was worked in Excel to filter those records that contained a mixture of words (invasive species, Panama alien species, invasive species in Panama, invasive exotic species in Panama, Panama invasive species). Additionally, a thorough search was carried out in specialized journals on the subject, such as *BioInvasions Records* (Reabic); *Biological Invasions* (Springer); *Revista Bioinvasiones; Check List—The Journal of Biodiversity Data* (PENSOFT); as well as the journals published in Panama that are found on the ABC Platforms (SENACYT), such as *National Resources* and *SIBIUP* of the University of Panama.

The search was done using the keywords invasive species, exotic species, invasive species in Panama, invasive exotic species in Central America, invasive species of insects in Panama, invasive species in America, new records + Panamá, pest + Panama, introduced species + Panamá, and "alien species" and filtered by the words Panama, new records + Panama, pest + Panama, and introduced species + Panama.

Due to the limited literature published on the subject for Panama, the country's specialists from the different groups under study were asked for a list of the species considered invasive in the country. We visited the following organizations and reviewed their bibliographies on the subject: Environment Ministry, the Aquatic Resources Authority of Panama, the Panama Canal Authority, the Herbarium of the Panama University, and the Directorate of Plant Health of the Ministry of Agricultural Development.

Other sources of information used were *Fifth National Report of Panama to the Convention on Biological Diversity* [13]; *Invasives in Mesoamerica and the Caribbean* [14]; *National Biodiversity Strategy and Action Plan 2018–2050* [15], and the Ministerio de Desarrollo Agropecuario database of invasive species 2019.

The information obtained was organized in tables for easier understanding. A database of the species was created (in cases where the information was obtained) with the following variables: report date (year), province, altitude, place of origin. This allowed us to answer the research question and achieve one of the objectives. Also, to make easy the access to information on invasive species of the phyla Arthropoda and Chordate, another two tables were created, one including the taxonomic identification of the groups and the reference according to the article or database that was obtained for the species and the other containing limitations on the distribution of pets under official control in the Panama Republic.

3. Results

Through the collection of information, the results indicated that approximately 141 invasive exotic species of the Arthropoda and Chordata phyla have been reported in Panama. Of the 141 species, 50 species belonged to the Arthropoda phylum and 91 species belong to the Chordate phylum, as indicated in taxonomy Table 1.

Table 1. A taxonomic list of invasive species recorded for the Panama Republic.

Species	Family	Order	Class	Ecological Status	Associated Organism or Habitat	Ref.
		ARTHROPODA				
Hypothenemus hampei Ferrari, 1867	Curculionidae	Coleoptera	Insecta	Pest species	Coffee	[16]
Coccotrypes advena Blandford 1894	Curculionidae	Coleoptera	Insecta	Pest species	Coffee	[17]
Ambrosiodmus obliquus LeConte, 1878	Curculionidae	Coleoptera	Insecta	Pathogen transmitter	Wood/forestry trees	[17]
Coptoborus ricini Eggers, 1932	Curculionidae	Coleoptera	Insecta	Pest species	Wood/forestry trees	[17]
Xyleborus bispinatus Eichhoff, 1864	Curculionidae	Coleoptera	Insecta	Fungal pathogen transmitter	Wood/forestry trees	[17]
Xylosandrus morigerus Blandford, 1894	Curculionidae	Coleoptera	Insecta	Pest species	Wood/forestry trees	[17]
Xylosandrus compactus Eichhinoff, 1876	Curculionidae	Coleoptera	Insecta	Pest species	Wood/forestry trees	[17]
Cryptocarenus seriatus Eggers, 1933	Curculionidae	Coleoptera	Insecta	Pest species	Wood	[17]
Coccotrypes vulgaris Eggers, 1923	Curculionidae	Coleoptera	Insecta	Pest species	Wood/trees	[17]
Dendrocranulus tardus Blandford, 1896	Curculionidae	Coleoptera	Insecta	Pest species	Wood/forestry trees	[17]
Scolytopsis puncticollis Blandford, 1896	Curculionidae	Coleoptera	Insecta	Pest species	Stem borer	[17]
Xylosandrus crassiusculus Motschulsky, 1866	Curculionidae	Coleoptera	Insecta	Pest species	Wood/forestry trees	[18]
Xyleborinus exiguus Walker, 1859	Curculionidae	Coleoptera	Insecta	Pest species	Tree polyphagous	[17]
Rhynchophorus palmarum Linneus, 1758	Curculionidae	Coleoptera	Insecta	Vector of the red ring (RR)	Palms	[19]
Euoniticellus intermedius Reiche, 1849	Scarabaeidae	Coleoptera	Insecta	Beneficial	Pasturelands	[20]
Harmonia axyridis Pallas, 1773	Coccinellidae	Coleoptera	Insecta	Pest species	Native species displace	[21]
Callosobruchus phaseoli Gyllenhal, 1833	Chrysomelidae	Coleoptera	Insecta	Pest species	*Cajanus cajan*	[22]
Saperda candida Fabricus, 1787	Cerambycidae	Coleoptera	Insecta	Pest species	Wood	[23]
Brachyplatys subaeneus Westwood, 1837	Plastaspidae	Hemiptera	Insecta	Pest species	*Cajanus cajan*	[24]
Diaphorina citri Kuwayama, 1908	Psyllidae	Hemiptera	Insecta	Pest species	Citrus plants	[25]
Pseudacysta perseae Heideman, 1908	Tingidae	Hemiptera	Insecta	Pest species	*Persea americana*	[26]
Aulacaspis yasumatsui Takagi, 1977	Diaspididae	Hemiptera	Insecta	Pest species	*Cycas* spp.	[27]
Aphis spiraecola Patch, 1914	Aphididae	Hemiptera	Insecta	Pest species	*Citrus*	[28]
Thrips palmi Karny, 1990	Thripidae	Thysanoptera	Insecta	Pest species	Cucurbitaceae Solanaceae	[29]
Apis mellifera scutellata Linneus, 1758	Apidae	Hymenoptera	Insecta	Pest species	Human	[30]
Tapinoma melanocephalum Fabricius, 1793	Formicidae	Hymenoptera	Insecta	Pest species	Human, native species displace	[23]
Paratrechina longicornis Motshulsky, 1863	Formicidae	Hymenoptera	Insecta	Pest species	Environmental damage	[23]
Monomorium floricola Jerdon, 1851	Formicidae	Hymenoptera	Insecta	Pest species	Native species displace	[23]
Quadrastichus erythrinae Kim, 2004	Eulophidae	Hymenoptera	Insecta	Pest species	*Erytrina* spp.	[31]
Nylanderia fulva Mayr, 1862	Formicidae	Hymenoptera	Insecta	Pest species	Crops	[32]
Ceratitis capitata Weidemann, 1824	Tephritidae	Diptera	Insecta	Pest species	Fruits	[33]
Anastrepha obliqua Macquart, 1835	Tephritidae	Diptera	Insecta	Pest species	Fruits	[34]
Anastrepha grandis Macquart, 1846	Tephritidae	Diptera	Insecta	Pest species	Cucurbitacea	[35]
Aedes aegypti Linnaeus, 1762	Culicidae	Diptera	Insecta	Disease vector	Human	[36]

Table 1. *Cont.*

Species	Family	Order	Class	Ecological Status	Associated Organism or Habitat	Ref.
		ARTHROPODA				
Aedes albopictus Skuse, 1895	Culicidae	Diptera	Insecta	Disease vector	Human	[36]
Tuta absoluta Meyrick, 1917	Gelechiidae	Lepidoptera	Insecta	Pest species	Tomato and potato	[12]
Blattella germanica Linnaeus, 1767	Ectobiidae	Blattodea	Insecta	Pest species	Human	[37]
Brevipalpus phoenicis Geijskes, 1936	Tenuipalpidae	Acarida	Arachnida	Pest species	Citrus	[38]
Brevipalpus californicus Banks, 1904	Tenuipalpidae	Acarida	Arachnida	Pest species CiLV vector	Citrus	[39]
Dermanyssus gallinae De Geer, 1778	Dermanyssidae	Mesostigmata	Arachnida	Hematophagous ectoparasite	Human and other animals	[40]
Rhithropanopeus harrisii Gould, 1841	Panopeidae	Decapoda	Malacostraca	Pest species	Aquatic biome	[41]
Elamenopsis kempi Chopra y Das, 1930	Hymenosomatidae	Decapoda	Malacostraca	Pest species	Aquatic biome	[42]
Procambarus clarkia Girard, 1852	Cambaridae	Decapoda	Malacostraca	Pest species	Aquatic biome	[35]
Penaeus monodon Fabriciud, 1798	Penaeidae	Decapoda	Malacostraca	Pest species	Aquatic biome	[35]
Cherax tenuimanus Smith, 1912	Astacidae	Decapoda	Malacostraca	Pest species	Aquatic biome	[35]
Macrobrachium amazonicum Heller, 1862	Palaemonidae	Decapoda	Malacostraca	Pest species	Aquatic biome	[43]
Macrobrachium rosenbergii De Man, 1879	Palaemonidae	Decapoda	Malacostraca	Pest species	Aquatic biome	[44]
Goniopsis cruentata Latrelle, 1803	Grapsidae	Decapoda	Malacostraca	Pest species	Aquatic biome	[43]
Pachygrapsyus gracilis de Saussure, 1857	Grapsidae	Decapoda	Malacostraca	Pest species	Aquatic biome	[43]
Raillietiella frenatus Riley & Self, 1981	Cephalobaenidae	Cephalobaenida	Maxillopoda	Parasite	Native toads	[45]
		CHORDATA				
Mus musculus Linnaeus, 1758	Muridae	Rodentia	Mammalia	Pest species	Human	[23]
Rattus norvegicus Berkenhout, 1769	Muridae	Rodentia	Mammalia	Pest species	Human	[23]
Herpestes auropunctatus Hodgson, 1836	Herpestidae	Carnivora	Mammalia	Pest species	Health biodiversity	[23]
Herpestes javanicus Hilaire, 1818	Herpestidae	Carnivora	Mammalia	Pest species	Native species affect	[23]
Columba livia Gmelin, 1789	Columbidae	Columbiformes	Aves	Pest species	Human	[23]
Bubulcus ibis Linnaeus, 1758	Ardeidae	Pelecaniformes	Aves	Pest species	Human	[23]
Mimus gilvus Viellot, 1808	Mimidae	Passeriformes	Aves			[23]
Passer domesticus Linnaeus, 1758	Passeridae	Passeriformes	Aves	Pest species	Human	[46]
Sicalis flaveola Linnaeus, 1766	Thraupidae	Passeriformes	Aves	Pest species		[46]
Rachycentron canadum Linnaeus, 1766	Rachycentridae	Carangiformes	Actinopterygii	Incidental record	Marine biome	[47]
Oreochromis niloticus Linnaeus, 1758	Cichlidae	Perciformes	Actinopterygii	Pest species	Aquatic biome	[48,49]
Oreochromis aureus Steindachner, 1864	Cichlidae	Perciformes	Actinopterygii	Pest species	Aquatic biome	[50]
Oreochromis mossambicus Peters, 1852	Cichlidae	Perciformes	Actinopterygii	Pest species	Aquatic biome	[50]
Oreochromis urolepis Norman, 1912	Cichlidae	Perciformes	Actinopterygii	Pest species	Aquatic biome	[50]
Coptodon rendalli Boulenger, 1897	Cichlidae	Perciformes	Actinopterygii	Pest species	Aquatic biome	[50]
Cichla ocellaris Bloch & Schneider, 1801	Cichlidae	Perciformes	Actinopterygii	Pest species	Aquatic biome	[33]
Cichla monoculus Agassiz, 1831	Cichlidae	Perciformes	Actinopterygii	Pest species	Aquatic biome	[49,50]
Parachromis managuensis Günther, 1867	Cichlidae	Perciformes	Actinopterygii	Pest species	Aquatic biome	[49,51]
Astronotus ocellatus Agassiz, 1881	Cichlidae	Perciformes	Actinopterygii	Pest species	Aquatic biome	[49]
Mesonauta festivus Heckel, 1840	Cichlidae	Perciformes	Actinopterygii	Pest species	Aquatic biome	[49]
Vieja maculicauda Regan, 1905	Cichlidae	Perciformes	Actinopterygii	Pest species	Aquatic biome	[43]
Eleotris melanosoma Bleeker, 1853	Eleotridae	Gobiiformes	Actinopterygii	Pest species	Aquatic biome	[52]

Table 1. *Cont.*

Species	Family	Order	Class	Ecological Status	Associated Organism or Habitat	Ref.
		CHORDATA				
Butis koilomatodon Bleeker, 1849	Butidae	Gobiiformes	Actinopterygii	Pest species	Aquatic biome	[52]
Lepomis macrochirus Rafinesque, 1819	Centrarchidae	Perciformes	Actinopterygii	Pest species	Aquatic biome	[44]
Lepomis microlophus Günther, 1859	Centrarchidae	Perciformes	Actinopterygii	Pest species	Aquatic biome	[53]
Micropterus salmoides Lacepede, 1802	Centrarchidae	Perciformes	Actinopterygii	Pest species	Aquatic biome	[49,53]
Pomoxis annularis Rafinesque, 1818	Centrarchidae	Perciformes	Actinopterygii	Pest species	Aquatic biome	[52]
Pomoxis nigromaculatus Lesueur, 1829	Centrarchidae	Perciformes	Actinopterygii	Pest species	Aquatic biome	[49,52]
Lepomis humilis Girard, 1858	Centrarchidae	Perciformes	Actinopterygii	Pest species	Aquatic biome	[49]
Dormitator maculatus Meek and Hildebrand, 1923	Eleotridae	Perciformes	Actinopterygii	Pest species	Aquatic biome	[43]
Hypleurochilus pseudoaequipinnis Bath, 1994	Blenniidae	Perciformes	Actinopterygii	Pest species	Aquatic biome	[43]
Leptophilypnus fluviatilis Meek and Hildebrand, 1916	Eleotridae	Perciformes	Actinopterygii	Pest species	Aquatic biome	[43]
Lophogobius cyprinoides Pallas, 1770	Gobiidae	Perciformes	Actinopterygii	Pest species	Aquatic biome	[43]
Lupinoblennius vinctus Poey, 1867	Blenniidae	Perciformes	Actinopterygii	Pest species	Aquatic biome	[43]
Omobranchus punctatus Valennciennes, 1836	Blenniidae	Perciformes	Actinopterygii	Pest species	Aquatic biome	[52]
Sciaenops ocellatus Linnaeus, 1766	Sciaenidae	Acanthuriformes	Actinopterygii	Pest species	Aquatic biome	[35]
Oncorhynchus mykiss Walbaum, 1792	Salmonidae	Salmoniformes	Actinopterygii	Pest species	Aquatic biome	[23,49]
Salmo trutta Linnaeus, 1758	Salmonidae	Salmoniformes	Actinopterygii	Pest species	Aquatic biome	[53]
Pterois volitans Linnaeus, 1758	Scorpaenidae	Scorpaeniformes	Actinopterygii	Pest species	Aquatic biome	[54]
Cirrhinus molitorella Valenciennes, 1844	Cyprinidae	Cypriniformes	Actinopterygii	Pest species	Aquatic biome	[35]
Ctenopharyngodon Idella Valenciennes, 1844	Cyprinidae	Cypriniformes	Actinopterygii	Pest species	Aquatic biome	[52]
Cyprinus carpio Linnaeus, 1758	Cyprinidae	Cypriniformes	Actinopterygii	Pest species	Aquatic biome	[23,49]
Hypophthalmichthys molitrix Valenciennes, 1844	Cyprinidae	Cypriniformes	Actinopterygii	Pest species	Aquatic biome	[52]
Hypophthalmichthys nobilis Richardson, 1845	Cyprinidae	Cypriniformes	Actinopterygii	Pest species	Aquatic biome	[23]
Mylopharyngodon piceus Richardson, 1846	Cyprinidae	Cypriniformes	Actinopterygii	Pest species	Aquatic biome	[35]
Ictiobus cyprinellus Valenciennes, 1844	Catostomidae	Cypriniformes	Actinopterygii	Pest species	Aquatic biome	[35]
Gambusia holbrooki Girard, 1859	Poeciliidae	Cyprinodontiformes	Actinopterygii	Pest species	Aquatic biome	[49]
Xiphophorus helleri Heckel, 1848	Poeciliidae	Cyprinodontiformes	Actinopterygii	Pest species	Aquatic biome	[49,55]
Gambusia affinis Baird & Girard, 1853	Poeciliidae	Cyprinodontiformes	Actinopterygii	Pest species	Aquatic biome	[55]
Gambusia nicaraguensis Günther, 1866	Poeciliidae	Cyprinodontiformes	Actinopterygii	Pest species	Aquatic biome	[55]
Poecilia reticulata Peters, 1859	Poeciliidae	Cyprinodontiformes	Actinopterygii	Pest species	Aquatic biome	[55]
Xiphophorus variatus Meek, 1904	Poeciliidae	Cyprinodontiformes	Actinopterygii	Pest species	Aquatic biome	[55]
Ictalurus punctatus Rafinesque, 1818	Ictaluridae	Siluriformes	Actinopterygii	Pest species	Aquatic biome	[35]
Colossoma macropomum Cuvier, 1816	Serrasalmidae	Characiformes	Actinopterygii	Pest species	Aquatic biome	[49,52]
Piaractus brachypomus Cuvier, 1818	Serrasalmidae	Characiformes	Actinopterygii	Pest species	Aquatic biome	[35]
Piaractus mesopotamicus Holmberg, 1887	Serrasalmidae	Characiformes	Actinopterygii	Pest species	Aquatic biome	[35]
Microphis brachyurus lineatus Kaup, 1856	Syngnathidae	Syngnathiformes	Actinopterygii	Pest species	Aquatic biome	[43]
Megalops atlanticus Valenciennes, 1847	Megalopidae	Elopiformes	Actinopterygii	Pest species	Aquatic biome	[43]
Anchoa parva Meek & Hildebrand, 1923	Engraulidae	Clupeiformes	Actinopterygii	Pest species	Aquatic biome	[43]

Table 1. *Cont.*

Species	Family	Order	Class	Ecological Status	Associated Organism or Habitat	Ref.
		CHORDATA				
Hemidactylus brookii Gray, 1845	Gekkonidae	Squamata	Reptilia	Species with potential to displace native species	Terrestrial biome	[56]
Hemidactylus frenatus Dúmeril & Bibron, 1836	Gekkonidae	Squamata	Reptilia	Species with potential to displace native species	Terrestrial biome	[23,56]
Hemidactylus mabouia Moreau de Jonnes, 1818	Gekkonidae	Squamata	Reptilia	Scarcely documented species	Terrestrial biome	[23,56]
Hemidactylus turcicus Linnaeus, 1758	Gekkonidae	Squamata	Reptilia	Scarcely documented species	Terrestrial biome	[56,57]
Lepidodactylus lugubris Dúmeril & Bibron, 1836	Gekkonidae	Squamata	Reptilia	Species with potential to displace native species	Terrestrial biome	[56,58]
Anolis sagrei Cocteau in Dúmeril & Bibron, 1837	Polychrotidae	Squamata	Reptilia	Species with potential to displace native species	Terrestrial biome	[59]
Sphaerodactylus argus Gosse, 1850	Sphaerodactylidae	Squamata	Reptilia	Scarcely documented species	Terrestrial biome	[44,56]
Trachemys scripta elegans Wied, 1838	Emydidae	Testudines	Reptilia	Species with potential to displace native species	Semiaquatic	[23]
Eleutherodactylus antillensis Reinhardt & Lütken, 1863	Eleutherodactylidae	Anura	Amphibia	Scarcely documented species	Terrestrial biome	[60,61]
Eleutherodactylus johnstonei Barbour, 1914	Eleutherodactylidae	Anura	Amphibia	Scarcely documented species	Terrestrial biome	[60]
Eleutherodactylus planirostris Cope, 1862	Eleutherodactylidae	Anura	Amphibia	Scarcely documented species	Terrestrial biome	[62]
Rana catesbeiana Shaw, 1802	Ranidae	Anura	Amphibia	Scarcely documented species	Aquatic and semi-aquatic environments	[35]
Ascidia incrassata Heller, 1878	Ascidiidae	Phlebobranchia	Ascidiacea	Pest species	Aquatic biome	[63]
Ascidia sydneiensis Stimpson, 1855	Ascidiidae	Phlebobranchia	Ascidiacea	Pest species	Aquatic biome	[63]
Phallusia nigra Savigny, 1816	Ascidiidae	Phlebobranchia	Ascidiacea	Pest species	Aquatic biome	[63]
Botrylloides nigrum Herdman, 1886	Styelidae	Stolidobranchia	Ascidiacea	Pest species	Aquatic biome	[63]
Botryllus planus Van Name, 1902	Styelidae	Stolidobranchia	Ascidiacea	Pest species	Aquatic biome	[63]
Polyandrocarpa anguinea Sluiter, 1898	Styelidae	Stolidobranchia	Ascidiacea	Pest species	Aquatic biome	[63]
Polyandrocarpa sagamiensis Tokioka, 1953	Styelidae	Stolidobranchia	Ascidiacea	Pest species	Aquatic biome	[63]
Polyandrocarpa zorritensis Van Name, 1931	Styelidae	Stolidobranchia	Ascidiacea	Pest species	Aquatic biome	[63]
Styela canopus Savigny, 1816	Styelidae	Stolidobranchia	Ascidiacea	Pest species	Aquatic biome	[63]
Symplegma brakenhielmi Michaelsen, 1904	Styelidae	Stolidobranchia	Ascidiacea	Pest species	Aquatic biome	[63]
Symplegma rubra Monniot, 1972	Styelidae	Stolidobranchia	Ascidiacea	Pest species	Aquatic biome	[63]
Herdmania pallida Heller, 1878	Pyuridae	Stolidobranchia	Ascidiacea	Pest species	Aquatic biome	[63]
Microcosmus exasperatus Heller, 1878	Pyuridae	Stolidobranchia	Ascidiacea	Pest species	Aquatic biome	[63]
Pyura vittata Stimpson, 1852	Pyuridae	Stolidobranchia	Ascidiacea	Pest species	Aquatic biome	[63]
Didemnum perlucidum Monniot F, 1983	Didemnidae	Aplousobranchia	Ascidiacea	Pest species	Aquatic biome	[63]
Didemnum psammatodes Sluiter, 1895	Didemnidae	Aplousobranchia	Ascidiacea	Pest species	Aquatic biome	[63]
Diplosoma listerianum Milne Edwards, 1841	Didemnidae	Aplousobranchia	Ascidiacea	Pest species	Aquatic biome	[63]
Lissoclinum fragile Van Name, 1902	Didemnidae	Aplousobranchia	Ascidiacea	Pest species	Aquatic biome	[63]
Trididemnum orbiculatum Van Name, 1902	Didemnidae	Aplousobranchia	Ascidiacea	Pest species	Aquatic biome	[63]
Polyclinum constellatum Savigny, 1816	Polyclinidae	Aplousobranchia	Ascidiacea	Pest species	Aquatic biome	[63]

Of the 50 species belonging to the Arthropoda phylum, 37 species belonged to the Insecta class, three to the Arachnida class, nine to the Malacostraca class, and one to the Maxillopoda class. Of the 91 species of the Chordate phylum, 49 belonged to the Actinopterygii class, 20 to the Ascidiacea class, eight species to the Reptilia class, five to the bird class, and four to the Amphibia and Mammalia class.

In the Arthropoda Phylum, 18 species belonged to the Coleoptera order, being the most abundant of the groups in this phylum, followed by Decapoda with nine species;

Hymenoptera with six species; Diptera and Hemiptera with five each; Acarida with two species; and one species each in the Thysanoptera, Lepidoptera, Blattodea, Mesostigmata, and Cephalobaenida orders.

In the Chordate phylum, 24 species belonged to the Perciforme order; 11 belonged to the Stolidobranchia order; seven species each belonged to the Squamata and Cypriniformes orders; six species each belonged to the Cyprinodontiformes and Aplousobranchia orders; four to the Anura order; three each to the Passeriformes, Characiformes, and Phlebobranchia orders; two each to the Rodentia, Carnivora, Salmoniformes, and Gobiiformes orders; and one species each to the Columbiformes, Pelecaniformes, Carangiformes, Acanthuriformes, Scorpaeniformes, Elopiformes, Clupeiformes, Syngnathiformes, Siluriformes, and Testudines orders.

The three most abundant families of the Arthropoda phylum were Curculionidae (14 spp.), Formicidae (4 spp.), and Tephritidae (3 spp.). For the Chordata phylum, the three most abundant families were Cichlidae (11 spp.); Styelidae (8 spp.); and Cyprinidae, Poeciliidae, and Centrarchidae (6 spp.).

We labeled one case, *Rachycentron canadum* (Linnaeus, 1766) [9,13], as "under scrutiny", as the data located it in the Pacific zone of the Panama Republic, but it has never been seen in other Panamanian aquatic ecosystems.

Based on the information obtained, we can say that the province with the highest record of invasive species introduced in the Panama Republic is the Panama province (111 records), followed by Colon (21 records), Chiriqui province (19 records), Veraguas (15 records), Cocle province (13 records), West Panama (12 records), Herrera (10 records), Bocas del Toro (7 records), Los Santos and Darien (10 records), and Kuna Yala (1 record) (Table 2 & Figure 1).

Table 2. Bionomic data about the invasive species recorded for the Panama Republic.

Species	Province	District	Site	Introduction Year	Continent of Origin
Hypothenemus hampei [1]	Chiriqui	Chiriqui	Border with Costa Rica	2007	Africa
Coccotrypes advena [1]	Panama, Colon	Panama, Colón	Panama Canal	2007	Asia
Ambrosiodmus obliquus [2]	Panama				
Coptoborus ricini [2]	Panama				
Xyleborus bispinatus [2]	Panama				
Xylosandrus morigerus [1]	Panama				
Xylosandrus compactus [1]	Panama				
Cryptocarenus seriatus [2]	Panama				
Coccotrypes vulgaris [2]	Panama				
Dendrocranulus tardus [2]	Panama				
Scolytopsis puncticollis [1]	Panama				
Xylosandrus crassiusculus [1]	Panama	Gatun	Barro Colorado	2004	Asia
Xyleborinus exiguus [1]	Panama	Gatun	Barro Colorado	2019	Asia
Rhynchophorus palmarum [2]	Panama	Panama, Colon	Panama		
Euoniticellus intermedius [1]	Chiriqui	Renacimiento	Chiriqui	2015	Africa
Harmonia axyridis [1]	Colon	Colon	Centro Regional Colon	2014	Europe
Callosobruchus phaseoli [2]	Panama	Arraijan	Las villas de Arraijan	2022	Neotropics
Saperda candida [2]					
Brachyplatys subaeneus [1]	Herrera	Chitre	Costa Pacífica	2016	Asia
Diaphorina citri [1]	Bocas del Toro	Bocas del Toro	Bocas del Toro	2016	Asia
Pseudacysta perseae [1]	Los Santos	Santa Maria	Peninsula de Azuero	2017	Europe
Aulacaspis yasumatsui [1]	Panama	Panama	Balboa	2011–2021	Asia
Aphis spiraecola [2]	Cocle	Penonome, Ola	Cocle	2002	South America
Thrips palmi [1]	Herrera	Chitre	Los Chicharrones	2006	Asia
Apis mellifera scutellata [1]	Panama	Panama	Panama Republic		Africa
Tapinoma melanocephalum [1]					
Paratrechina longicornis [1]					
Monomorium floricola [1]					
Quadrastichus erythrinae [1]	Panama	Panama	Panama Canal	2018	Asia
Nylanderia fulva [2]	Chiriqui	Chiriqui	Chiriqui		
Ceratitis capitata [1]	Chiriqui	Chiriqui	Chiriqui	1963	South America
Anastrepha obliqua [2]	Cocle	Cocle	Cocle	2011	
Anastrepha grandis [2]	Darien	La Palma	Darien	2009	South America
Aedes aegypti [1]	Panama	Panama	Panama	1985	Africa
Aedes albopictus [1]	Panama	Panama	Panama	2003	Asia
Tuta absoluta [1]	Panama	Panama	Panama		

Table 2. *Cont.*

Species	Province	District	Site	Introduction Year	Continent of Origin
Blattella germanica [1]	Panama	Panama	Panama		Africa
Brevipalpus phoenicis [1]					
Brevipalpus californicus [1]					
Dermanyssus gallinae [1]	Herrera	Ocu	Las Guabas	2018	
Rhithropanopeus harrisii [1]	Panama, Colon		Gatun Lake	1969	North America
Elamenopsis kempi [1]	Panama, Colon		Gatun Lake	2007	Africa
Procambarus clarkii [1]	Panama	Panama		1987	
Penaeus monodon [1]	Panama	Panama	Panama		North America
Cherax tenuimanus [1]	Panama	Panama	Panama		Oceania
Macrobrachium amazonicum [2]	Panama, Colon		Gatun Lake		
Macrobrachium rosenbergii [1]	Panama			1981	Asia
Goniopsis cruentata [1]	Panama, Colon		Gatun Lake		
Pachygrapsyus gracilis [1]	Panama, Colon		Gatun Lake		
Raillietiella frenatus [1]	Panama				
Mus musculus [1]	Panama	Panama	Panama	1973	Europe, Asia
Rattus norvegicus [1]	Panama	Panama	Panama		Asia
Herpestes auropunctatus [1]	Panama	Panama	Panama		Asia
Herpestes javanicus [1]					
Columba livia [1]	Panama	Panama	Panama	1608	Europe, Africa, Asia
Bubulcus ibis [1]	Panama				
Mimus gilvus [2]	Panama				
Passer domesticus [1]	Panama	Panama	Panama	2019	Europe
Sicalis flaveola [1]	Panama	Panama	Panama		South America
Rachycentron canadum [1]	Panama		Gulf of Panama	2015	South America
Oreochromis niloticus [1]	Chiriqui, Veraguas, Panama, Colon		Fortuna, La Yeguada, Gatun, Alajuela, Bayano and rivers	1976	Africa
Oreochromis aureus [1]	Panama	Panama	Panama Canal	1976	Northwest Africa
Oreochromis mossambicus [1]	Panama	Gatún, Alajuela	Panama Canal	1940	Southeast Africa
Oreochromis urolepis [1]	Panama	Panama	Panama Canal	1979	Southeast Africa
Coptodon rendalli [1]	Panama	Panama	Panama Canal	1977	Africa, Middle East
Cichla ocellaris [1]	Panama, Colon		Gatun Lake	1967	South America
Cichla monoculus [2]	Panama, Colon, Veraguas		Gatun, Alajuela y La Yeguada	1969	South America
Parachromis managuensis [1]	Chirqui, Veraguas, Panama, Colon		Fortuna, La Yeguada, Gatun, Alajuela and rivers	1973	Central America (Atlantic Slope)
Astronotus ocellatus [1]	Panama, Colon		Gatun Lake, Alajuela	1991	South America
Mesonauta festivus [2]	Panama, Colon		Gatun Lake	1998–1999	South America
					Africa, Asia, Indo-Pacific
Vieja maculicauda [2]	Chiriqui, Veraguas, Los santos, Herrera, Cocle, Panama Oeste, Panama, Darien			1939	
Eleotris melanosoma [1]	Panama	Panama	Panama		
Butis koilomatodon [1]	Panama	Panama	Miraflores Locks	1973	Occidental Indo-Pacific
Lepomis macrochirus [1]	Chiriqui, West Panama, Panama		Lagunas de Volcan y San Carlos, Las Cumbres	1955	North America
Lepomis microlophus [2]	Panama	Panama	Panama		North America
Micropterus salmoides [1]	Chiriqui		Lagunas de Volcan	1955	North America
Pomoxis annularis [2]	Panama	Panama	Panama		North America
Pomoxis nigromaculatus [2]	West Panama	La Laguna	Laguna de San Carlos	1925–1935	North America
Lepomis humilis [2]	West Panama	La Laguna	Laguna de San Carlos	1925–1935	North America
Dormitator maculatus [2]	Panama	Panama	Panama Canal	1937	
Hypleurochilus pseudoaequipinnis [2]	Panama	Panama	Panama Canal	1971	
Leptophilypnus fluviatilis [2]	Panama	Panama	Panama Canal	1937	
Lophogobius cyprinoides [2]	Panama	Panama	Panama Canal	1937	
Lupinoblennius vinctus [2]	Panama	Panama	Panama Canal	1967	
Omobranchus punctatus [1]	Panama	Panama	Panama Canal	2011	East Asia
Sciaenops ocellatus [2]	Cocle	Aguadulce	El Salado	1987	North America
Oncorhynchus mykiss [1]	Chiriqui		Chiriqui Viejo, Caldera, Boquete	1925	North America
Salmo trutta [1]	Panama	Panama	Panama		Europe
Pterois volitans [1]	Colon	Colon	Panamanian caribbean	2008	Pacific Ocean, Asia
Butis koilomatodon [1]	Panama	Panama	Miraflores locks	1973	Occidental Indo-Pacific
Cirrhinus molitorella [1]	Panama	Panama	Panama	1980	Asia
Ctenopharyngodon idella [1]	Panama	Panama	Panama	1978	Asia
Cyprinus carpio [1]	Chiriqui, Veraguas		Fortuna, La Yeguada	1976	Europe, Asia
Hypophthalmichthys molitrix [1]	Panama	Panama	Panama	1978	Asia
Hypophthalmichthys nobilis [1]	Panama	Panama	Panama	1978	Asia

Table 2. *Cont.*

Species	Province	District	Site	Introduction Year	Continent of Origin
Mylopharyngodon piceus [1]	Panama	Panama	Panama		East Asia
Ictiobus cyprinellus [2]	Veraguas	Santiago	Santiago	1987	Cuba
Gambusia holbrooki [1]	Panama, Colon		Gatun Lake		North America
Xiphophorus helleri [1]	Cocle, Panama		Las Cumbres, Anton Valley	1950's y 1970's	Central America (Atlantic Slope)
Gambusia affinis [1]	Panama	Panama	Panama		
Gambusia nicaraguensis [2]	Panama	Panama	Panama		
Poecilia reticulata [1]	Veraguas, Panama		Santiago, Pacora, Chame, Las Cumbres, Cerro Azul	1910–1912	South America
Xiphophorus variatus [2]	Panama	Panama	Panama		
Ictalurus punctatus [1]	Panama	Panama	Panama	1981	North America
Colossoma macropomum [2]	Panama, Chiriqui, Veraguas, Colon		La Yeguada, río Sereno y Alajuela	1980	South America
Piaractus brachypomus [2]	Panama	Panama	Panama	1980	South America
Piaractus mesopotamicus [2]	Panama	Panama	Panama	1986–1987	South America
Microphis brachyurus lineatus [2]	Panama	Panama	Panama Canal	1971	Tropical region, Indo-Pacific
Megalops atlanticus [2]	Panama	Panama	Panama Canal	1937	
Anchoa parva [2]	Panama	Panama	Panama Canal	1937	
Hemidactylus brookii [1]	Panama	Panama	Tocumen		
Hemidactylus frenatus [1]	Panama	Panama	Chepo, Tocumen	2019	Tropical region, Indo-Pacific
Hemidactylus mabouia [1]	Panama	Panama	Canal zone	1994	Sub-Saharan Africa
Hemidactylus turcicus [1]	Panama	Panama	Canal Zone	1914	Mediterranean
Lepidodactylus lugubris [1]	Not specific			2004	Sri Lanka, Africa
Anolis sagrei [2]	Panama	Panama	Albrook	2018	Cuba
Sphaerodactylus argus [2]	Bocas del Toro, San Blas	Bocas del Toro	Bocas del Toro, San Blas	2002	Jamaica, Caribbean Islands
Trachemys scripta elegans [1]	Panama	Panama	Panama		United States
Eleutherodactylus antillensis [1]	Panama	Panama	Panama City	1950–1960	Puerto Rico, Virgin Islands
Eleutherodactylus johnstonei [2]	Panama	Panama	Panama	1980	
Eleutherodactylus planirostris [2]	Panama	Panama	Panama		Cuba, Bahamas
Rana catesbeianus [2]	Panama	Panama	Panama		North America
Ascidia incrassata [1]	Panama	Panama	Panama Canal	2009	Indo-Pacific
Ascidia sydneiensis [1]	Panama	Panama	Panama		Australia-Oceania
Phallusia nigra [1]	Panama	Panama	Panama		Indo-Pacific
Botrylloides nigrum [1]	Panama	Panama	Panama Canal	2002	South America
Botryllus planus [2]	Panama	Panama	Panama		Indo-Pacific
Polyandrocarpa anguinea [1]	Panama	Panama	Panama		Africa
Polyandrocarpa sagamiensis [2]	Panama	Panama	Panama		South America
Polyandrocarpa zorritensis [1]	Panama	Panama	Panama		South America
Styela canopus [1]	Colon	Colon	Panama Canal		Indo-Pacific
Symplegma brakenhielmi [1]	Panama	Panama	Panama		
Symplegma rubra [1]	Panama	Panama	Taboguilla		Indo-Pacific
Herdmania pallida [1]	Bocas del Toro	Colon Island	Bocas del Toro		South America
Microcosmus exasperatus [1]	Panama	Panama	Panama		Colombia, South America
Pyura vittata [1]	Panama	Panama	Panama		
Didemnum perlucidum [1]	Panama	Panama	Panama		South America
Didemnum psammatodes [1]	Panama	Panama	Panama		
Diplosoma listerianum [1]	Panama	Panama	Panama		
Lissoclinum fragile [1]	Panama	Panama	Panama		
Trididemnum orbiculatum [2]	Panama	Panama	Panama		South America
Polyclinum constellatum [1]	Panama	Panama	Panama		

Species with number: 1 = global invaders. 2 = local invaders [33].

The four continents with the most record of invasive species origins that have been reported in Panama were Asia (22 records), South America (21 records), Africa (15 records), and North America (14 records) (Table 2). The results indicate that the majority of registered invasive species in Panamá are global invasive species (Table 2).

Among the species reported as invasive for the Panama Republic, nine species are under official control, all belonging to the insect class, for causing damage to cucurbit crops, tomato, citrus, coffee, and species of the Arecaceae family (Table 3).

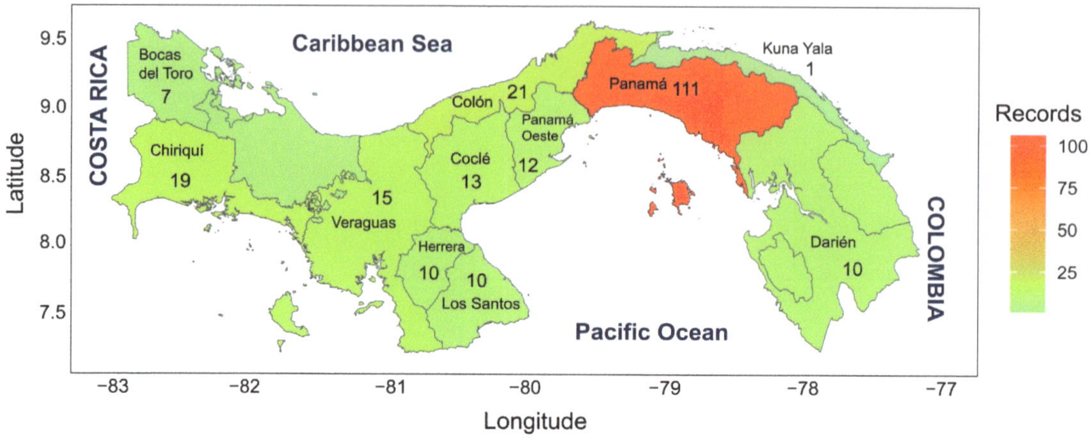

Figure 1. Map of the Republic of Panama indicating the number of invasive species by province.

Table 3. The invasive species recorded for the Panama Republic for the first time (class Insecta).

Pests	Species	Host	Ecological Status
South American cucurbit fly	*Anastrepha grandis* Macquart, 1846	Cucurbitaceae	Pest species
Tomato moth	*Tuta absoluta* Meyrick, 1917	Tomato	Pest species
Huanglongbing (HLB)	*Diaphorina citri* Kuwayama, 1908	Citrus	Pest species
Mediterranean fly	*Ceratitis capitata* Wiedemann, 1824	Fruit plants (citrus, cucurbitaceae)	Pest species
Oriental *Trips palmi*	*Thrips palmi* Karny, 1990	Cucurbitaceae	Pest species
Coffee berry borer	*Hypothenemus hampei* Ferrari, 1867	Coffee	Pest species
Citrus leprosis virus	*Brevipalpus californicus* Banks, 1904	Citrus	Pest species
	Brevipalpus phoenicis Geijskes, 1936	Citrus	Pest species
Red ring disease	*Rhynchophorus palmarum* Linnaeus, 1758	Arecaceae	Pest species

4. Discussion

Since its emergence and closure, Panama's Isthmus has been a transit node in global invasion flows. Due to the closure of Panama's Isthmus, natural events such as the great American biotic exchange [8] have developed, and later during Panamá's colonial times, due to its geographical position, Panama's Isthmus was used to transfer merchandise that promoted accidental or intentional species exchange. Later, with the construction of the Panama Canal Railway, the movement of all types of cargo from all over the world increased, making Panama one of the countries with the most species movement globally. Likewise, the presence for almost 100 years of the so-called Panama Canal Zone (area around the Panama Canal under the jurisdiction of the United States) led to the introduction of exotic species to the country without any type of control, many of which became invasive [64].

The results indicate that the main Panamanian ecosystem occupied by invasive exotic species is the aquatic ecosystem, with fish being the most dominant. Most of these species were introduced intentionally and, in many cases, are used as a food source in both farming systems and reservoirs. In terrestrial ecosystems, insects are the largest recorded group of invasive species. The invasive species of this group have become pests in agricultural crops or forest plantations, reducing crop yields and affecting wood commercialization (Table 3). Other species of insects affect ornamental plants, in which case, the damage has been less quantified. In this group, important vectors of human viral diseases (dengue, zika, and chikungunya) are also registered.

In the case of amphibians and reptiles, documentation of the ecological importance of invasive species is scarce. Most records are associated with urban areas, and the extent of their distribution depends greatly on anthropogenic activities [62–65]. For example,

Eleutherodactylus johnstonei uses the vegetation associated with the herbaceous substrate as places to vocalize or perch [66]. *Eleutherodactylus planirostris* [62] is an introduced frog that has most likely moved to disturbed locations, such as forest edges and outside of residential home gardens, that represent its known habitat [65]. On the other hand, *Trachemys scripta elegans* is a species introduced for the pet trade [44], and it prefers calm waters with soft and muddy bottoms, aquatic vegetation, and suitable places for sunbathing. Within their native range, red-eared sliders occupy an ecological niche as predators and prey. They are resistant and, outside their native area, they occupy the same ecological niche with great adaptability [67]. *A. sagrei*, for its part, mainly occupies open areas, is highly territorial, and can ecologically displace native species. It is well adapted to finding food and avoiding predators in newly colonized habitats. Therefore, its capacity to displace other native species may be high [59]. Some species have only been reported, but no studies have been carried out on their current status and ecological impact. Some examples are *Rana catesbeiana* [44], *Sphaerodactylus argus* [68], and *Eleutherodactylus antillensis* [44]. Authors have suggested that some of these invasive amphibians and reptiles can be categorized as pests [44]; however, there is no evidence to support this, so we prefer to categorize them as "Species with potential to displace native species" or "Scarcely documented species".

In the case of mammals, despite being from only a few species, the effect on public health is immense, as is the case of *Mus musculus* and *Rattus norvegicus*. For their part, the two species of the Herpestes genus are aggressive predators that cause imbalances in the local fauna.

According to our results from the Arthropoda and Chordate phyla, there are an estimated 141 invasive species for the Panama Republic. Compared with the information from other Iberoamerican countries, we found that for the countries that are part of the Andean Community (Colombia, Ecuador, Venezuela, Peru, and Bolivia), 227 invasive exotic species have been identified, mostly plants (92 species), insect pests (61 species), and vertebrates (30 species) [48]. In Costa Rica, 235 invasive species have been reported from all groups of organisms [69]. For the Dominican Republic, 192 species have been reported, of which 38 species were fish, 4 were amphibians, 8 were reptiles, 13 were birds, and 13 were mammals [70]. While in other latitudes, such as the Iberian Peninsula (Spain, Portugal, and Andora), 100 invasive species have been recorded [71].

When comparing the number of invasive species based on the surface area of each country (Andean Community and Dominican Republic) where the study was conducted, we found that the number is higher for the Panama Republic (141 species) compared to the Andean Community (166 species) and Dominican Republic (138 species).

In the Andean Community, the area of Panama is 75,416.6 km^2; the area of Colombia is 1142 million km^2; the area of Ecuador is 256,370 km^2; the area of Venezuela is 916,445 km^2; the area of Peru is 1285 million km^2; the area of Bolivia is 1099 million km^2; and the area of the Dominican Republic is 48,442 km^2. This indicates that the Panama Republic, even though it is a small country compared to those previously mentioned, has quite a considerable number of invasive species in the territory.

The economic activity of the Dominican Republic is based on beaches and tourism [72]; Colombia on the free market (exchange of goods and services between people and companies through monetary transactions) [73]; Ecuador and Venezuela on oil revenues [74,75]; Peru on the exploitation, processing, and export of natural resources, mainly mining, agriculture, and fishing [76]; and Bolivia on natural gas [77]. Since the Panama Republic is a country that bases its economy on logistics and transportation, commerce, and finance, we can assume that this is the reason why the constant influx of invasive species is so high compared to the others countries.

The results clearly indicate that the Panama Province is where the greatest number of exotic invasive species have been recorded. This is because the Panama Province is where the largest movement of merchandise operations takes place through the Panama Canal, the Panama Railway, and two of the main seaports.

Invasive exotic species are one of the main threats to countries' economies and local diversity [1,2]. In Panama, in the case of species that cause economic activity damage, the state permanently monitors the containers of fresh products that enter through different ports and airports of the country [78]. The surveillance programs of the Ministry of Agricultural Development include a list of quarantine species that are monitored for detention in customs facilities. This surveillance is carried out by highly qualified personnel to diagnose species considered of quarantine interest that may affect different crops [78]. Once the invasive exotic species evades the detection line and is introduced into the country, the state develops constant sampling programs in the different provinces to restrict the local distribution of the species (Table 3). An example of this is the Fruit Fly Monitoring Program, which through maximum trapping, has prevented species such as *Anastrepha grandis* or *Ceratitis capitata* from spreading throughout the country. The Panamanian state also invests resources in the control and monitoring of exotic species that are vectors of viral diseases such as dengue, zika, and chikungunya. In recent times, these public health programs developed by the Ministry of Health have extended to the border with the Colombian Republic due mainly to the massive migratory phenomenon across the Colombian–Panamanian border in recent years.

In reference to invasive exotic species that affect natural ecosystems, the situation is different. Due to its variety of natural ecosystems, Panama offers an innumerable number of habitats that invasive exotic species can occupy. According to our results, Panama City is the main gateway for invasive exotic species to enter the country. Panama City is surrounded by urban forests with different states of conservation and management, which offers habitats that are highly degraded or contaminated by exotic species. Likewise, the Panama Province and the Colon Province, which is where the Panama Canal is located and which separate the Pacific Ocean from the Caribbean Sea by only 82 km, offer a variety of aquatic habitats for the species that come on ships from different parts of the world. To address this situation and lack of information, the Panamanian state, through the Ministry of the Environment, has established the *National Biodiversity Strategy and Action Plan 2018–2030*, which addresses the problem of invasive species in the country [15].

5. Conclusions

The Panama Republic, due to its geographical position and commercial activity, is an important point of global biological invasion. Panamanian aquatic ecosystems are those that have been the most occupied by invasive species. However, invasive species of the insect group quantitatively cause the most damage to humans and the environment.

When comparing the number of invasive species based on the surface area of each country in which this study was conducted, we found that the number was highest for the Panama Republic.

Due to the limitations of this work, it is possible that some species were not included due to a scarcity of information, or that they have not been considered invasive for some researchers because they may correspond to natural distribution patterns, like the case of the coyote.

Author Contributions: Conceptualization, E.M. and D.R.-G.; methodology, E.M. and D.R.-G.; research, E.M., H.A.G.B., R.F., L.R.-S., Y.A., O.G.L.-C. and D.R.-G.; Writing—preparation of original draft, E.M. and D.R.-G.; Writing-revisions and editing, E.M., H.A.G.B., R.F., L.R.-S., Y.A., O.G.L.-C. and D.R.-G.; acquisition of funds, E.M. All authors have read and agreed to the published version of the manuscript.

Funding: This research was funded by the University of Panama, the Panamanian National Research System (SNI), and through SENACYT DDCCT economic subsidy contract No. 004-2023.

Institutional Review Board Statement: Not applicable.

Informed Consent Statement: Not applicable.

Data Availability Statement: The data are included in the paper.

Acknowledgments: We thank Darío Luque, who provided us with the database of invasive exotic species of MIAMBIENTE; Mgtr. Mirna Samaniego from STRI, who looked for information in the Web of Science; Pedro Méndez, Ricardo Moreno, Jorge García, and Chelina Bastista for assisting with the names of the invasive species for their respective groups; Digna in the MIAMBIENTE documentation section for providing books on invasive species; and the Marine Resources Authority of Panama for offering information on fish introduced in Panama.

Conflicts of Interest: The authors declare no conflicts of interest.

References

1. Capdevila-Argüelles, L.; Zilletti, B.; Suárez Álvarez, V.Á. Causas de La Pérdida de Biodiversidad: Especies Exóticas Invasoras. *Mem. Real Soc. Española Hist. Nat.* **2013**, *2*, 55–76.
2. Carballo, G.O. Invasiones Biológicas. In *Ecosistemas*; CSIC: Madrid, Spain, 2009; p. 215.
3. Lockwood, J.; Hoopes, M.; Marchetti, M. Invasion Ecology. In *Invasion Ecology*; Blackwell Publishing: Malden, MA, USA, 2007; p. 304.
4. Pauchard, A.; Quiroz, C.; Garcia, R.; Anderson, C.; Arroyo, M. Invasiones Biológicas en América Latina y el Caribe: Tendencias en Investigación para la Conservac. In *Conservación Biológica: Perspectivas desde América Latina*; Editorial Universitaria: Santiago, Chile, 2011; pp. 79–94.
5. Waits, L.P.; Paetkau, D. Noninvasive Genetic Sampling Tools for Wildlife Biologists: A Review of Applications and Recommendations for Accurate Data Collection. *J. Wildl. Soc.* **2005**, *69*, 1419–1433. [CrossRef]
6. Ribera, I.; Melic, A.; Torralba, A. Introducción y guía visual de los artrópodos. IDE@. 2015, pp. 1–30. Available online: http://sea-entomologia.org/IDE@/revista_2.pdf (accessed on 10 January 2023).
7. Parques Naturales de la Comunitat Valenciana. PHYLUM CHORDATA. 2023. Available online: https://parquesnaturales.gva.es/es/web/acuarium-virtual-ifac/phylum-chordata (accessed on 15 July 2023).
8. Autoridad Nacnional del Ambiente (ANAM). *Atlas Ambiental de la República de Panamá*; Novo Art, S.A., Ed.; Autoridad Nacnional del Ambiente (ANAM): Ancón, Panama, 2012.
9. Autoridad Nacional de Ambiente. *Informe Sobre Especies Exóticas de Panamá*; Autoridad Nacional de Ambiente de Panamá: Ancón, Panama, 2010.
10. Muirhead, J.R.; Minton, M.S.; Miller, W.A.; Ruiz, G.M. Projected Effects of the Panama Canal Expansion on Shipping Traffic and Biological Invasions. *Divers. Distrib.* **2014**, *21*, 75–87. [CrossRef]
11. Mundial, B. Panamá, Panorama General. Available online: https://www.bancomundial.org/es/country/panama/overview (accessed on 15 July 2023).
12. European Environment Agency. *Pathways of Introduction of Invasive Species, Their Prioritization and Management*; Convention on Biological Diversity: Montreal, QC, Canada, 2014.
13. Autoridad Nacional de Ambiente. *Quinto Informe Nacional de Biodiversidad de Panamá Ante el Convenio Sobre Diversidad Biológica*; Autoridad Nacional de Ambiente: Ancón, Panama, 2014.
14. Hernández, G.; Lahmann, E.J.; Pérez-Gil, R. *Invasores en Mesoamérica y el Caribe: Resultados del Taller Sobre Especies Invasoras: Ante los Retos de su Presencia en Mesoamérica y el Caribe*; IUCN: San José, Costa Rica, 2002; p. 54.
15. Ministerio de Ambiente. *Estrategia y Plan de Acción Nacional de Biodiversidad (EPANB) (2018–2050)*; Ministerio de Ambiente: Panama City, Panama, 2018.
16. Pérez, H. *Manejo de la Broca del Café en la República de Panamá*; Sociedad Mexicana de Entomología, A.C.: Texcoco, Mexico, 2007; pp. 33–36. Available online: https://www.cabi.org/wp-content/uploads/Perez-2006-Coffee-berry-borer.pdf (accessed on 15 July 2023).
17. Salgado Lizardo, C. Especies de Scolytinae (Coleoptera: Curculionidae) asociados a puertos y recintos aduaneros en las provincias de Panamá y Colón. *Scientia* **2022**, *32*, 64–86. Available online: http://up-rid.up.ac.pa/6129/ (accessed on 15 July 2023).
18. Department for Enviroment Food & Rural Affairs. Rapid Pest Risk Analysis for: *Xylosandrus crassiusculus*. 2015, pp. 1–3. Available online: https://efsa.onlinelibrary.wiley.com/doi/pdf/10.2903/sp.efsa.2020.EN-1903 (accessed on 15 July 2023).
19. FAO. Declara de Control Oficial la Enfermedad Conocida Como Anillo Rojo de las Palmáceas. 2009. N° 26.213: Resuelto N° 71. Available online: https://www.fao.org/faolex/results/details/fr/c/LEX-FAOC086829/ (accessed on 15 July 2023).
20. Solís, A.; Cambra, R.A.; Gonzales, M. Primer registro para Panamá de la especie invasiva *Euoniticellus intermedius*. *Tecnociencia* **2015**, *17*, 39–43. Available online: https://www.researchgate.net/publication/306291580_PRIMER_REGISTRO_PARA_PANAMA_DE_LA_ESPECIE_INVASIVA_Euoniticellus_intermedius_Reiche_1849_COLEOPTERA_SCARABAEIDAE_SCARABAEINAE (accessed on 15 July 2023).
21. Thomas, H.; Haelewaters, D. A case of silent invasion: Citizen science confirms the presence of *Harmonia axyridis* (Coleoptera, Coccinellide) in Central America. *PLoS ONE* **2019**, *14*, e0220082. [CrossRef]

22. Abrego, J.; Santos-Murgas, A.; Rivera, J.; Vargas, C. Primer Reporte de *Eupelmus pulchriceps* (Hymenoptera: Eupelmidae) parasitando *Callosobruchus phaseoli* (Coleoptera: Chrysomelidae) plaga de *Cajanus cajan* en Panamá. *La Técnica Rev. De Agrocienc.* **2022**, *12*, 1–9. Available online: https://www.researchgate.net/publication/366055332_Primer_reporte_de_Eupelmus_pulchriceps_HymenopteraEupelmidae_parasitando_Callosobruchus_phaseoli_ColeopteraChrysomelidae_plaga_de_Cajanus_cajan_en_Panama (accessed on 15 July 2023).
23. Global Invasive Database. Available online: https://www.iucngisd.org/gisd/ (accessed on 15 July 2023).
24. Gálvez, D.; Murcia-Moreno, D.; Añino, Y.; Ramos, C. The Asian heminopteran *Brachyplatys subaeneus* (Westwood, 1837) (Hemiptera: Heteroptera: Plataspidae) in Protected are in Panama. *BioInvasions Rec.* **2022**, *11*, 95–100. Available online: https://www.reabic.net/journals/bir/2022/1/BIR_2022_Galvez_etal.pdf (accessed on 15 July 2023). [CrossRef]
25. Atencio-Valdespino, R. Distribución de *Diaphorina citri* Kuwayama (Hemiptera: Liviidae) en zonas de producción citrícola de Panamá. *Agron. Mesoam.* **2023**, *34*, 51106. Available online: https://www.scielo.sa.cr/scielo.php?pid=S1659-13212023000200018&script=sci_abstract&tlng=es (accessed on 15 July 2023). [CrossRef]
26. Anovel, C.; Aguilera, V.; Herrera, J.A. Presencia de *Pseudacysta perseae* (Heidemann, 1908) (Insecta: Hemiptera: Tingidae) en Panamá. *IDESIA* **2020**, *38*, 123–127. Available online: https://www.scielo.cl/scielo.php?script=sci_arttext&pid=S0718-34292020000300123 (accessed on 15 July 2023).
27. Castillo-Gómez, M.; Chang-Pérez, R.; Jiménez-Puello, J.; Santos Murgas, A.; Medianero, E. Knowledge of cycad Aulacaspis scale (Hemiptera: Diaspididae) in Panama. *BioInvasion Rec.* **2021**, *10*, 1015–1021. Available online: https://www.reabic.net/journals/bir/2021/4/BIR_2021_Castillo-Gomez_etal.pdf (accessed on 15 July 2023). [CrossRef]
28. Emmen, D.; Quiros, D.; Vargas, A. Enemigos naturales de áfidos (Hemiptera: Aphididae) en plantaciones de cítricos de la provincia de Coclé, Panamá. *Tecnociencia* **2012**, *14*, 133–148. Available online: https://www.researchgate.net/publication/328841685_ENEMIGOS_NATURALES_DE_AFIDOS_HEMIPTERA_APHIDIDAE_EN_PLANTACIONES_DE_CITRICOS_DE_LA_PROVINCIA_DE_COCLE_PANAMA (accessed on 15 July 2023).
29. José-Ángel, H.V.; Anovel-Amet, B.A. Identificación de *Thrips palmi* Karny (Thysanoptera: Thripidae) en cultivos de cucurbitáceas en Panamá. *Agron. Mesoam.* **2013**, *24*, 47–55.
30. Moritz, R.F.A.; Härtel, S.; Neumann, P. Global invasions of the western honeybee (*Apis mellifera*) and the consequences for biodiversity. *Écoscience* **2005**, *12*, 289–301. [CrossRef]
31. Medianero, E.; Zachrisson, B. Erythrina gall wasp, *Quadrastichus erythrinae* (Kim, 2004) (Hymenoptera: Eulophidae: Tetrastichinae): A new pest in Central America. *BioInvasions Rec.* **2019**, *8*, 452–456. Available online: https://www.reabic.net/journals/bir/2019/2/BIR_2019_Medianero_Zachrisson.pdf (accessed on 15 July 2023). [CrossRef]
32. Murgas, I.; Pitti, C.; Cambra, M.; Cambra, R. First report of the invasive ant *Nylanderia fulva* (Mayr, 1862) (Hymenoptera: Formicidae) in Panama. *BioInvsion Rec.* **2022**, *12*, 78–85. Available online: https://www.reabic.net/journals/Bir/2023/1/BIR_2023_Murgas_etal.pdf (accessed on 15 July 2023). [CrossRef]
33. Saavedra, R. Demografia de Moscas del Género *Anastrepha* (Diptera: Tephritidae), e Identificación de Poblaciones Ocasionales o Establecidas de *Ceratitis capitata* (Diptera: Tephritidae), en la Zona Media-Alta de Chicá y el Parque Nacional Campana, Distrito de Chame, Provincia de Panamá. Master's Thesis, Universidad de Panamá, Panama City, Panama, 2004. Available online: http://up-rid.up.ac.pa/4678/ (accessed on 15 July 2023).
34. Weems, H.V.; Heppner, J.B.; Steck, G.J. *West Indian Fruit Fly Scientific Name: Anasthrepha obliqua (Macquart) (Insecta: Diptera: Tephritidae)*; DPI Entomology: Gainesville, FL, USA, 2001; EENY-198. Available online: https://edis.ifas.ufl.edu/publication/IN355 (accessed on 26 June 2024).
35. FAO. Introduced Species Fact Sheets-2023. Available online: https://www.fao.org/fishery/en/introsp/3993/en (accessed on 15 July 2023).
36. Eskildsen, G.A.; Rovira, J.R.; Smith, O.; Miller, M.J.; Bennett, K.L.; McMillan, W.O.; Loaiza, J. Maternal invasion history of *Aedes aegypti* and *Aedes albopictus* into the Isthmus of Panama: Implications for the control of emergent viral disease agents. *PLoS ONE* **2018**, *13*, e0194874. [CrossRef] [PubMed]
37. Iturralde, D. Diversidad y Distribución de Cucarachas (Insecta: Blatodea: Blattaria) en la República de Panamá. Bachelor's Thesis, Universidad de Panamá, Panama City, Panama, 2023. Available online: http://up-rid.up.ac.pa/6357/ (accessed on 15 July 2023).
38. Saavedra-Dominguez, F.; Bernal, A.; Childers, C.C.; Kitajima, E.W. First report of Citrus leprosis in Panamá. *Pant Dis.* **2001**, *85*, 228. Available online: https://www.researchgate.net/publication/240595025_First_Report_of_Citrus_leprosis_virus_in_Panama (accessed on 15 July 2023).
39. Rujano, R.E. Demografía de Ácaros en los Cítricos. Master's Thesis, Universidad de Panamá, Panama City, Panama, 2002; 108p. Available online: http://up%E2%80%93rid.up.ac.pa/4370/ (accessed on 15 July 2023).
40. González, M.; Jiménez, Y.; Marín, J.; Tuñon, Y.; Him, J. Presencia del *Dermanyssus gallinae* en aves de corral en Las Guabas de Ocú, Panamá. *Guacamaya* **2018**, *3*, 25–29. Available online: https://www.researchgate.net/publication/328615585_Presencia_del_Dermanyssus_gallinae_en_aves_de_corral_en_Las_Guabas_de_Ocu_Herrera_Panama (accessed on 15 July 2023).
41. Dominique, R.; Mark, T. Established population of the North America Harris mud crab, *Rhithropanopeus harrisii* (Gould 1841) (Crustacea: Brachyura: Xanthidae) in the Panama Canal. *Aquat. Invasions* **2007**, *2*, 155–161.
42. Kam, Y.; Schloder, C.; Roche, D.; Torchin, M. The Iraqi crab, *Elamenopsis kempi* in the Panama Canal: Distribution, abundance and interactions with the exotic Noth America crab, *Rhithropanopeus harrisii. Aquat. Invasions* **2011**, *6*, 439–445. Available online: http://www.aquaticinvasions.net/2011/AI_2011_6_3_Kam_etal.pdf (accessed on 15 July 2023). [CrossRef]

43. Smithsonian Tropical Research Institute. Available online: https://serc.si.edu/labs/marine-invasions-research (accessed on 20 July 2023).
44. Valdés, V.V. Impactos positivos y negativos de la introducción de animales exóticos en Panamá. *Tecnol. Marcha* **2009**, *22*, 91–97. Available online: https://www.researchgate.net/publication/279640630_Impactos_positivos_y_negativos_de_la_introduccion_de_animales_exoticos_en_Panama (accessed on 15 July 2023).
45. Kelehear, C.; Saltonstall, K.; Torchin, M. An introduced pentastomid parasite (Raillietiella frenata) infects native cane toads (Rhinella marina) in Panama. *Parasitology* **2015**, *142*, 675–679. Available online: https://repository.si.edu/bitstream/handle/10088/22662/stri_Kelehear_etal_2014_introduced_pentastomes_in_native_cane_toads.pdf (accessed on 15 July 2023). [CrossRef] [PubMed]
46. Biota Panamá. Base de Datos de Aves de Panamá-2007. Available online: https://biota.wordpress.com/2007/12/24/base-de-datos-de-aves-de-panama/ (accessed on 15 August 2023).
47. Vega, A.; Vergara, Y.; Robles, Y.A. Primer registro de la cobia, *Rachycentron canadum* Linnaeus (Pisces: RACHYCENTRIDAE) en el Pacífico panameño. *Tecnociencia* **2016**, *18*, 13–19. Available online: http://up-rid.up.ac.pa/188/ (accessed on 15 July 2023).
48. Gutiérrez Bonilla, F.d.P. *Estado de Conocimiento de Especies Invasoras: Propuesta de Lineamientos Para El Control de Los Impactos*; Instituto de Investigación de Recursos Biológicos Alexander Von Humboldt: Bogotá, Colombia, 2006.
49. Díaz-Ferguson, E. Tras la Pista de las especies invasoras. *Imagina-SENACYT* **2022**, *13*, 51–53. Available online: https://imagina.senacyt.gob.pa/wp%E2%80%93content/uploads/2021/10/IMAGINA-13-print.pdf (accessed on 15 July 2023).
50. González-Gutiérrez, R. *Los Principales Peces de los Lagos y Embalses Panameños*, 7th ed.; Smithsonian Libraries: Panama City, Panama, 2016; pp. 56–114.
51. Autoridad de los Recursos Acuáticos de Panamá. Available online: https://arap.gob.pa/ (accessed on 5 May 2022).
52. FishBase. List of Freshwater Fishes Reported from Panama. Available online: https://www.fishbase.se/country/CountryChecklist.php?resultPage=2&what=list&trpp=50&c_code=591&cpresence=Reported&sortby=alpha2&ext_CL=on&ext_pic=on&vhabitat=fresh (accessed on 5 May 2022).
53. Welcomme, R.L. *International Introductions of Inland Aquatic Species*; FAO Fisheries Technical Paper 294; FAO Fisheries Department: Rome, Italy, 1988; pp. 294–318.
54. Schofield, P.J. Update on geographic spread of invasive lionsfishes (*Pterois volitans* Linnaeus, 1758 and *P. miles* Bennett, 1828) in the western North Atlantic Ocean, Caribbean Sea and Gulf of Mexico. *Aquat. Invasions* **2010**, *5*, S117–S122. Available online: http://www.aquaticinvasions.net/2010/Supplement/AI_2010_5_S1_Schofield.pdf (accessed on 15 July 2023). [CrossRef]
55. Valdés-Díaz, S. *Parivivos de la familia Poeciliidae de Panamá Guia de Identificación*, 1st ed.; Parque Recreativo Omar: Panamá City, Panama, 2021; p. 144.
56. Auth, D.L. *Checklist and Bibliography of the Amphibians and Reptiles of Panama*; Smithsonian Herpetological Information Service: Washington, DC, USA, 1994; pp. 2–98.
57. Farallo, V.; Swanson, R.; Hood, G.; Troy, J.; Forstner, M. New county records for the Mediterranean house gecko (*Hemidactylus turcicus*) in Central Texas with comments on human-mediated dispersal. *Appl. Herpetol.* **2009**, *6*, 196–198. Available online: https://brill.com/view/journals/ah/6/2/article-p196.xml?ebody=previewpdf-96220 (accessed on 15 July 2023).
58. Díaz-Pérez, J. Primer registro de *Lepidodactylus lugubris* (Sauria: Gekkonidae) para el departamento de Sucre, Colombia. *Rev. Colomb. Cienc. Anim. RECIA* **2012**, *4*, 163–167. Available online: https://www.researchgate.net/publication/323916731_Primer_registro_de_Lepidodactylus_lugubris_Sauria_Gekkonidae_para_el_departamento_de_Sucre_Colombia (accessed on 15 July 2023).
59. Batista, A.; Ponce, M.; Garcés, O.; Lassitier, E.; Miranda, M. Silent pirates: *Anolis sagrei* Duméril & Bibron, 1837 (Squamata, Dactyloidae) taking over Panama City, Panama. *Check List* **2019**, *15*, 455–459. Available online: https://checklist.pensoft.net/article/34178/ (accessed on 15 July 2023).
60. Ibáñez, D.; Roberto, R. *Los Anfibios del Monumento Natural Barro Colorado, Parque Nacional Soberanía y Áreas Adyacentes*; Fundación Natura, Círculo Herpetológico de Panamá; Smithsonian Tropical Research Institute: Panama City, Panama, 1999; p. 192.
61. Jaramillo, A.C.; Wilson, L.D.; Ibáñez, R.; Jaramillo, F. The herpetofauna of Panama: Distribution and conservation status. In *Conservation of Mesoamerican Amphibians and Reptiles*; Wilson, L.D., Townsend, J.H., Johnson, J.D., Eds.; Eagle Mountain Publishing: Eagle Mountain, UT, USA, 2010; pp. 604–671.
62. Crawford, A.; Alonso, R.; Jaramillo, C.; Sucre, S.; Ibañez, R. Los códigos de barras de ADN identifican una tercera especie invasora de Eleutherodactylus (Anura: Eleutherodactylidae) en la Ciudad de Panamá, Panamá. *Zootaxa* **2011**, *2890*, 65–67.
63. Carman, M.; Bullard, S.; Rocha, R.; Lambert, G. Ascidians at the Pacific and Atlantic entrances to the Panama Canal. *Aquat. Invasions* **2011**, *6*, 371–380. Available online: https://www.researchgate.net/publication/228491724_Ascidians_at_the_Pacific_and_Atlantic_entrances_to_the_Panama_Canal (accessed on 15 July 2023). [CrossRef]
64. Digna, R.G. Una Aproximación por Metaanálisis al Número de Especies Exóticas Invasoras y sus Áreas Más Frecuentes de Introducción en la República de Panamá. Bachelor's Thesis, The Panama University, Panama City, Panama, 2022.
65. Antúnez-Fonseca, C.; Juarez-Peña, C.; Sosa-Bartuano, Á.; Alvarado-Larios, R.; Sánchez-Trejo, L.; Vega-Rodriguez, H. First records in El Salvador and new distribution records in Honduras for *Eleutherodactylus planirostris* Cope, 1862 (Anura, Eleutherodactylidae), with comments on its dispersal and natural history. *Caribb. J. Sci.* **2021**, *51*, 37–43. [CrossRef]
66. Ortega, J.E.; Serrano-Cardozo, V.H.; Pinilla, M.P.R. Diet composition and microhabitat of *Eleutherodactylus johnstonei* in an introduced population at Bucaramanga City, Colombia. *Herpetol. Rev.* **2005**, *36*, 238–241.

67. Burger, J. *Red-Eared Slider Turtles (Trachemys scripta elegans)*; Freshwater Ecology and Conservation Lab, University of Washington: Seattle, WA, USA, 2009.
68. Harris, D.M.; Kluge, A.G. *The Sphaerodactylus (Sauria: Gekkonidae) of Middle America*; University of Michigan: Ann Arbor, MI, USA, 1984.
69. Universidad del Atlántico. Especies Invasoras Acuáticas y Salud. In Proceedings of the Memorias II Seminario de La Red Temática InvaWet, Barranquilla, Colombia, 27–28 October 2016.
70. Ministerio de Medio Ambiente y Recursos Naturales. *Especies Invasoras en República Dominicana*; Ministerio de Medio Ambiente y Recursos Naturales: Santo Domingo, Dominican Republic, 2023.
71. Casals, F.; Sánchez-González, J. *Guía de Las Especies Exóticas e Invasoras de Los Ríos, Lagos y Estuarios de La Península Ibérica*; Proyecto LIFE INVASAQUA; Sociedad Ibérica de Ictiología: Navarra, Spain, 2020.
72. Dominican Republic Economy. Economy 2020. Available online: https://www.republicadominicana.org.br/rp/economia/ (accessed on 15 July 2023).
73. Cámara de Comercio de Colombia. Available online: https://www.ccb.org.co/blog/colombia-un-pais-de-regiones-e-industrias-que-impulsan-la-economia-nacional (accessed on 15 July 2023).
74. Instituto de Investigación Económicas y Sociales-Universidad Católica Andrés Bello. *Informe de Coyuntura Venezuela*; Instituto de Investigación Económicas y Sociales-Universidad Católica Andrés Bello: Caracas, Venezuela, 2023. [CrossRef]
75. Instituto Nacional de Estadística y Censos. *Clasificación Nacional de Actividades Económicas (CIIU Rev. 4.0)*; Instituto Nacional de Estadística y Censos: Quito, Ecuador, 2012.
76. Varona-Castillo, L.; Gonzales-Castillo, J.R. Crecimiento Económico y Distribución Del Ingreso En Perú. Problemas Del Desarrollo. *Rev. Latinoam. Econ.* **2021**, *52*, 79–107. [CrossRef]
77. Instituto Nacional de Estadística. *Clasificación de Actividades Económicas de Bolivia*; Instituto Nacional de Estadística: Madrid, Spain, 2022.
78. Janeth, S.; Nicolás, P.H.; Enrique, M. Áfidos (Hemiptera: Sternorrhyncha: Aphididae) interceptados en vegetales frescos en los puertos de Manzanillo International Terminal y Panama Ports Company (Colón Panamá). *IDESIA (Chile)* **2022**, *40*, 137–143.

biology

MDPI

Article

Introduced Spiders in Panama: Species Distributions and New Records

Daniel Murcia-Moreno [1] **and Dumas Gálvez** [1,2,3,*]

1 Smithsonian Tropical Research Institute, Ciudad de Panamá 0843-03092, Panama; daniel.murcia@up.ac.pa
2 Estación Científica Coiba, Coiba AIP, Calle Gustavo Lara, Bld. 145B, Clayton 0843-01853, Panama
3 Programa Centroamericano de Maestría en Entomología, Universidad de Panamá, Estafeta Universitaria, Avenida Simón Bolívar, Ciudad de Panamá 0824, Panama
* Correspondence: dgalvez@coiba.org.pa

Simple Summary: Organisms in nature are often not restricted to a single geographical location and are able to disperse to new areas. However, humans have provided means for many species to disperse over much larger distances (e.g., ships, planes) than those they would normally reach by natural means. Introduced species often can have negative effects for native organisms and local economies (known as invasive species) and for this reason a first step is to quantify the number of introduced species in a particular region. Although arachnids are a group that has been extensively studied in terms of invasive species, some countries lack information on this issue, which is the case for Panama. Here we provide the first list of introduced spiders in Panama. By reviewing scientific literature, museum collections, online databases and fieldwork, we identified 31 species of introduced spiders in the country and six of those were identified by this study. Although none of the species seem to generate negative effects for local ecosystems, more research is needed to evaluate any potential effects to native species. Similarly, monitoring policies should consider spiders as a group of organisms with potential risk for local ecosystems.

Abstract: Biological invasions occur when organisms are moved from their native range and introduced into new areas, where they can spread and become a potential risk for native organisms. Invasive species are well recorded for vertebrates, plants and a number of invertebrates. A taxa for which there is a lack of information in some countries are arachnids, and this is the case of reports on introduced species in Panama. Here, we report the first list of introduced spider species in the country as a result of reviewing scientific literature, online databases, museum collections and carrying out field collections. We report six new introduced species out of a total of 31 species, belonging to 10 families and 25 genera. A large proportion of these species originated from Asia. Moreover, most of the species, with historic data available, have expanded their distribution in the country. We also present a list of species that might be already present in the country or may be introduced in the future. Our work highlights the capability of spiders to invade and spread into new territories. Therefore, monitoring policies and control strategies should also consider them as potential risk for local ecosystems.

Keywords: biological invasion; exotic species; Central America; neotropics

1. Introduction

The dispersal of organisms is essential for the distribution of life on Earth, it plays a critical role for the maintenance of biodiversity, contributing to the strengthening of diverse

https://doi.org/10.3390/biology14010004

forms and functions in living beings [1]. However, the arrival of exotic species may impose new challenges for native species, like new competitors and diseases, predation and habitat alteration, among others [2–6]. Expansion in the distribution ranges of species between continents is mainly driven by human activities, and it is experiencing an increase in recent decades [1,7,8]. Introduced non-native species to a particular ecosystem that can cause or are likely to cause harm to the economy, environment or human health are known as invasive species [9]. In that line, some introduced species may not acquire the status of invasive since they do not seem to generate a clear problem [10].

Globally, most research on invasive species has focused on plants, vertebrates, and several groups of invertebrates, including spiders [7,8,11,12]. However, it is crucial to recognize that invasive spiders can have a significant impact on ecosystems due to their ability to reach high densities in both natural and urban environments, potentially affecting native species diversity [7,13]. In addition, some of these invasive spiders also have medical relevance, as their bites can have potentially lethal consequences [14–16].

Panama is recognized for its rich biological diversity and biological invasions have received continued attention. Over the years, a variety of invasive taxa have been reported, including plants, birds, fish, reptiles, amphibians and invertebrates [17–26]. The introduction of these species to Panama is the result of multiple activities like pest control, promotion of agriculture and aquaculture, exotic pet trade and accidental introductions due to human activity. Unfortunately, these introductions, instead of generating benefits, have given rise to significant problems in terms of economic and ecological damages [26–28]. In spite of those efforts, invasive spider species has received little attention in government reports [29] or scientific studies [26].

In the country, the first list of spider species by Nentwig [30] registered 1223 species, which currently reaches approximately 1240 species (Murcia-Moreno, unpubl. data). However, the proportion of introduced species within this figure remains unknown. Therefore, here we present a list of introduced species resulting from a review of scientific literature, review of a museum collection, review of available databases and opportunistic sampling. Our study provides a complement to a recent work by Rodríguez-Gavilanes et al. [26] that listed invasive species of Chordates and Arthropods in Panama but it did not include spiders in their report. Moreover, the present work provides basic information on the diversity and distribution of introduced spider species in the country. Although further work is required to decipher potential ecological or economic costs imposed by these species, we assumed an invasive status for many of them, given their wide distribution and long establishment in the country.

2. Materials and Methods

This study was carried out from 2022–2024 and consisted of an extensive bibliographic review of scientific articles and online databases: (1) World Spider Catalog [31]; (2) The Global Biodiversity Information Facility [32] and (3) iNaturalist [33]. We used the list of species provided by Nentwig [30] as the main guide to check for introduced species. None of the most recent reviews on biological invasions in Panama provided information on introduced spiders [26,29].

We also used Web of Science (WOS: all databases) to search for articles on introduced spider species in Panama. We performed independent searches with the terms 'invasive species', 'invasive spider', 'exotic spider', 'especie invasora' and 'especie exotica'. Each of these searches was refined with the term 'Panama' (Table S1).

Our field work was based on an unsystematic sampling of sites visited during 2022–2024 during both dry and rainy seasons. The spiders that we sampled were opportunistically collected from visits to multiple sites that included natural and anthropogenic habitats like ports, buildings in urban and rural areas, green areas in urban areas, protected areas and tourist sites. We collected spiders by means of forceps or aspirators and were placed into 70% ethanol. Collections were made under the permit 0033-2021 (Ministry of Environment). Finally, we reviewed specimens from the Invertebrate Museum of the University of Panama (MIUP).

From the review and visits to the MIUP, we obtained historical records (first time report) and distribution data for all species. We considered species as introduced or cosmopolitan following the taxonomic nomenclature of the World Spider Catalog [31]. We identified collected spiders by morphological characteristics, by means of a stereomicroscope (Nikon SMZ 800, Tokyo, Japan). We confirmed identifications following published guides in scientific articles, books and websites. Samples are stored in the laboratory of Dumas Gálvez at the Central American Masters in Entomology, University of Panama. All samples are freely available.

3. Results

Our WOS searches did not generate documents of which we were not already aware (Table S1). Here, we report 31 species of introduced spiders to Panama, belonging to 10 families and 25 genera (Table 1). The families with the highest number of representative species were Theridiidae (7 spp.), Oonopidae (7 spp.), Araneidae (4 spp.) and Pholcidae (4 spp.) (Table 1, Figure 1). Species from Asia seem to be the most common introduced species (Table 1). We also report six new introduced species to the country: *Crossopriza lyoni* (Blackwall, 1867) (Figure 2), *Cyrtophora citricola* (Forsskål, 1775) (Figure 3), *Neoscona adianta* (Walckenaer, 1802), *Parasteatoda tepidariorum* (C. L. Koch, 1841), *Steatoda grossa* (C. L. Koch, 1838) and *Theridion melanostictum* (O. Pickard-Cambridge, 1876) (Figure 4, Tables 1, 2 and S2).

Table 1. List of first-time reports of introduced spider species in Panama.

Species	Area of Origin	First Record	Distribution	Reference
Araneidae				
Cyrtophora citricola (Forsskål, 1775)	Southern Europe, Africa, Middle East, Asia (Pakistan, India, China, Japan).	this work		
Gea heptagon (Hentz, 1850)	Pacific Is, Australia.	1929	Panamá Oeste, Panama	[30,34,35]
Neoscona adianta (Walckenaer, 1802)	Europe, North Africa, Asia.	this work		
Neoscona nautica (L. Koch, 1875)	Asia, Pacific Is.	1885	Bocas del Toro, Chiriquí, Panamá Oeste, Colón	[30,36–42]
Cithaeronidae				
Cithaeron praedonius (O. Pickard-Cambridge, 1872)	North Africa, Greece, Cyprus, Turkey, Middle East to India, Malaysia	2024	Panamá Oeste	[31,43]
Filistatidae				
Labahitha marginata (Kishida, 1936)	Taiwan, Philippines, Papua New Guinea, Pacifi Is.	2022	Chiriquí, Veraguas	[44,45]
Ochyroceratidae				
Theotima minutissima (Petrunkevitch, 1929)	Tropical Asia	1995	Panamá Oeste	[46]

Table 1. *Cont.*

Species	Area of Origin	First Record	Distribution	Reference
Oonopidae				
Brignolia parumpunctata (Simon, 1893)	Asia (Pakistan, India, Sri Lanka, China (Yongxing Is. = Woody Is.), Philippines, Indonesia (Sulawesi, Banda Is.).	1951	Panamá Oeste, Panamá, Colón	[30,47–49]
Ischnothyreus peltifer (Simon, 1892)	Tropical Asia	1951	Los Santos, Panamá Oeste, Panamá, Colón	[30,47,48,50]
Ischnothyreus velox (Jackson, 1908)	Tropical Asia	2012	Panamá, Colón	[50]
Opopaea apicalis (Simon, 1893)	Asia (China (Yongxing Is. = Woody Is.), Thailand, Indonesia, Philippines).	2009	Panamá Oeste, Panamá, Colón	[51]
Opopaea deserticola (Simon, 1892)	Southeast Asia	1951	Panamá Oeste, Panamá, Colón	[30,47,51]
Triaeris stenaspis (Simon, 1892)	Africa	1938	Bocas del Toro, Chiriquí, Coclé, Panamá Oeste, Panamá, Colón	[30,47,48,50,52,53]
Xestaspis parmata (Thorell, 1890)	Southeast Asia	2019	no data	[31]
Pholcidae				
Artema atlanta (Walckenaer, 1837)	North Africa, Middle East	1925	Colón	[30,54]
Crossopriza lyoni (Blackwall, 1867)	Probably native to Africa and/or Asia	this work		
Micropholcus fauroti (Simon, 1887)	Temperate Asia	1996	Panamá Oeste	[55,56]
Physocyclus globosus (Taczanowski, 1874)	North America	1925	Chiriquí, Veraguas, Coclé, Panamá Oeste, Panamá, Colón	[30,54,56–59]
Salticidae				
Hasarius adansoni (Audouin, 1826)	Africa, Middle East	1925	Chiriquí, Panamá	[30,54,60]
Menemerus bivittatus (Dufour, 1831)	Africa	1901	Chiriquí, Panamá Oeste, Panamá, Colón	[30,35,54,57,61,62]
Plexippus paykulli (Audouin, 1826)	Africa	1929	Panamá Oeste, Panamá, Colón	[30,35,57,61,63]
Sparassidae				
Heteropoda venatoria (Linnaeus, 1767)	Tropical Asia	1936	Veraguas, Coclé, Panamá Oeste, Panamá, Colón	[64–66]
Tetragnathidae				
Tetragnatha nitens (Audouin, 1826)	Tropical and subtropical Asia	1925	Chiriquí, Coclé, Panamá Oeste, Colón	[30,54,67]
Tetragnatha vermiformis (Emerton, 1884)	Temperate and tropical Asia	1981	Panamá Oeste	[30,68,69]

Table 1. *Cont.*

Species	Area of Origin	First Record	Distribution	Reference
Theridiidae				
Latrodectus geometricus (C.L. Koch, 1841)	Africa	1959	Panamá, Colón	[70,71]
Latrodectus mactans (Fabricius, 1775)	Probably native to North America only.	1902	Chiriquí, Veraguas, Panamá Oeste	[30,38,54,62,70]
Meotipa pulcherrima (Mello-Leitão, 1917)	East Asia, Papua New Guinea, Pacific Is.	1936	Chiriquí, Veraguas, Panamá Oeste, Panamá, Colón	[30,57,64,72–74]
Parasteatoda tepidariorum (C.L. Koch, 1841)	Asia	this work		
Steatoda erigoniformis (O. Pickard-Cambridge, 1872)	East Mediterranean, Middle East, Caucasus, India, East Asia	1957	Veraguas, Panamá, Colón	[30,64,75,76]
Steatoda grossa (C. L. Koch, 1838)	Europe, Turkey, Russia (Europe to Far East), Caucasus, Iran, Kazakhstan, Central Asia, China, Korea, Japan	this work		
Theridion melanostictum (O. Pickard-Cambridge, 1876)	Macaronesia, Mediterranean to Egypt, India, Central Asia, China, Japan	this work		

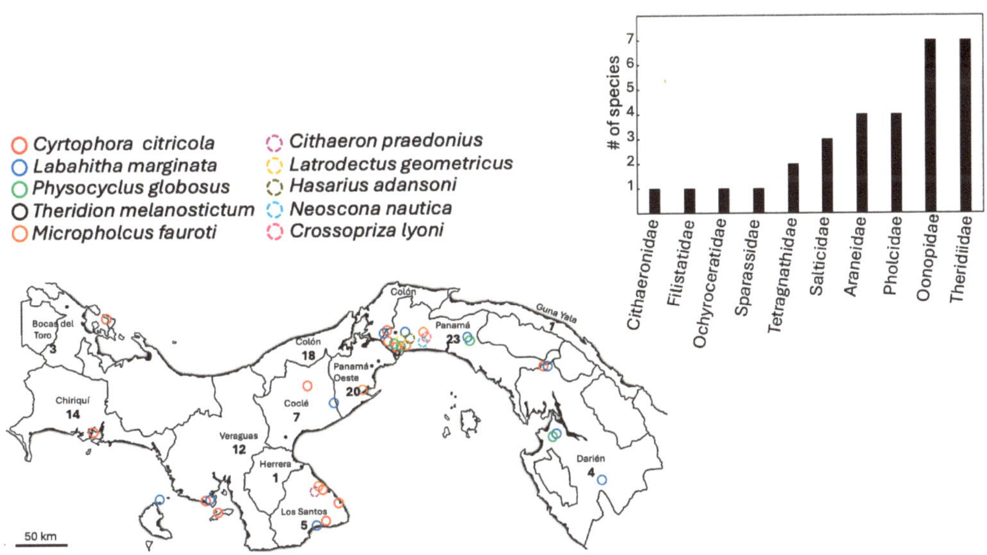

Figure 1. New records of introduced species based on collected individuals from different locations in Panama, and the total number of invasive species per province. Box shows the number of species per family in the country.

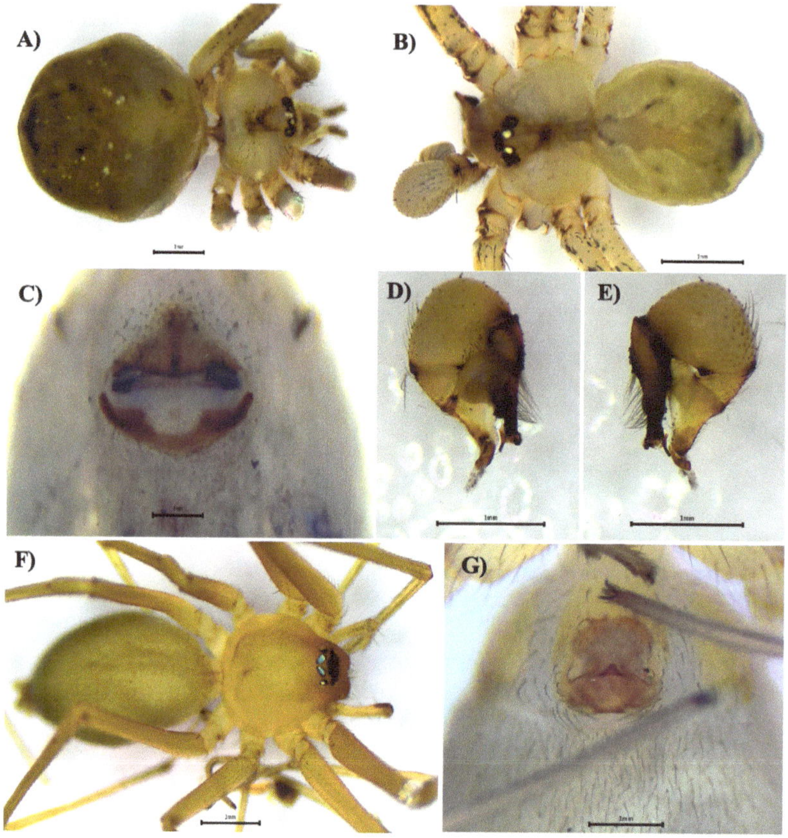

Figure 2. *Crossopriza lyoni*, female, dorsal view (**A**). Male, dorsal view (**B**). Female epigynum, ventral view (**C**). Male left palp, prolateral view (**D**), retrolateral view (**E**). *Cithaeron praedonius*, female, dorsal view (**F**). Epigynum, ventral view (**G**).

We found evidence from the museum collection and our sampling that some species have expanded their distribution in the country (Table 2) as compared to the historic data (Table 1). We also report new location records for most of the species; for instance, *Cithaeron praedonius* (O. Pickard-Cambridge, 1872) (Figure 2, Table 2), *Labahitha marginata* (Kishida, 1936) (Figure 5, Table 2) and *Micropholcus fauroti* (Simon, 1887) (Figure 6, Table 2). Moreover, rooted on a presence/absence evaluation per province, *Cyrtophora citricola* (all provinces), *Heteropoda venatoria* (Linnaeus, 1767), *Labahitha marginata* and *Physocyclus globosus* (Taczanowski, 1874) seem to be the most widespread species in the country (Figure 1, Tables 1, 2 and S3). Overall, the provinces of Panama, Panamá Oeste, Colón and Chiriquí seem to have received the largest numbers of introduced species and widest distributions locally (Tables 2, 3 and S3).

We also present data on species that could be present in Panama, given their presence in the two neighboring countries (Costa Rica and Colombia), but have not yet been reported, as well as records that required confirmation due to insufficient evidence (Tables 3 and S4).

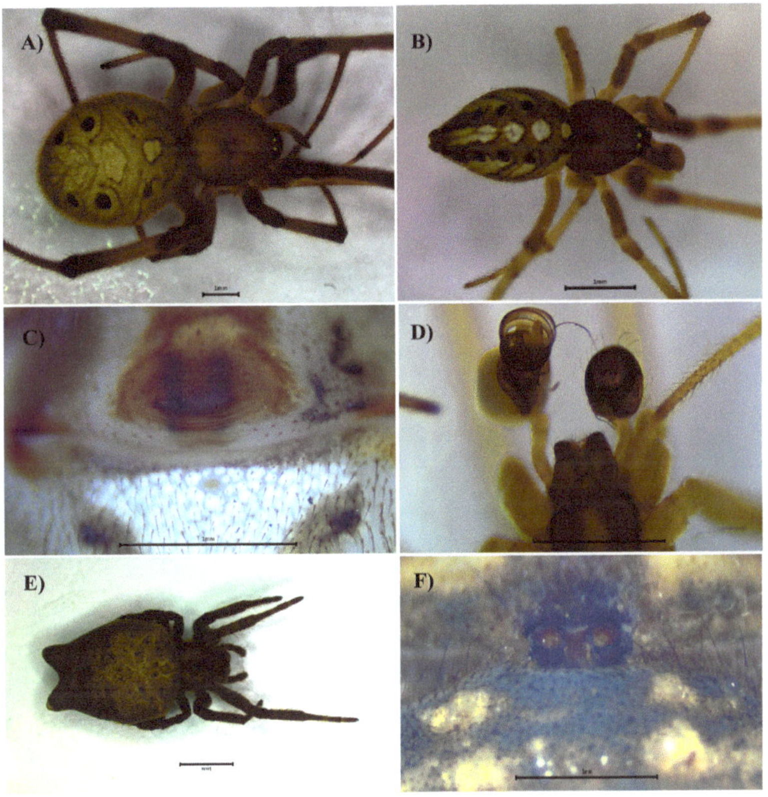

Figure 3. *Latrodectus geometricus*, female, dorsal view (**A**). Male, dorsal view (**B**). Female epigynum, ventral view (**C**). Male palps, ventral view (**D**). *Cyrtophora citricola*, female, dorsal view (**E**). Epigynum, ventral view (**F**).

Figure 4. *Theridion melanostictum*, female, dorsal view (**A**). Female epigynum, ventral view (**B**). Male, dorsal view (**C**). Male right palp, ventral view (**D**).

Table 2. List of introduced spider species, with new data distribution from collection and databases review. * literature review.

Species	Location	Basis of Record	Reference
Araneidae			
Cyrtophora citricola (Forsskål, 1775)	Bocas del Toro, Chiriquí, Veraguas, Herrera, Los Santos, Coclé, Panamá Oeste, Panamá, Colón, Comarca Kuna Yala, Darien	Collected specimen, Inaturalist	[77]
Gea heptagon (Hentz, 1850)	Panamá Oeste, Panamá	MIUP Collection	
Neoscona adianta (Walckenaer, 1802)	Chiriquí	Gbif database	[78]
Neoscona nautica (L. Koch, 1875)	Bocas del Toro, Chiriquí, Veraguas, Panamá Oeste, Panamá	MIUP Collection, Inaturalist, Collected specimen	[42]
Cithaeronidae			
Cithaeron praedonius (O. Pickard-Cambridge, 1872)	Veraguas, Los Santos, Panamá	Collected specimen, Inaturalist	[79]
Filistatidae			
Labahitha marginata (Kishida, 1936)	Chiriquí, Veraguas *, Los Santos, Coclé, Panamá, Colón, Darien	Collected specimen	[45]
Oonopidae			
Ischnothyreus peltifer (Simon, 1892)	Los Santos *	MIUP Collection	[50]
Opopaea deserticola (Simon, 1892)	Panamá	MIUP Collection	
Xestaspis parmata (Thorell, 1890)	Panamá Oeste, Panamá, Colón	Gbif database	[80]
Pholcidae			
Crossopriza lyoni (Blackwall, 1867)	Panamá	Collected specimen	
Micropholcus fauroti (Simon, 1887)	Los Santos, Panamá	Collected specimen	
Physocyclus globosus (Taczanowski, 1874)	Panamá, Colón, Darien	Collected specimen	
Crossopriza lyoni (Blackwall, 1867)	Panamá	Collected specimen	
Salticidae			
Hasarius adansoni (Audouin, 1826)	Veraguas, Coclé, Panamá, Colón	MIUP Collection, Collected specimen, Inaturalist	[81]
Menemerus bivittatus (Dufour, 1831)	Bocas del Toro, Chiriquí, Veraguas, Los Santos, Herrera, Coclé, Panamá Oeste, Panamá, Colón	MIUP Collection, Collected specimen, Inaturalist	[82]
Plexippus paykulli (Audouin, 1826)	Veraguas, Panamá Oeste, Panamá	MIUP Collection, Collected specimen, Inaturalist	[83]
Sparassidae			
Heteropoda venatoria (Linnaeus, 1767)	Chiriquí, Veraguas, Panamá Oeste, Panamá, Comarca Kuna Yala, Darien	Inaturalist	[84]
Tetragnathidae			
Tetragnatha nitens (Audouin, 1826)	Panamá	MIUP Collection	
Theridiidae			
Latrodectus geometricus (C.L. Koch, 1841)	Panamá	MIUP Collection, Collected specimen	
Parasteatoda tepidariorum (C.L. Koch, 1841)	Chiriquí	Gbif database, Inaturalist	[85,86]
Steatoda grossa (C.L. Koch, 1838)	Chiriquí, Veraguas	Inaturalist	[87]
Theridion melanostictum (O. Pickard-Cambridge, 1876)	Panamá Oeste, Panamá, Colón	Gbif database, Collected specimen	[88]

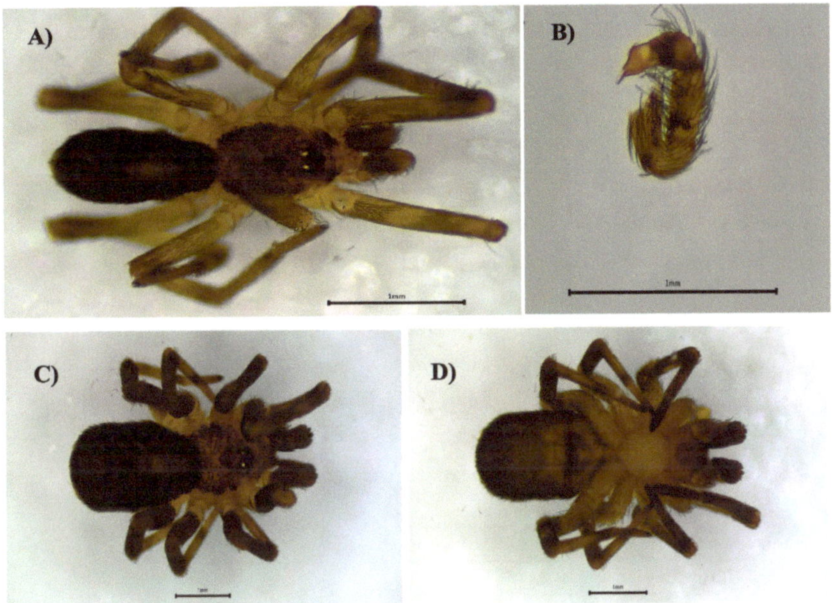

Figure 5. *Labahitha marginata*, male, dorsal view (**A**). Male right palp, retrolateral view (**B**). *Labahitha marginata*, female, dorsal view (**C**). Female, ventral view (**D**).

Figure 6. *Micropholcus fauroti*, male, dorsal view (**A**). Female, dorsal view (**B**). Male right palp, prolateral view (**C**), retrolateral view (**D**). Female epigynum, ventral view (**E**).

Table 3. Putative introduced spider species or with the potential for introduction in Panama.

Species	Location	Basis of Record	Reference
Agelenidae			
Tegenaria domestica (Clerck, 1757)	not reported	Costa Rica, Colombia	[62,89,90]
Tegenaria pagana (C.L. Koch, 1840)	not reported	Colombia	[90,91]
Araneidae			
Neoscona moreli (Vinson, 1863)	not reported	Colombia	[31,40,92]
Nephilingis cruentata (Fabricius, 1775)	not reported	Colombia	[31,90,92,93]
Gnaphosidae			
Marinarozelotes kulczynskii (Bösenberg, 1902)	not reported	Colombia	[31,90,92,94]
Urozelotes rusticus (Fabricius, 1775)	not reported	Colombia	[90]
Oecobiidae			
Oecobius navus (Blackwall, 1859)	to be confirmed	Colombia	[64,90,92,95]
Oonopidae			
Opopaea concolor (Blackwall, 1859)	not reported	Costa Rica, Colombia	[51,92]
Pholcidae			
Pholcus phalangioides (Fuesslin, 1775)	to be confirmed	Costa Rica	[96,97]
Smeringopus pallidus (Blackwall, 1858)	not reported	Costa Rica	[56,96,98]
Theridiidae			
Steatoda nobilis (Thorell, 1875)	not reported	Colombia	[31,90,92,99]

4. Discussion

The introduced spiders in Panama present a diverse and complex panorama in terms of their distribution and origin. Most of the species recorded belonged to the families Theridiidae, Oonopidae, Araneidae and Pholcidae, in comparison to regions like South America, where the most common introduced species belong to Oonopidae, Theridiidae, Salticidae and Gnaphosidae [92] or in Europe, where Pholcidae and Theridiidae contain a large proportion of the introduced species [7]. The dominance of families such as Theridiidae, Pholcidae, Oonopidae and Araneidae could be related to their morphological size, the similarity of their climatic niches with those of their place of origin and their ability to adapt to microclimatic changes (e.g., anthropogenic factors). These families include numerous species well adapted to dry niches in their areas of origin, which facilitates their survival especially in urban environments, which is the case in for instance in Europe [100]. Importantly, future work with molecular techniques such as DNA barcoding could improve the accuracy on the detection of introduced species and establish the point of origin.

Although it is difficult to establish the origin of the introduced species, our results suggest that species from Asia are the most common invaders in Panama. However, certainty of the invasion origin of some species is low, given that they could come from a location that is not necessarily their site of origin (e.g., *B. parumpunctata*, *O. deserticola*) [92]. Still, spider introductions in Panama could be partially explained by the constant traffic through the Panama Canal which allows the transfer of organism by ships [101] and the fact that the Atlantic—Pacific route involving Asia is a major traffic route through the canal [102]. This hypothesis is supported by the fact that provinces around the canal (Panama, Panamá Oeste and Colón) showed the largest proportion of introduced species. In line with our results, Rodríguez-Gavilanes et al. [26] found the highest number of invasive taxa for the province of Panama. Another likely entry point is the transport network [103] connection between the Chiriquí province and Costa Rica, and this province ranks fourth in the proportion of introduced species recorded, in line with this idea. Also supporting this idea is the fact that the other province that borders Costa Rica (Bocas del Toro), but with significantly less traffic, show relatively low number of introduced species, as well as Darién with no road connections with Colombia. Still, there is the possibility that these figures are the result of biased sampling in each province.

Understanding whether sampling bias is an issue to understand introduced species distribution is important because it can shed some light also on whether these species can naturally disappear in the sites where they manage to invade and establish. For instance, *Latrodectus mactans* and *Artema atlanta* have not been reported in almost a century [30,54]. Thus, if those species have truly disappeared then which factors are responsible? Related to this, more work is needed to understand whether there is some correlation between habitat type and the likelihood of establishment by introduced species [104]. For instance, the apparent most common introduced species in Panama, *Cyrtophora citricola*, has been able to occupy a novel set of abiotic conditions in the Americas, which are not present in their native range of distribution [8]. Still, most of the introduced species in Panama come from warm regions such as Africa, Asia and the Mediterranean (Table 1), which suggests that they find favorable similar abiotic conditions in the country.

The distribution pattern can also provide estimation on colonization dates. A study from Chile suggests that residence time is a determining factor in predicting the invasive success and range size of introduced spiders. Species with more than 100 years in the country have shown significant range expansions [105] and other studies have found a positive relationship between residence time and invasion success [106,107]. Based on this hypothesis, species such as *C. citricola*, *L. marginata* and *M. bivittatus* may have established in Panama a long time ago and be among the first species to arrive, which could explain their wide distribution range.

New distribution ranges that include Panama are also expected for introduced spiders that are currently present in the two neighboring countries, Costa Rica and Colombia. It is likely that there are introduced spider species in Panama that have not yet been confirmed, such as *Tegenaria domestica* and *Steatoda nobilis*, in addition to those that require more evidence, such as *Oecobius navus* and *Pholcus phalangioides*. Quintero [64] mentioned the presence of *O. navus* in the country, but did not provide references or collection data. We found no historical or current evidence to support its presence in Panama and the specimens that we have collected correspond to *Oecobius cocinnus*, a native and synanthropic species. Although we did not collect specimens of *P. phalangioides* in this study, we identified a previous record probably from Panama. However, the associated coordinates are not reliable, and the date of collection is unknown. This record was identified by Pholcidae expert Hubert Bernhard [97]. The possibility that other unconfirmed species are in the process of becoming established in the country is feasible. While some introduced species are known to form populations quickly, others require more time and studies to determine if they can maintain sustainable populations in the new ecosystem [92].

Once introduced species establish in those new locations, it remains unknown whether some of these species could be biologically controlled [108] or whether they could provide ecological services [109]. Moreover, further ecological work is required to understand whether some of these introduced species pose a risk [110] or even benefits [111] for native species, and how their presence may influence local ecosystems with potential cascading effects [112] or interacting effects [4]. For example, *C. citricola* webs can interfere with plant growth in agricultural systems, possibly resulting in a lower yield [113] and pollination success, which would make this species a true invasive species. On the other hand, species like *L. mactans* and *L. geometricus* could pose a health risk for humans and animals due to their potentially risky venoms [105,107].

There have not been studies aimed at understanding the impacts of these introduced species in Panama, either related to ecological or economics effects. However, based on our experience and trends observed in the literature for other invasive spider species [7], we estimate that their overall impact is low. Perhaps one of the species that possess the highest economical risk for coffee plantations is the most widespread, *C. citricola*, as reported in Colombia [113], but this does not seem the case in Panama. Regarding monitoring policies, the Panamanian government have traditionally implemented policies for monitoring arthropod species of risk in entry points of the country, for example, species that represent a threat for agriculture [114]. Unfortunately, there has not been monitoring programs for other taxa, like spiders; however, recent policies are targeted to cope with the issue of invasive species in general [34]. Our work highlights the fact that spiders are part of the taxa that can easily establish and spread in the country, which calls for attention of the monitoring policies and potentially control strategies.

5. Conclusions

Our work has identified a complex scenery for introduced spiders in Panama, originating from different geographic regions, which may be driven by multiple factors such as international trade, transport and environmental change. The new records of introduced spiders expand the knowledge of Panama's biodiversity and call for a strict monitoring of their distribution and potential ecological impacts, which includes their interaction with native organisms, and long-term impacts. Our work updates the list of introduced animal species in the country to 172, and arachnids represent 19% of the animal species recorded. Future efforts should focus on monitoring new introductions, validating historical records, and establishing policies that minimize the ecological and health risks associated with these species.

Supplementary Materials: The following supporting information can be downloaded at: https://www.mdpi.com/article/10.3390/biology14010004/s1, Table S1: Results of searches carried in Web of Sciences; Table S2: Coordinates of collected individuals of invasive spiders in Panama; Table S3: Summary of presence/absence (1 and 0, respectively) of invasive spider species per regions of Panama. Table S4: Details on some unconfirmed introduced spider species in Panama [115].

Author Contributions: Conceptualization: D.M.-M. Data Collection: D.M.-M. and D.G. Data Analysis: D.M.-M. and D.G. Drafting of manuscript: D.M.-M. Editing of manuscript: D.G. All authors have read and agreed to the published version of the manuscript.

Funding: This research was funded by SENACYT, grant number FID22-034 and 'Sistema Nacional de Investigación', grant number SNI-II.

Institutional Review Board Statement: Ethical review and approval were waived for this study due to that the Ethics Committee of the University of Panama only warrants permits for studies on vertebrates.

Informed Consent Statement: Not applicable.

Data Availability Statement: All data supporting this manuscript is available as Supplementary Materials.

Acknowledgments: We thank Roderick Valdés, Carlos Vega and Braulio Bonilla for support in the field. We thank the PCMENT for the laboratory space and use of equipment. We also thank the staff of Invertebrate Museum of the University of Panama (MIUP) for access to the arachnological collection. Finaly, we thank Dra. Sabrina Amador who facilitated the use of her laboratory for taking photographs.

Conflicts of Interest: The authors declare no conflicts of interest.

References

1. Nentwig, W. Biological Invasions: Why It Matters. In *Biological Invasions*; Ecological Studies Series; Springer: Berlin/Heidelberg, Germany, 2008.
2. Vilà, M.; Weiner, J. Are Invasive Plant Species Better Competitors than Native Plant Species? Evidence from Pair-Wise Experiments. *Oikos* **2004**, *105*, 229–238. [CrossRef]
3. Marbuah, G.; Gren, I.; Mckie, B. Economics of Harmful Invasive Species: A Review. *Diversity* **2014**, *6*, 500–523. [CrossRef]
4. Didham, R.K.; Tylianakis, J.M.; Gemmell, N.J.; Rand, T.A.; Ewers, R.M. Interactive Effects of Habitat Modification and Species Invasion on Native Species Decline. *Trends Ecol. Evol.* **2007**, *22*, 489–496. [CrossRef] [PubMed]
5. Young, H.S.; Parker, I.M.; Gilbert, G.S.; Guerra, A.; Nunn, C.L. Introduced Species, Disease Ecology, and Biodiversity—Disease Relationships. *Trends Ecol. Evol.* **2017**, *32*, 41–54. [CrossRef]
6. Xie, J. Invasive Species Threatening Ecosystems: Challenges and Strategies. *Ukranian J. Ecol.* **2023**, *13*, 32–34. [CrossRef]
7. Nentwig, W. Introduction, Establishment Rate, Pathways and Impact of Spiders Alien to Europe. *Biol. Invasions* **2015**, *17*, 2757–2778. [CrossRef]
8. Segura-Hernández, L.; Barrantes, G.; Chacón-Madrigal, E.; Rodríguez-García, A. Species Distribution Models and Climatic Niche Comparisons Provide Clues on the Geographic Origin of a Spider Invasion in the Americas. *Biol. Invasions* **2023**, *25*, 251–265. [CrossRef]
9. Iannone, B.; Carnevale, S.; Main, M.B.; Hill, J.E.; McConnell, J. Invasive Species Terminology: Standardizing for Stakeholder Education. *J. Ext.* **2020**, *58*, 27. [CrossRef]
10. Lockwood, J.; Hoopes, M.; Marchetti, M. *Invasion Ecology*; Blackwell Publishing: Malden, MA, USA, 2007; 304p.
11. Mollot, G.; Pantel, J.H.; Romanuk, T.N. The Effects of Invasive Species on the Decline in Species Richness: A Global Meta-Analysis. In *Advances in Ecological Research*; Bohan, D., Dumbrell, A., Massol, F., Eds.; Academic Press: Oxford, UK, 2017; pp. 61–83.
12. Řezáč, M.; Růžička, V.; Hula, V.; Dolanský, J.; Machač, O.; Roušar, A. Spiders Newly Observed in Czechia in Recent Years—Overlooked or Invasive Species. *Bioinvasion Rec.* **2021**, *10*, 555–566. [CrossRef]
13. Hann, S. Evidence for the Displacement of an Endemic New Zealand Spider, *Latrodectus katipo* Powell by the South African Species *Steatoda capensis* Hann (Araneae: Theridiidae). *N. Zealand J. Zool.* **1990**, *17*, 295–307. [CrossRef]
14. Nentwig, W.; Kuhn-Nentwing, L. Spider Venoms Potentially Lethal to Humans. In *Spider Ecophysiology*; Nentwig, W., Ed.; Springer: Berlin/Heidelberg, Germany, 2012; pp. 253–264.
15. Nentwig, W.; Gloor, D.; Kropf, C. Spider Taxonomists Catch Data on Web. *Nature* **2015**, *528*, 479. [CrossRef] [PubMed]
16. Nentwig, W.; Pantini, P.; Vetter, R.S. Distribution and Medical Aspects of Loxosceles Rufescens, One of the Most Invasive Spiders of the World (Araneae: Sicariidae). *Toxicon* **2017**, *132*, 19–28. [CrossRef] [PubMed]
17. Gonzalez, R. Estado de Los Peces Exóticos Introducidos En Las Aguas Continentales de Panama. *Brenesia* **1995**, *43–44*, 55–59.
18. Svenning, J. Non-Native Ornamental Palms Invade a Secondary Tropical Forest in Panama. *Palms* **2002**, *46*, 81–86.
19. Kirkendall, L.; Odegaard, F. Ongoing Invasions of Old-Growth Tropical Forests: Establishment of Three Incestuous Beetle Species in Central America. *Zootaxa* **2007**, *1588*, 53–62. [CrossRef]
20. Angehr, G.; Dean, R. *The Birds of Panama: A Field Guide*; Zona Tropical Publications: San José, CA, USA, 2010.
21. Holway, D.A.; Lach, L.; Suarez, A.V.; Tsutsui, N.D.; Case, T.J. The Causes and Consequences of Ant Invasions. *Annu. Rev. Ecol. Syst.* **2002**, *33*, 181–233. [CrossRef]
22. Lopez, O. Introduced Alien Plant Species in the Neotropics: The Panama Case. *Open Ecol. J.* **2012**, *5*, 84–89. [CrossRef]
23. Ros, M.; Ashton, G.V.; Lacerda, M.B.; Carlton, J.T.; Vázquez-Luis, M.; Guerra-García, J.M.; Ruiz, G.M. The Panama Canal and the Transoceanic Dispersal of Marine Invertebrates: Evaluation of the Introduced Amphipod Paracaprella Pusilla Mayer, 1890 in the Pacific Ocean. *Mar. Environ. Res.* **2014**, *99*, 204–211. [CrossRef]
24. Batista, A.; Ponce, M.; Garcés, O.; Lassiter, E.; Miranda, M. Silent Pirates: Anolis Sagrei Duméril & Bibron, 1837 (Squamata, Dactyloidae) Taking over Panama City, Panama. *Check List* **2019**, *15*, 455–459.
25. Gálvez, D.; Murcia-Moreno, D.; Añino, Y.; Ramos, C. The Asian Hemipteran Brachyplatys Subaeneus (Westwood, 1837) (Hemiptera: Heteroptera: Plataspidae) in a Protected Area in Panama. *Bioinvasion Rec.* **2022**, *11*, 95–100. [CrossRef]
26. Rodríguez-Gavilanes, D.; Garcés Botacio, H.; Fuentes, R.; Rodriguez-Scott, L.; Añino, Y.; López-Chong, O.; Medianero, E. An Annotated Checklist of Invasive Species of the Phyla Artrhropods and Chordates in Panama. *Biology* **2024**, *13*, 571. [CrossRef] [PubMed]
27. Perez, J.; Alfonsi, C.; Nirchio, M.; Muñoz, C.; Gómez, J. The Introduction of Exotic Species in Aquaculture: A Solution or Part of the Problem? *Interciencia* **2003**, *28*, 234–238.

28. Carpio, A.J.; Álvarez, Y.; Oteros, J.; León, F.; Tortosa, F.S. Intentional Introduction Pathways of Alien Birds and Mammals in Latin America. *Glob. Ecol. Conserv.* **2020**, *22*, e00949. [CrossRef]
29. Autoridad Nacional del Ambiente (ANAM). *Informe Sobre Especies Exóticas de Panamá*; Autoridad Nacional del Ambiente: Ancón, Panama, 2010.
30. Nentwig, W. Spiders of Panama. In *Flora and Fauna Handbook*; Sandhill Crane Press, Inc.: Leiden, The Netherlands, 1993.
31. World Spider Catalog. Available online: http://wsc.nmbe.ch (accessed on 16 October 2024).
32. GBIF. The Global Biodiversity Information Facility (Year) What Is GBIF? Available online: https://www.gbif.org/what-is-gbif (accessed on 16 October 2024).
33. INaturalist. Available online: https://www.inaturalist.org (accessed on 16 October 2024).
34. Ministerio de Ambiente. *Estrategia y Plan de Acción Nacional de Biodiversidad (EPANB) (2018–2050)*; Ministerio de Ambiente: Ancón, Panama, 2018.
35. Banks, N. Spiders from Panama. *Bull. Mus. Comp. Zool.* **1929**, *69*, 53–96.
36. Levi, H. The Spider Genera Gea and Argiope in America (Araneae: Araneidae). *Bull. Mus. Comp. Zool.* **1968**, *136*, 319–352.
37. Keyserling, E. Neue Spinnen Aus America. VI. In *Verhandlungen der Kaiserlich-Königlichen Zoologisch-Botanischen Gesellschaft in Wien*; BHL: Washington, DC, USA, 1885; Volume 34, pp. 489–534.
38. Pickard-Cambridge, F.O. Arachnida-Araneida and Opiliones. In *Biologia Centrali-Americana, Zoology*; BHL: London, UK, 1904; pp. 465–560.
39. Berman, J.D.; Levi, H.W. The Orb Weaver Genus Neoscona in North America (Araneae: Araneidae). *Bull. Mus. Comp. Zool.* **1971**, *141*, 465–500.
40. Levi, H. American Neoscona and Corrections to Previous Revisions of Neotropical Orb-Weavers (Araneae: Araneidae). *Psyche* **1993**, *99*, 221–239. [CrossRef]
41. Caira, J.; Barbeau, E.; Carl, W.; Marian, E. Rettenmeyer Army Ant Guest Collection. University of Connecticut. Available online: https://www.gbif.org/occurrence/2332530583 (accessed on 13 November 2023).
42. Grant, S.; Webbink, K.; Turcatel, M.; Shuman, R. Field Museum of Natural History (Zoology) Insect, Arachnid and Myriapod Collection. Version 12.56. Field Museum. *Neoscona nautica*. Available online: https://www.gbif.org/dataset/7931dcab-94f1-46ce-8092-56e4335423de (accessed on 13 November 2023).
43. Miranda, R.J.; Pérez-pérez, J.I.; Bermúdez, C.S.E. Primer Registro *Cithaeron praedonius* O. Pickard-Cambridge, 1872 (Araneae: Cithaeronidae) En Panamá. *Rev. Ibérica Arachnol.* **2024**, *44*, 140–142.
44. Magalhaes, I.L.F.; Berry, J.W.; Koh, J.K.H.; Gray, M.R. Labahitha Spiders (Arachnida: Araneae: Filistatidae) from Islands in the Indian and Pacific Oceans. *Eur. J. Taxon.* **2022**, *805*, 1–51. [CrossRef]
45. Murcia-Moreno, D.; Gálvez, D. Preliminary Checklist of Spiders (Araneae) from Coiba National Park, Panama. *Biodivers. Data J.* **2024**, *12*, e117642. [CrossRef]
46. Deeleman-Reinhold, C.L. The Ochyroceratidae of the Indo-Pacific Region (Araneae). *Raffles Bull. Zool. Suppl.* **1995**, *2*, 43.
47. Chickering, A.M. The Oonopidae of Panama. *Bull. Mus. Comp. Zool.* **1951**, *106*, 205–245.
48. Chickering, A. The Genus Triaeris Simon (Araneae, Oonopidae) in Central America and the West Indies. *Psyche* **1969**, *75*, 351–359. [CrossRef]
49. Platnick, N.I.; Dupérré, N.; Ott, R.; Kranz-Baltensperger, Y. The Goblin Spider Genus Brignolia (Araneae, Oonopidae). *Bull. Am. Mus. Nat. Hist.* **2011**, *349*, 1–131. [CrossRef]
50. Platnick, N.I.; Berniker, L.; Kranz-Baltensperger, Y. The Goblin Spider Genus Ischnothyreus (Araneae, Oonopidae) in the New World. *Am. Mus. Novit.* **2012**, *3759*, 1–32. [CrossRef]
51. Platnick, N.I.; Dupérré, N. The Goblin Spider Genera Opopaea and Epectris (Araneae, Oonopidae) in the New World. *Am. Mus. Novit.* **2009**, *3649*, 1–43. [CrossRef]
52. Grant, S.; Webbink, K.; Turcatel, M.; Shuman, R. Field Museum of Natural History (Zoology) Insect, Arachnid and Myriapod Collection. Version 12.56. Field Museum. *Triaeris stenaspis*. Available online: https://www.gbif.org/occurrence/1142390241 (accessed on 14 November 2023).
53. Morris, P.J. Museum of Comparative Zoology, Harvard University. Version 162.397. Museum of Comparative Zoology, Harvard University. Occurrence Dataset. *Triaeris stenaspis*. Available online: https://www.gbif.org/occurrence/477963419 (accessed on 14 January 2024).
54. Petrunkevitch, A. Arachnida from Panama. *Trans. Connect. Acad. Arts Sci.* **1925**, *27*, 51–248.
55. Penney, D. *Micropholcus fauroti* (Simon, 1887), a New Spider Record for Panama. *Newsl. Br. Arachnol. Soc.* **1996**, *77*, 11.
56. Huber, B. Pholcidae—The Longest Legs in the Web. Available online: http://www.pholcidae.de/ (accessed on 30 August 2024).
57. Chickering, A. Additions to the List of Known Species of Spiders from Barro Colorado Island, Panama. *Trans. Am. Microsc. Soc.* **1936**, *55*, 449–456. [CrossRef]
58. Eberhard, W. Notes on the Ecology and Behaviour of *Physocyclus globosus* (Araneae, Pholcidae). *Bull. Br. Arachnol. Soc.* **1992**, *9*, 38–42.

59. Valdez-Mondragón, A. Revisión Taxonómica de Physocyclus Simon, 1893 (Araneae: Pholcidae), Con La Descripción de Especies Nuevas de México. *Rev. Ibérica Arachnol.* **2010**, *18*, 3–80.
60. Natural History Museum. Natural History Museum (London) Collection Specimens. Occurrence Dataset. Available online: https://www.gbif.org/occurrence/1057060364 (accessed on 14 November 2023).
61. Petrunkevitch, A. The Spider Fauna of Panama and Its Central American Affiliation. *Am. Nat.* **1929**, *63*, 455–469. [CrossRef]
62. Pickard-Cambridge, F.O. Arachnida-Araneida and Opiliones. In *Biologia Centrali-Americana, Zoology*; BHL: London, UK, 1902; pp. 313–424.
63. European Bioinformatics Institute (EMBL-EBI). GBIF Helpdesk. INSDC Sequences. Version 1.58. European Nucleotide Archive (EMBL-EBI). Available online: https://www.gbif.org/occurrence/3818336894 (accessed on 14 November 2023).
64. Quintero, A.D., Jr. Preliminary Biodiversity Assessment and Notes on the Biology of the Arachnids (Arachnida: Scorpriones, Amplypygi and Araneae) of Bahía Honda Region (Veraguas, Panama). In *Studies on the Biodiversity of the Bahía Honda Region (Veraguas, Panama)*; Castroviejo, S., Ibáñez, A., Eds.; Consejo Superior de Investigaciones Científicas Madrid: Madrid, Spain, 2005; pp. 363–491.
65. Harvard University; Morris, P. Museum of Comparative Zoology, Harvard University. Version 162.397. Museum of Comparative Zoology, Harvard University. Occurrence Dataset. *Heteropoda venatoria*. Available online: https://www.gbif.org/occurrence/477904768 (accessed on 14 January 2024).
66. Harvard University; Morris, P. Museum of Comparative Zoology, Harvard University. Version 162.397. Museum of Comparative Zoology, Harvard University. Occurrence Dataset. *Heteropoda venatoria*. Available online: https://www.gbif.org/occurrence/477967886 (accessed on 14 January 2024).
67. Chickering, A. The Genus Tetragnatha (Araneae, Argiopidae) in Panama. *Bull. Mus. Comp. Zool.* **1957**, *116*, 299–354.
68. Levi, H. The American Orb-Weaver Genera Dolichognatha and Tetragnatha North of Mexico (Araneae: Araneidae, Tetragnathinae). *Bull. Mus. Comp. Zool.* **1981**, *149*, 271–318.
69. Senckenberg Collection Arachnology SMF. Available online: https://www.gbif.org/occurrence/208068132 (accessed on 14 November 2023).
70. Levi, H. The Spider Genus Latrodectus (Araneae, Theridiidae). *Trans. Am. Microsc. Soc.* **1959**, *78*, 7–43. [CrossRef]
71. Garb, J.E.; González, A.; Gillespie, R.G. The Black Widow Spider Genus Latrodectus (Araneae: Theridiidae): Phylogeny, Biogeography, and Invasion History. *Mol. Phylogenet. Evol.* **2004**, *31*, 1127–1142. [CrossRef]
72. Harrison, J. Some Spiders Associated with Banana Plants in Panama. *Ann. Entomol. Soc. Am.* **1968**, *61*, 878–884. [CrossRef]
73. Levi, H. More American Spiders of the Genus Chrysso (Araneae, Theridiidae). *Psyche* **1962**, *69*, 209–237. [CrossRef]
74. Grant, S.; Webbink, K.; Turcatel, M.; Shuman, R. Field Museum of Natural History (Zoology) Insect, Arachnid and Myriapod Collection. Version 12.56. Field Museum. *Chrysso pulcherrima*. Available online: https://www.gbif.org/occurrence/1142346962 (accessed on 15 November 2023).
75. Levi, H. The Spider Genera Crustulina and Steatoda in North America, Central America, and the West Indies. *Bull. Mus. Comp. Zool.* **1957**, *117*, 367–424.
76. Levi, H. The Spider Genera Steatoda and Enoplognatha in America (Araneae, Theridiidae). *Psyche* **1962**, *69*, 11–36. [CrossRef]
77. iNaturalist Contributors. INaturalist Research-Grade Observations. *C. citricola*. Available online: https://www.inaturalist.org/observations?place_id=7003&quality_grade=research&subview=map&taxon_id=130082&verifiable=any (accessed on 18 February 2024).
78. Grant, S.; Webbink, K.; Turcatel, M.; Shuman, R. Field Museum of Natural History (Zoology) Insect, Arachnid and Myriapod Collection. Version 12.56. Field Museum. *Neoscona adianta*. Available online: https://www.gbif.org/occurrence/1142345101 (accessed on 15 November 2023).
79. iNaturalist Contributors. INaturalist Research-Grade Observations. *C. praedonius*. Available online: https://www.inaturalist.org/observations?place_id=7003&subview=map&taxon_id=556721 (accessed on 18 August 2024).
80. Morris, P.J. Museum of Comparative Zoology, Harvard University. Version 162.413. Museum of Comparative Zoology, Harvard University. Occurrence Dataset. *Xestaspis parmata*.
81. iNaturalist Contributors. INaturalist Research-Grade Observations. *H. adansoni*. Available online: https://www.inaturalist.org/observations?place_id=7003&quality_grade=research&subview=map&taxon_id=347886&verifiable=any (accessed on 18 February 2024).
82. iNaturalist Contributors. INaturalist Research-Grade Observations. *M. bivittatus*. Available online: https://www.inaturalist.org/observations?place_id=7003&quality_grade=research&subview=map&taxon_id=68062&verifiable=any (accessed on 18 February 2024).
83. iNaturalist Contributors. INaturalist Research-Grade Observations. *P. paykulli*. Available online: https://www.inaturalist.org/posts/39072-research-grade (accessed on 18 February 2024).

84. iNaturalist Contributors. INaturalist Research-Grade Observations. *H. venatoria*. Available online: https://www.inaturalist.org/observations?place_id=7003&quality_grade=research&subview=map&taxon_id=48765&verifiable=any (accessed on 18 February 2024).
85. Grant, S.; Webbink, K.; Turcatel, M.; Shuman, R. *Field Museum of Natural History (Zoology) Insect, Arachnid and Myriapod Collection. Version 12.56. Field Museum. Parasteatoda tepidariorum*; FAO: Rome, Italy, 2019.
86. iNaturalist Contributors. INaturalist Research-Grade Observations. *P. tepidarioum*. Available online: https://www.inaturalist.org/observations?place_id=7003&quality_grade=research&subview=map&taxon_id=120583&verifiable=any (accessed on 18 February 2024).
87. iNaturalist Contributors. INaturalist Research-Grade Observations. *S. grossa*. Available online: https://www.inaturalist.org/observations?place_id=7003&subview=map&taxon_id=61997 (accessed on 18 February 2024).
88. International Barcode of Life Project (iBOL). International Barcode of Life Project (IBOL). Occurrence Dataset. Available online: https://www.gbif.org/occurrence/4916435761 (accessed on 7 September 2024).
89. Ramírez, M.; Grismado, C.; Blick, T. Notes on the Spider Family Agelenidae in Southern South America (Arachnida: Araneae). *Rev. Ibérica Aracnol.* **2004**, *9*, 179–182.
90. ArachnoTrAC. Arachnids from Tropical Andean Countries. Arañas Tropicales—ArachnoTrAC Initiative (W. Galvis, Ed. 2020–2024) and Research Group in Arachnology and Myriapodology (GAM-UN)—Universidad Nacional de Colombia, Bogotá. Available online: https://Sites.google.com/view/arachnotrac (accessed on 30 August 2024).
91. Roth, V. A Review of the South American Spiders of the Family Agelenidae (Arachnida, Araneae). *Bull. Am. Mus. Nat. Hist.* **1967**, *134*, 297–345.
92. Dupérré, N. Araneae (Spiders) of South America: A Synopsis of Current Knowledge. *N. Zealand J. Zool.* **2022**, *50*, 3–117. [CrossRef]
93. Kuntner, M. A Monograph of Nephilengys, the Pantropical "hermit Spiders" (Araneae, Nephilidae, Nephilinae). *Syst. Entomol.* **2007**, *32*, 95–135. [CrossRef]
94. Müller, H. Spiders from Colombia XVII. Records and Range Extensions of Palpimanidae and Gnaphosidae from Middle and South America (Arachnida: Araneae). *Faun. Abhandl. Dresd.* **1994**, *19*, 141–144.
95. Santos, A.J.; Gonzaga, M.O. On the Spider Genus Oecobius Lucas, 1846 in South America (Araneae, Oecobiidae). *J. Nat. Hist.* **2003**, *37*, 239–252. [CrossRef]
96. Víquez, C.; Longhorn, S. Arachnids of Central America. Available online: http://arachnids.myspecies.info (accessed on 31 August 2024).
97. Harvard University; Morris, P. Museum of Comparative Zoology, Harvard University. Version 162.441. Museum of Comparative Zoology, Harvard University. Occurrence Dataset. *Pholcus phalangioides*. Available online: https://www.gbif.org/occurrence/477899765 (accessed on 31 August 2024).
98. Huber, B. The Pholcid Spiders of Costa Rica (Araneae: Pholcidae). *Rev. Biol. Trop.* **1998**, *45*, 1583–1634.
99. Faúndez, E.I.; Carvajal, M.A.; Darquea-Schettini, D.; González-Cano, E. Nuevos Registros de Steatoda Nobilis (Thorell, 1875) (Araneae: Theridiidae) de Sudamérica. *Rev. Ibérica Aracnol.* **2018**, *33*, 52–54.
100. Kobelt, M.; Nentwig, W. Alien Spider Introductions to Europe Supported by Global Trade. *Divers. Distrib.* **2008**, *14*, 273–280. [CrossRef]
101. Ruiz, G.; Lorda, J.; Arwine, A.; Lion, K. Shipping Patterns Associated with the Panama Canal: Effects on Biotic Exchange? In *Bridging Divides*; Gollasch, S., Galil, B., Cohen, A., Eds.; Springer: Dordrecht, The Netherlands, 2006; p. 315. ISBN 9781402050466.
102. Llacer Montero, F.J. The Panama Canal: Operations and Traffic. *Mar. Policy* **2005**, *29*, 223–234. [CrossRef]
103. Hulme, P.E. Trade, Transport and Trouble: Managing Invasive Species Pathways in an Era of Globalization. *J. Appl. Ecol.* **2009**, *46*, 10–18. [CrossRef]
104. Narimanov, N.; Kempel, A.; Van Kleunen, M.; Entling, M. Unexpected Sensitivity of the Highly Invasive Spider *Mermessus trilobatus* to Soil Disturbance in Grasslands. *Biol. Invasions* **2021**, *23*, 1–6. [CrossRef] [PubMed]
105. Taucare-Ríos, A.; Bizama, G.; Bustamante, R.O. Using Global and Regional Species Distribution Models (SDM) to Infer the Invasive Stage of *Latrodectus geometricus* (Araneae: Theridiidae) in the Americas. *Environ. Entomol.* **2016**, *45*, 1379–1385. [CrossRef] [PubMed]
106. Dawson, W.; Moser, D.; Van Kleunen, M.; Kreft, H.; Pergl, J.; Pyšek, P.; Essl, F. Global Hotspots and Correlates of Alien Species Richness across Taxonomic Groups. *Nat. Ecol. Evol.* **2017**, *1*, 0186. [CrossRef]
107. Mazza, G.; Tricarico, E. *Invasive Species and Human Health*; CABI: Wallingford, UK, 2018.
108. Hoddle, M.S. A New Paradigm: Proactive Biological Control of Invasive Insect Pests. *BioControl* **2024**, *69*, 321–334. [CrossRef]
109. Ewel, J.J.; Putz, F.E. A Place for Alien Species in Ecosystem Restoration. *Front. Ecol.* **2004**, *2*, 354–360. [CrossRef]
110. Houser, J.D.; Ginsberg, H.; Jakob, E.M. Competition between Introduced and Native Spiders (Araneae: Linyphiidae). *Biol. Invasions* **2014**, *16*, 2479–2488. [CrossRef]
111. Burger, J.C.; Patten, M.A.; Prentice, T.R.; Redak, R.A. Evidence for Spider Community Resilience to Invasion by Non-Native Spiders. *Biol. Conserv.* **2001**, *98*, 241–249. [CrossRef]

112. Hogg, B.; Daane, K. Impacts of Exotic Spider Spillover on Resident Arthropod. *Ecol. Entomol.* **2015**, *40*, 69–77. [CrossRef]
113. Cárdenas-Murillo, R.; Posada-Flórez, F.; Bustillo Pardey, A. Daños Causados Por Arañas En Los Cafetales. *Av. Técnicos Cenicafé* **1997**, *242*, 1–4.
114. Sánchez, J.; Pérez-Hidalgo, N.; Medianero, E. Áfidos (Hemiptera: Sternorrhyncha: Aphididae) Interceptados En Vegetales Frescos En Los Puertos de Manzanillo International Terminal y Panama Ports Company (Colón Panamá). *IDESIA* **2022**, *40*, 137–143. [CrossRef]
115. Nentwig, W.; Blick, T.; Bosmans, R.; Gloor, D.; Hänggi, A.; Kropf, C. Spiders of Europe. Available online: https://www.araneae.nmbe.ch (accessed on 2 February 2024).

biology

MDPI

Article

Risk Perception: Chemical Stimuli in Predator Detection and Feeding Behaviour of the Invasive Round Goby *Neogobius melanostomus*

Natalia Z. Szydłowska *, Pavel Franta, Marek Let, Vendula Mikšovská, Miloš Buřič and Bořek Drozd

South Bohemian Research Center of Aquaculture and Biodiversity of Hydrocenoses, Faculty of Fisheries and Protection of Waters, University of South Bohemia in České Budějovice, Zátiší 728/II, 389 25 Vodňany, Czech Republic; pfranta@frov.jcu.cz (P.F.); mlet@frov.jcu.cz (M.L.); vmiksovska@frov.jcu.cz (V.M.); buric@frov.jcu.cz (M.B.); drozd@frov.jcu.cz (B.D.)

* Correspondence: nszydlowska@frov.jcu.cz

Simple Summary: The round goby is considered one of the most widespread Ponto-Caspian invasive species, which poses a threat to ecosystems and food web alterations. While it is known for direct predation, its interactions with native higher predators and its feeding behaviour upon their presence remain unclear. Therefore, we investigated how the exposure to chemical alarm cues emitted by conspecifics and the odour of a native predator, the European eel, affects the round goby's feeding activity. Surprisingly, the gobies did not exhibit any pronounced threat sensitivity to both types of chemical stimuli, which was observed in unchanged food consumption probability and gut evacuation rates. The obtained outcomes suggest boldness, great efficiency in food processing, and a potential competitive advantage over native counterparts when colonising new ecosystems, irrespective of local higher predator presence.

Abstract: The round goby *Neogobius melanostomus* is a notoriously invasive fish originating from the Ponto-Caspian region that in recent decades has successfully spread across the globe. One of its primary impacts is direct predation; in addition, when entering new ecosystems, the round goby is likely to become a food resource for many higher native predators. However, little is known either about the indirect effects of predators on the round goby as prey or its feeding behaviour and activity. The non-consumptive effect of the presence of higher native predators presumably plays an important role in mitigating the impact of non-native round gobies as mesopredators on benthic invertebrate communities, especially when both higher- and mesopredators occupy the same habitat. We tested the food consumption probability and gut evacuation rates in round gobies in response to chemical signals from a higher predator, the European eel *Anguilla anguilla*. Gobies were placed individually in experimental arenas equipped with shelters and exposed to water from a tank in which (a) the higher predator had actively preyed on a heterospecific prey, earthworms *Lumbricus* sp. (the heterospecific treatment; HS); (b) the higher predator had fed on round gobies (the conspecific treatment; CS); or (c) the water was provided as a control treatment (C). To ensure exposure to the chemical stimuli, this study incorporated the application of skin extracts containing damaged-released alarm cues from the CS treatment; distilled water was used for the remaining treatments. No significant differences were observed in either the food consumption probability or gut evacuation rate in the tested treatments. Despite the lack of reaction to the chemical stimuli, round gobies did exhibit high evacuation rates ($R = 0.2323 \pm 0.011\ h^{-1}$; mean $\pm$ SE) in which complete gut clearance occurred within 16 h regardless of the applied treatment. This rapid food processing suggests high efficiency and great pressure on resources regardless of the presence or not of a higher predator. These findings hint at the boldness of round gobies, which did not exhibit any pronounced threat sensitivity. This would seem to suggest great efficiency in food processing and a potential competitive advantage over local native species when colonising new ecosystems, irrespective of the presence of native predators. Our study did not detect any non-consumptive effect attributable to the higher predator, given that the feeding activity of the invasive round goby was not altered.

Keywords: alarm cues; *shreckstoff*; food consumption; gut evacuation rate; predation efficiency; aquatic invasions; non-native species

1. Introduction

Alien invasive species continue to pose a significant threat to biodiversity [1]. Their rising impacts may be attributable in many cases to globalisation [2,3] and progressive global climate changes [4,5] and have serious ecological and economic consequences [6–8]. In recent decades, the Ponto-Caspian region has emerged as a notable source of fresh species introductions throughout Europe [9], which include a significant number of representatives from the Gobiidae (Actinopterygii) family [10]. At least five species have established populations in European inland freshwater ecosystems [11], of which the most widespread and well-known is the round goby *Neogobius melanostomus* (Pallas, 1814). This fish has successfully expanded its range worldwide [12] and now serves as a model species for studying invasions in freshwater ecosystems [13].

As an invasive species, the round goby exhibits numerous specific traits that have facilitated its successful colonisation of new freshwater and marine ecosystems on both sides of the Atlantic Ocean [14–17]. Key among the characteristics facilitating its establishment is its opportunistic feeding strategy [18,19] that guarantees effective food resource exploitation. However, since many introduced species are not only predators but also prey for native species when they enter new ecosystems, the round goby's ability to manage risks posed by unfamiliar predators might also be contributing to its invasive success [20,21].

When entering new ecosystems, the round goby notably disrupts community dynamics and food webs [22], primarily through direct predation on benthic invertebrate communities [23,24], as well as on fish fry and eggs [25,26]. Hence, depending on the round goby's diet shifts during ontogeny [27], as well as seasonal and/or environmental conditions [28–30], different macroinvertebrate assemblages will experience intense predatory pressure leading to significant declines [31–33]. Additionally, intensified foraging on particular taxa and competition for habitat and food resources will further contribute to the depletion of native fish abundance and richness [34–36], thereby leading potentially to their replacement [37]. Moreover, besides its commonly attributed role as a predator, the round goby has become an essential dietary item for numerous native higher-predator fish species [38–40] and in some cases is even the main food item due to its long-term presence in the ecosystem [41].

These predator–prey interactions are crucial drivers of trophic dynamics and structural and functional changes within ecosystems [42]. Recent studies have shown that predators can exert threats and affect prey populations not solely via direct consumption but also indirectly [43,44]. Non-consumptive effects are observed when mesopredators face the trade-off between predation risk and other activities such as, for instance, resource acquisition [45]. Compared to the direct foraging effect, non-consumptive effects appear to be equally or even more significant [46], since they can affect the entire prey population by inducing adaptations of defensive strategies [47]. This can lead to modifications in fish morphology, physiology, and behaviour where, depending on the length of exposure to the risk (acute or chronic), the response of fish will differ [48].

Following that, the early detection of a local predation risk presence represents a keystone ability for prey survival [49,50]. Foragers gather information by perceiving and relying on a variety of stimuli [44,51], with chemical communication believed to be fundamental for risk assessment and prey avoidance [52–55] irrespective of environmental conditions [56]. These stimuli include stress-inducing cues resulting from prey–predator interactions [57]. Both predator and prey specimens can actively release these cues: predator-borne cues include kairomones and digestion-released (dietary) cues recognisable by conspecifics of the consumed prey [58,59], while prey-borne cues derived from epidermic injuries or active foraging serve as pre-attack alarm alerts that warn conspecifics of the threat of a potential attack [53,60–62]. As a result, all signals may lead to both behavioural [63–65] and physiological [59,66,67] anti-predatory responses, which are often specific to a given species [60,67]. Different fish species appear to respond differently to identical chemical pictures [68] or to the type of chemical signal to which they are exposed [69]. All of the abovementioned comprise adaptations that significantly contribute to increasing the prey survival rate [70].

Although significant advances have been made in studying predation on the round goby [71], our understanding of the indirect effects of the presence of higher predators on mesopredators (i.e., the round goby) remains limited. Therefore, in this study, we aimed to investigate whether or not round gobies respond to chemical stimuli and whether or not the presence of the local higher predator, the European eel *Anguilla anguilla* (Linnaeus, 1758), affects gobies' feeding and digestion efficiency. We assessed this by counting the number of prey items consumed and by analysing evacuation rates (R), i.e., the mass of food expelled from the digestive tract per unit of time [72]. This evacuation rate parameter is commonly used in direct in situ methods to quantify food consumption, e.g., [73–75], since it regulates feeding rates [76].

Foragers have to confront trade-offs between predator avoidance and other fitness-related activities [61,77]. This can manifest itself in altered feeding behaviour [78] including shifts in feeding-hiding ratios [63,79], a reduction in food intake in fish exposed to the chemical cues provided by higher predators [80–84], or even complete feeding latency [65,69,85]. Thus, we expected to observe similar reactions in round gobies. In addition to recognising predator odours, we predicted that the round goby would respond sharply to alarm cues derived from conspecific individuals. Therefore, we hypothesised that exposure to chemical cues, i.e., predator odour and conspecific alarm cues, would significantly contribute to a specific anti-predatory response. This response was expected to manifest itself in the form of decreased foraging activity and slower gut evacuation rates due potentially to lower consumption rates [86].

2. Materials and Methods

2.1. Collection and Maintenance of Study Organisms

N. melanostomus (a non-native mesopredator) individuals in a size range 55–75 mm were collected in the river Elbe (Dolní Žleb, North Bohemia, Czech Republic GPS: 50.50335 N, 14.13019 E) by electrofishing (a backpack pulsed-DC electrofishing unit FEG 1500, EFKO, Leutkirch, Germany) (~0.5 m water depth; 0.02–0.49 m s^{-1} water velocity). Fish capture was conducted in a recently established population first documented by Buřič et al. [87]. Fish were transported to the experimental facilities of the Institute of Aquaculture and Protection of Waters (FFPW USB) in České Budějovice where they were acclimatised in a recirculating aquaculture system (RAS) with a total volume of 1600 L for at least one month before the experiments were conducted. Selected physical–chemical parameters (T, O_2, pH) were monitored using daily measurements conducted with an HQ40d digital multimeter (Hach Lange GmbH, Düsseldorf, Germany).

A. anguilla (a native higher predator) were captured at another locality on the river Elbe (Lovosice, North Bohemia, Czech Republic; 50.5165686N, 14.0595781E), where the co-occurrence of these two species—as well as eel predation on round gobies—has been confirmed (obs. pers.). Given the shared habitat [88] and the similar temperature conditions required for growth and consumption [89–91], the European eel was chosen as the higher predator for our experiment. Eels sized 540–840 mm (known to feed on small-size fish species including round gobies; [92,93]) were transported to the facility and stocked in the two separate recirculating systems (RAS; 2 × 1600 L) and were subsequently used in the experiment. For at least one month before the experiment began, the eels were fed on round gobies (first group, representing the conspecific treatment; CS) or on common earthworms *Lumbricus terrestris* (Linnaeus, 1758) (second group, representing the heterospecific treatment; HS) to acclimatise them to the food to be used during the experiment and to ensure adequate prey intake. All the abovementioned physical–chemical parameters were controlled daily.

The experimental food provided for the round gobies included *Chironomus* sp. larvae (frequently observed in their diet, [94,95]) purchased from a pet store and *Gammarus fossarum* (Koch in Panzer, 1835), selected as an alternative dietary source that is easily distinguishable from the tested food even after digestion. The *Gammarus fossarum* individuals (successfully preyed upon by round goby, [96]) were collected from Dobrovodský potok (Třebotovice, South Bohemia, Czech Republic; 48.9507 N, 14.5676 E), frozen on-site using dry ice, and subsequently stored at −80 °C until the beginning of the experiment.

2.2. Preparation of the Alarm Cues

The preparation of the alarm cues used in this study followed the methods described by Wisenden et al. [61]. Since it was unfeasible to guarantee a predation event at a specific time, to ensure the efficacy of the experiment and to strengthen the effects of the alarm cues, we created an extract from round goby skin containing alarm pheromones released from epidermal cells while fish are attacked/injured [97]. This allowed us to simulate the release of alarm signals without relying solely on spontaneous attacks by higher predators. The 30 donor round gobies in a size range of 71–90 mm (81.3 mm ± 5.5 mm, TL ± SD) were sacrificed by a blow to the head to prevent any potential interference in the release of chemical signals that could occur if anaesthetics were used. A skin fillet was carefully removed from both sides of each fish to obtain approximately 130.5 cm^2 of skin devoid of muscle tissue. All skin samples were promptly placed in 200 mL of distilled water, kept chilled on ice to prevent any decomposition until further processing, and homogenised. Subsequently, the homogenised samples were filtered through a glass-wool filter twice to remove any remaining particles (e.g., scales) and then diluted with distilled water to form a solution with a final concentration of 0.1 cm^2 of skin.ml^{-1} [61]. Subsequently, syringes with a volume of 5 mL were filled with the prepared extract and stored in the freezer at −29 °C before use in experiments.

2.3. Experimental Design

Based on the modified methodology described by Richter et al. [98], as well as our own previous experiences, we set up a series of five experimental trials (see below) to examine the gastric evacuation rate and food consumption probability under three different treatments. These treatments involved exposing tested fish to different conditions: (a) conspecific alarm cues and kairomones released by a higher predator fed on round gobies (conspecific treatment; CS); (b) chemical cues released by a higher predator fed on earthworms (heterospecific treatment; HS); and (c) clear water devoid of any chemical stimulants (control treatment; C).

Each treatment was tested in a parallel trial (five repetitions) within a separate recirculation system comprising four vertically arranged tanks, each with a capacity of 400 L (RAS; 4 × 400 L). The uppermost and lowermost tanks were fitted with biological and mechanical filters in each system, respectively. Depending on the treatment, the upper tank was stocked either with seven European eels fed on round gobies (whose numbers were controlled by the end of each experimental trial; CS treatment), seven European eels fed on earthworms (HS treatment), or no European eels (chemical cue-free water; C treatment). The water containing specific chemical cues flowed into the second lower tank, where 11 experimental arenas per treatment were placed, i.e., 33 arenas for each experimental trial were used at the same time. Each experimental arena consisted of an aquarium (total volume of 6.5 L) divided into two parts by a glass wall with a lower steel mesh to maintain the water flow (Figure 1). The tested fish was placed in the larger section (23.0 × 19.5 × 20 cm) equipped with a half-cut pipe serving as a shelter and an upper outlet for water flow. The smaller section (10 × 23 × 20 cm) had an inlet for flowing water (Figure 1). Each aquarium was covered with a matte translucent film to minimise stress and disturbance and to prevent visual interactions between gobies.

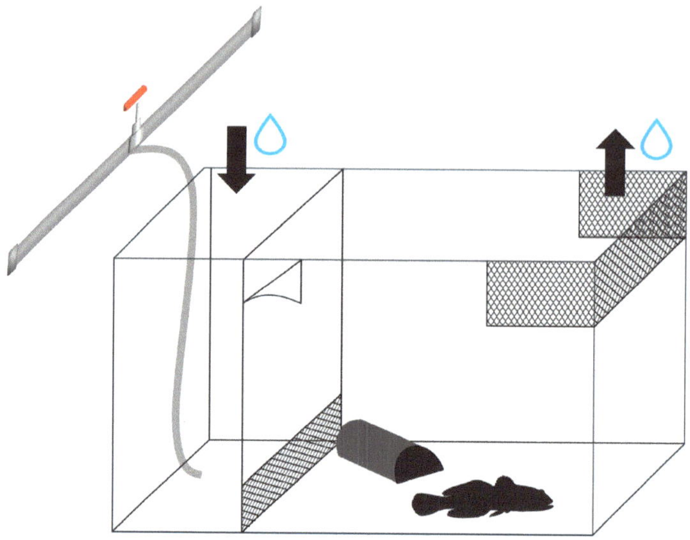

Figure 1. Experimental arena. Black arrows mark water inflow and outflow, respectively.

Before each trial, a 36-h fasting period was imposed on the fish to ensure complete gut clearance and preclude any potential influence of the remains of the acclimatisation food on the experiment outcomes. To prevent possible cannibalism provoked by starvation and to ensure that a particular individual consumed a given amount of food, all fish were placed separately in experimental arenas. The light regime was set to 12 h light–12 h dark, including dusk and dawn simulations. The temperature was maintained at a constant rate throughout the experiment at 21.48 ± 0.026 °C; mean $\pm$ SE. Automatic dataloggers Minikin T (EMS, Brno, Czech Republic) monitored the temperature every two hours.

To obtain the conspecific cues, 15 round gobies were introduced into the tanks where the eels had been feeding before the start of each trial. After each trial, tanks were replenished with gobies to ensure an equal number at the start of each trial. Following a 36-h starvation period, the round gobies were fed with the tested food (Food A) consisting of 12 *Chironomus* sp. larvae with an average whole portion weight of 0.074 g $\pm$ 0.023 g, (W $\pm$ SD) (see Figure 2), which were used to observe the evacuation rate of the fish. After two hours, the tested food remains were removed from each aquarium, which marked the start of the main experimental phase and sampling. Fish that did not consume any test food within this period were excluded from the remaining part of the experiment. At this time, easily distinguishable food, marked as Food B, was given to the fish to observe the evacuation rate while continuous feeding was applied [99]. Simultaneously, the external stimulant, i.e., the predator odour from the upper tank and prepared alarm cues or distilled water, were added to the treatments to observe the consumption probability of food B and the effects of the chemical cues on the evacuation rate of food A. Thus, four individuals of *Gammarus fossarum* (food B) with an average whole portion weight of 0.069 g $\pm$ 0.062 g (W $\pm$ SD) were provided to all experimental round gobies. Tubes were placed in the aquariums to allow clean water or water containing chemicals from the eel tanks to flow in throughout the experiment. Simultaneously, the first dose (5 mL) of alarm cues (for the conspecific treatment) or distilled water (for the heterospecific and control treatments) was released. To ensure continuous exposure to the chemical signals, which have been shown to persist in aquatic environments for up to six hours [100], the prepared solutions were released repeatedly every four hours (Figure 2). The experimental arenas were checked simultaneously to ensure that constant food availability was being maintained.

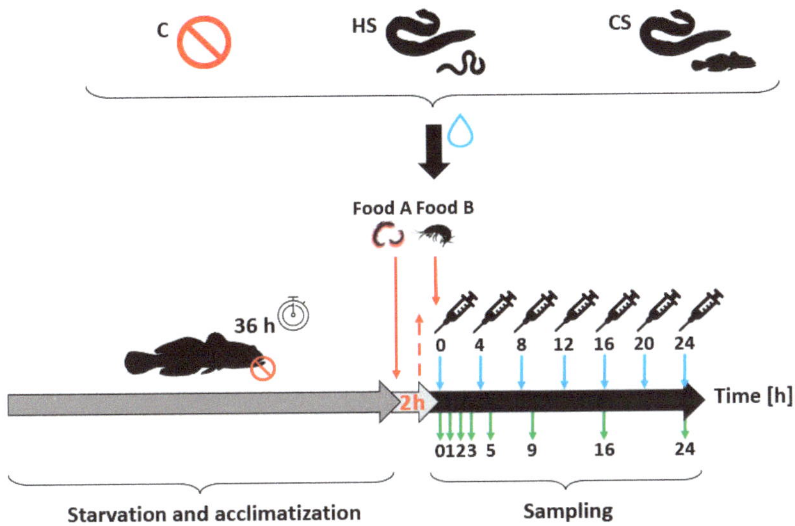

Figure 2. Experimental setup. C, control treatment (supplied by clear water without any chemical cues derived from predator or prey), HS, heterospecific treatment (supplied by water from a tank with predator fed by common earthworms), CS, conspecific treatment (supplied by water from a tank with predator fed by round gobies); solid red lines depict the application of food A (*Chironomus* sp. larvae) and food B (*Gammarus fossarum*); the dashed line represents the removal of the remains of food A; blue arrows represent the hours at which the chemical alarm cues (skin extract from the round goby in CS treatment) or distilled water (in C and HS treatments), depending on the treatment, were applied; green arrows denote the time intervals after feeding with food B at which the tested fish were sampled.

Fish samples were taken at eight time intervals: 0, 1, 2, 3, 5, 9, 16, and 24 h after feeding with food B (Figure 2). In total, we tested 144 fish specimens across five trials of the experiment, with six replications at each time point. When sampling each round goby, the number of prey individuals remaining (food B) in the experimental arena was counted to estimate the total amount of food consumed by the round goby individual over a given time. The collected fish were killed using the anaesthetic MS 222 and kept in a freezer at −80 °C until further analysis, conducted within two weeks of completing the experiment.

2.4. Sample Processing

Before determining the gut contents of the round goby individuals, several measurements were taken, including total length (TL; mm), standard length (SL; mm), body weight with a weighing accuracy of 0.01 g (W; g) (Kern and Sohn GmbH, Balingen, Germany), and sex determination [15]. Subsequently, each fish was dissected, with the gut extraction following the methodology outlined by Manko [101]. The extracted gut/intestine and its contents were weighed to an accuracy of 0.1 mg (EX224M analytical balance, OHaus Corporation, Parsippany, NJ, USA). The length of the gut was measured, and the position of each food type (foods A and B) was observed. Next, the gut contents (food A and B separately) were placed on glass microfibre discs (type 691, 516–0862; VWR International, Radnor, PA, USA) that had been preheated in a laboratory chamber furnace (LAC-Ht40A1; LAC s.r.o., Židlochovice, Czech Republic) to completely decompose all organic carbon molecules (30 min at 500 °C). All the microfibre discs were weighed (Mettler-Toledo Excellence Plus XP6 microbalance; Greifensee, Switzerland) before the gut content was placed upon them. The empty intestine was also weighed and subsequently placed on pre-numbered and weighed aluminium foil. The body of the de-gutted fish (portioned to ease the drying process) was weighed (M3P, Sartorius, Goettingen, Germany) and placed on aluminium foil alongside the empty gut. Finally, samples of the gut contents (food A and B), the fish carcasses, and the empty dissected gut were dried overnight at 105 °C and stored in a desiccator to determine the dry weight (DW) using microbalances.

2.5. Data Analysis

All data analyses and graphical procedures were performed using R software (R Development Core Team, v. 4.1.1, Vienna, Austria).

2.5.1. Assumptions of Observation Independence

We tested the assumption that all 24 design-based groups represented by round goby individuals sampled at eight individual sampling time points for each treatment did not differ either in standard length (SL) or wet weight (WW). Generalised linear models (GLMs) were employed with Gamma distribution and a 'log' link function. In addition, the differences between these design-based groups in the counts of the quantity of food A consumed were performed using a GLM with Poisson distribution and the 'log' link function. Due to the underdispersion detected, a quasi-likelihood estimation method was applied.

2.5.2. Gut Evacuation Rate

The evacuation rate was estimated with statistical modelling of the weight of the prey remaining in the gut over time, whereby the relative gut content (GI) was calculated as a ratio of the dry weight (DW) of the tested food (food A) at the time of sampling vs. fish dry body weight. Subsequently, based on (GI), we computed the evacuation rate (R). This parameter can be described using linear and/or exponential functions [72,74]. Hence, given the uncertainty presented within the literature, we fitted our data to both linear Equation (1) and exponential Equation (2) models, represented by the following:

$$GI_t = GI_0 - Rt \tag{1}$$

$$GI_t = GI_0 e^{-Rt} \tag{2}$$

where GI_t is relative gut content at time t, GI_0 is the initial intestine content at postprandial time t_0 (h), and R is the gastric evacuation rate [74,102].

We fitted the equations using GLMs with Gaussian distributions; 'identity' and 'log' link functions were applied. We tested the relationship between the relative gut content GI used as a response variable and the explanatory variables: (I) 'time' (continuous numerical variable) and (II) 'treatment' (three-level factor). All zero values for the relative gut content were replaced by 0.0014 mg to represent half of the overall minimum gut content mass. The use of a Gaussian distribution was appropriate given that the histogram of model residuals was found to be close to the Gaussian; the assumption of homoscedasticity was also fulfilled. The quality of the model fits was expressed using the McFadden coefficient of determination (R^2). Regression coefficients for each treatment yielded by the model with the higher R^2 were taken as the individual evacuation rates.

2.5.3. Consumption Probability

To test differences between treatments at all time points in the consumption probability of offered prey (food B), we employed a GLM with a binomial distribution with the logistic link function. The proportions of eaten and offered prey were used as a response, and observations were weighted by counts of offered prey. Due to the detected overdispersion, a quasi-likelihood estimation method was applied.

3. Results

A total of 144 round goby individuals (SL = 55.2 ± 2.51 mm; wet weight WW = 3.79 ± 0.56 g; mean ± SD) were tested during five experimental trials. Depending on the treatment, the round gobies were observed to eat 0.023 (C), 0.033 (CS), and 0.036 (HS) g of provided food B, which corresponds to 2.31% (C), 2.12% (CS), and 1.94% (HS) of the wet fish body weight.

3.1. Assumptions of Observation Independence

All 24 design-based groups were completely standardised in terms of round goby SL ($F_{23,120}$ = 0.93, p = 0.56), WW ($F_{23,120}$ = 0.88, p = 0.63), and amount of food A consumed ($F_{23,120}$ = 1.30, p = 0.18).

3.2. Gut Evacuation Rate

The fit of the exponential GLM Equation (2) was better than the fit of the linear GLM Equation (1) when comparing their R^2 values (exponential model: R^2 = 0.77; linear model: R^2 = 0.49). Both linear and exponential GLMs revealed that the interaction between 'treatment × time' (Model 1; Table 1) obtained by modelling the effects on *GI* was not significant (p = 0.61 and p = 0.96, respectively); thus, the evacuation rate was probably not influenced by the 'treatment'. A further exponential GLM with additive effects (Model 2; Table 1) revealed a significant effect for both 'treatment' ($F_{2,140}$ = 4.69, p = 0.01) and 'time' ($F_{1,140}$ = 453.90, p < 0.001) on *GI*. *GI* in CS was 23.7% higher than in C and 9.8% lower in C than in HS (Figure 3). Nevertheless, complete gut clearance in all treatments took 16 h (Figure 3). According to the 95% confidence interval overlap, the evacuation rate was the same (R = 0.2323 ± 0.011 g·g^{-1} fish·h^{-1}; mean ± SE given as an absolute value) in all treatments (Figure 4), which corresponds to the fitted decrease in *GI* per hour (20.7%). Since the linear GLM with additive effects did not reveal any significant effect of the 'treatment' ($F_{2,140}$ = 1.43, p = 0.24) and considering the low effect size of the 'treatment' in exponential GLM Model 2 (Table 1), it is doubtful that the 'treatment' had any impact on the *GI*.

Table 1. Relationships between relative gut content (*GI*) and the explanatory variables, i.e., 'treatment' and 'time'. A generalised linear model (GLM) with Gaussian distribution (with the log link function applied) was used to fit and test the relationships. The significances of all relationships were tested with partial *F*-tests. Asterisks following *p*-values denote significance: * 0.05, *** 0.001.

Model	Predictor	*df*	Deviance	*F* Value	*p*-Value	
Model 1: (*GI*) ~ Treatment × Time						
	Treatment	2	0.0015	3.13	0.047	*
	Time	1	0.0034	1.56	<0.001	***
	Treatment × Time	2	0.0014	0.04	0.964	
Residual *df*: = 138						
Model 2: (*GI*) ~ Treatment + Time						
	Treatment	2	0.0015	4.69	0.011	*
	Time	1	0.0061	453.91	<0.001	***
Residual *df*: = 140						

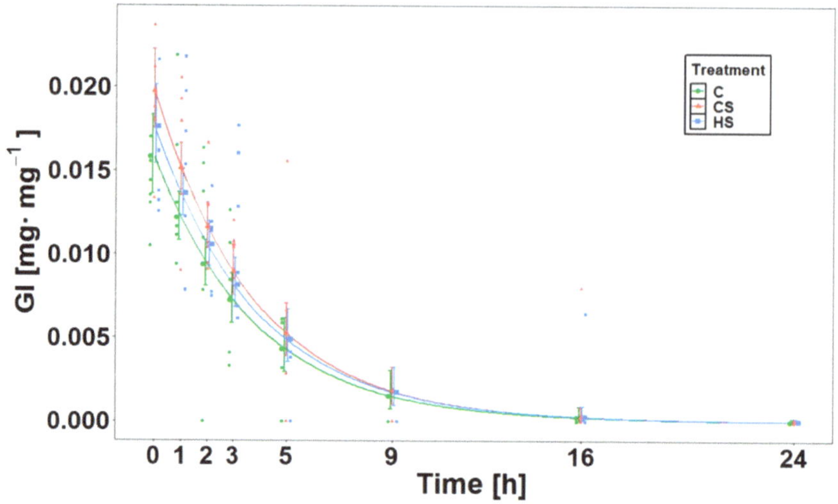

Figure 3. Relative gut content *GI* (computed as DW food A/DW fish) at the time after food application in the round goby (*Neogobius melanostomus*) under three experimental treatments: C, control treatment (no chemical cues); CS, conspecific treatment (conspecific alarm cues and kairomones released by a higher predator fed on round gobies); HS, heterospecific treatment (chemical cues released by a higher predator fed on earthworms).

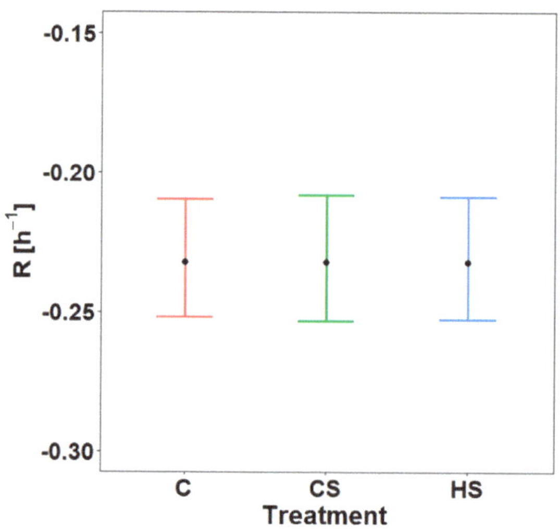

Figure 4. Overlap between the 95% confidence intervals of the gastric evacuation rates (R; mean ± S.D., n = 144) of the tested round gobies under three experimental treatments: C, control treatment (no chemical cues); CS, conspecific treatment (conspecific alarm cues and kairomones released by a higher predator fed on round goby individuals); HS, heterospecific treatment (chemical cues released by a higher predator fed on earthworms).

3.3. Consumption Probability

There were no significant effects of the interaction 'treatment × time' ($F_{12,105} = 0.59$, $p = 0.85$) (Table 2), as well as 'treatment' ($F_{2,117} = 1.86$, $p = 0.16$) or 'time' ($F_{6,117} = 1.20$, $p = 0.31$) (Table 2), indicating that the presence of the chemical stimulants did not significantly influence the probability of food B being consumed.

Table 2. Relationship between food consumption probability of food B (eaten and offered prey was used as a response, and observations were weighted by counts of offered prey) and explanatory variables, i.e., 'treatment' and 'time'. The table depicts two models: one includes the interaction between explanatory variables (Treatment × Time), while the second, additive effects of predictors (Treatment + Time). A generalised linear model (GLM) with quasibinomial distribution was used to fit and test the relationships. The significances of all relationships were tested with partial *F*-tests.

Model	Predictor	*df*	Deviance	*F* Value	*p*-Value
Food B consumption ~ Treatment × Time					
	Treatment	2	347.82	1.0178	0.3649
	Time	6	352.13	0.5607	0.7607
	Treatment × Time	12	364.15	0.5884	0.8473
Residual *df*: = 105					
Food B consumption ~ Treatment + Time					
	Treatment	2	375.73	1.8609	0.1601
	Time	6	386.55	1.1994	0.3115
Residual *df*: = 117					

4. Discussion

Our study revealed no discernible response in the invasive round goby to exposure to either type of chemical stimuli, and the feeding behaviour of the experimental groups subjected to different treatments was unaffected. The results of our experiment did not confirm the expected differences, including a decrease in the tested parameters due either to the presence of the predator's odour or the chemical alarm signals received from injured conspecifics.

Despite the slightly lower amount of food B eaten in the HS treatment compared to the CS treatment (and especially C), all tested round gobies consistently processed and evacuated prey at the same rates per unit of time in all treatments, which is a surprising finding that contradicts our expectations. Considering that food processing (here represented by the evacuation rate) can be influenced by a multitude of factors including food properties [103,104] and predator and prey biometric characteristics [72,105–107], as well as by environmental conditions [108–111], we anticipated some variation between treatments. For instance, the presence of predators and their direct chemical recognition can lead to physiological activity changes [58,112] including alterations in hormone levels such as cortisol [113,114], as occurs in the case of conspecific disturbance or alarm cues [62]. In the context of our study, these physiological responses could be attributable to food intake and gut evacuation rates that are modified as stress-induced behaviour. However, although they did not differ, the treatment and time did seem to have a significant impact on the relative gut content. Nevertheless, considering the (i) small size of the statistical effect of the treatment, (ii) the lack of differences in the counts of consumed prey and fish biometrics between treatments, and (iii) the observation that all tested fish experienced complete gut clearance within 16 h regardless of the applied cues, we consider that the biological effect of treatment on the *GI* was negligible.

As for the evacuation rate, no significant differences were observed in the counts of the consumed prey (food B) (as given by the tested consumption probability). This also contradicts our expectations, as previous studies have demonstrated suppressed feeding activity including decreases in foraging and consumption rates under predator-induced stress conditions [62,87,115,116], as well as feeding latency [86] or even a complete cessation of food intake, as reported in the studies conducted by Giaquinto and Hoffmann [117]. However, since the complete reduction of food intake observed in the pintado catfish *Pseudoplatystoma corruscans* (Spix and Agassiz, 1829) was a result of the exposure to chemosensory information from non-injured conspecifics [117], we should question which alarm signals—i.e., disturbance or *shreckstoff*—are most important in overall conspecific communication. On the other hand, Fraser and Huntingford [118] proposed that foragers may exhibit various responses to danger, with 'risk-reckless' behaviour being one such mechanism whereby individuals ignore exposed hazards while maintaining maximum feeding levels regardless of food abundance. Our study sheds light on the behaviour of round gobies within this framework and suggests that it is an adaptation aiding its invasiveness success [119]. This behaviour matches previous research on round goby boldness [120–122] and how this trait influences its foraging ability [123,124]. It also facilitates the effective exploitation of available food resources with a visible superiority of bolder individuals over shy ones when it comes to resource usage [125] when invading new areas. While bold behaviour is often observed in specimens at the invasion front and is more characteristic of new populations rather than established ones [126], it is notable that bold fish also exhibit a smaller size-at-age compared to shy fish [127], which agrees with the overall size distribution of our tested fish population. Nevertheless, the inability to effectively balance predator avoidance behaviour with other fitness-related activities such as foraging may compromise fish survival and overall fitness, particularly in established populations [61,77,128]. Therefore, it is crucial to undertake further studies examining additional behavioural responses, shelter usage patterns, and potential shifts in round goby activity when encountering predators.

Given the existing evidence—i.e., increased respiratory responses [129]—that round gobies are responsive to chemical stimuli (specifically, injury-released alarm signals from conspecifics), we should examine whether or not the properties of the chemical cues and the intensity of the applied stimulants in our study were sufficient to elicit a response from the tested organisms. Our stimulus preparation followed the methodology outlined by Wisenden et al. [61] that uses the whole skin subjected to mechanical processing and extraction. The issue is whether another method [130] involving skin incisions washed with water could be more effective, as found in responses primarily in minnows. Many previous studies have shown that the magnitude of reactions in fish increases with the concentration of cues—e.g., in goldfish *Carassius auratus* (Linnaeus, 1758) exposed to dietary cues from a predator [131]—where variations in concentrations may indicate different degrees of threat [132]. Moreover, it seems that the frequency of the application of the alarm substance might also influence outcomes if we compare our results to those from studies of the behavioural response of the monkey goby *Neogobius fluviatilis* (Pallas, 1814) [133]. Continuous dosing with damage-released alarm cues (skin extract) in these studies may have given more potent effects than in our experiment in which the extract was only applied every four hours. However, we believe that our application time interval was sufficient since Hazlett [100] has reported that alarm cues fade from water within six hours; other studies have shown that the biological activities of injury-released chemical alarm cues in minnows and amphipods last for 3–6 h at 18 °C [134]. Even so, the round goby could be a less sensitive species characterised by a higher threshold regarding concentrations of alarm cues and/or the frequency of stimuli required to activate an expected response. Hence, other characteristics such as the persistence of the stimulus should be taken into account since multiple environmental conditions may influence particle breakdown and the degradation of alarm cues (solar radiation, temperature, pH, or oxygen levels) [130]. However, these factors may not be applicable to our study since we conducted them under laboratory conditions in which all parameters were maintained constant throughout the experiment. Since Chivers et al. [130] observed a decrease in the strength of the reaction intensity after 30 min, it seems that the initial application of the stimulants is key and each further application could acclimatise fish further to the chemical signal.

The sensitivity of fish in perceiving the threat signal is thus probably key. Responses to olfactory cues require the ability to perceive the specific intensity (concentration) of the odorants that compose the stimuli [135]. The sensitivity of olfactory receptors to specific chemicals varies between species, with observed specific threshold concentrations for cues [68] and fish at individual levels [136]. Specific variation is closely related to the anatomy and morphology of the olfactory organ [137], and a histochemical study has confirmed the widespread presence of olfactory sensory neurons in the peripheral olfactory organ, which underlines the crucial role of chemosensory signalling in round goby survival [138]. Thus, we anticipated that our tested fish would respond to the odour of predators as well as to alarm cues. Nonetheless, the whole Gobiidae family (Actinopterygii), which is known for the diversified production of alarm substances [139], exhibits variations in responses to chemical alarm communications despite the presence of epidermal club cells associated with such responses [140,141], a rule that our research seems to confirm. Thus, we believe that along with the perception threshold, another possible explanation of our observations may relate to the fact that chemical cue recognition in round gobies requires more experience and learning, as has been observed in many other fish [81,142]. Utne-Palm [143] claims that the recognition of odour in two-spotted gobies *Gobiusculus flavescens* (Fabricius, 1779) requires experience and that predator identities can be rapidly learnt by associating the smell of a predator with those of a prey chemical cue [144].

Regardless of the production of alarm substances, we did still expect to observe a reaction to the predator odour in the heterospecific treatment since the used prey was not in any way phylogenetically related to or belonged to the same prey guild [70,145–158]. However, unlike many invertebrates and amphibians, most fish are innately unable to recognise their predators' kairomones, the exception being some groups such as salmonids [49,149]. Kairomones become associated with risk through direct exposure to alarm cues during a predatory attack or through the detection of dietary cues through passive [150], or even active [151], sampling behaviour during predator inspection. The ability of fish to respond to predators may not necessarily result from their ability to detect predator odours directly but rather from their ability to detect chemical alarm signals produced by conspecifics or heterospecifics in their diets. This has been confirmed in research on the pearl cichlid *Geophagus brasiliensis* (Quoy and Gaimard, 1824) by Arvigo et al. [69], which showed that predator odours did not induce any antipredator response. Similarly, the frillfin goby *Bathygobius soporator* (Valenciennes, 1837) is known to be indifferent to predator odours, although conspecific cues do affect its ventilation rate activity [66]. When exposed to predator odours and visual cues, the two-spotted goby likewise does not exhibit any innate anti-predator response [143].

On the other hand, research has demonstrated that even a single exposure to predator cues can enhance the learning process, thereby enabling fish to retain recognition of predators for extended periods and providing survival benefits. For instance, brook trout *Salvelinus fontinalis* (Mitchill, 1814) trained to recognise predators such as the chain pickerel *Esox niger* (Lesueur, 1818) have higher survival rates than untrained individuals [70]. Furthermore, learning could also play a critical role in acquiring experience with particular fish species and is required if the expected responses to predator cues are to be generated. However, given that the round goby and the European eel co-exist in the round goby source locality, we anticipate that a certain experience level would already have been established.

Our findings indicate that round gobies did not show any response to chemical stimuli from predatory odours combined with alarm substances, particularly those from conspecifics. This contradicts all expectations, as some fish species react to alarm substances produced by other species [152], with the most vital reactions typically provided by conspecific alarm substances [153]. However, little is known about fish anti-predator responses when predator and alarm cues are combined [69]. To date, anti-predator responses to the stimulants tested separately have been extensively studied in numerous fish species [60,65,154,155] showing that different types of chemical stimulants often induce different and specific responses [69]. Given this consideration, our approach using both chemical stimulants together may represent a limitation, so examining each stimulant separately could potentially provide a more complete understanding of the round goby's chemical perceptions since one of the chemical cues might mitigate the impact of the other (although in this case, we believe it to be unlikely). However, it is essential to acknowledge that under natural conditions, chemical signals consist of a complex mixture of compounds conveying specific risk information [136,140]. Signals transmitted by predators combined with cues released during the digestion process can provide further detailed knowledge regarding the abundance, location, and recent feeding habits of predators in their environment [56]. Therefore, our method produced robust results, which may agree well with field-related conditions.

Most importantly, although our study did not reveal any changes in the food processing consumed between the tested groups, it is essential to emphasise that round gobies maintained consistently high evacuation rates regardless of the presence of higher predators. Combined with a previous study [99], these findings confirm the great efficiency of round gobies in processing and digesting food when continuously fed, particularly on chironomids. Given that evacuation rates closely regulate feeding rates [76], the rapid evacuation of chironomid larvae suggests potentially high predatory pressure on these macrozoobenthos taxa, present in the round goby diet [31,95,96], which could lead to modifications in benthic community composition and reduced resource availability for other consumers [23]. This trait, allied with the boldness of the round goby—i.e., obviously low sensitivity to predation danger—might suggest a potential competitive advantage over other local native species when colonising new ecosystems. For instance, compared to the ruffe *Gymnocephalus cernua* (Linnaeus, 1758), whose population decline has been linked to the appearance of this goby [36,156] as well as to a temporal collapse in ruffe populations [35,157], the round goby evacuates chironomid larvae faster under similar thermal conditions as those used in our studies. Moreover, the high predator-threat sensitivity in ruffes [158] suggests that the round goby could outcompete other species for food resources when introduced into a new environment, irrespective of the threat from local top predators.

Our results indicate that other stimulants such as visual perception might be relevant in the sensory hierarchy of this fish and play the role of a starter in combination with chemical cues, as previously confirmed [143,159]. Interestingly, some studies conducted on other gobiids, e.g., two-spotted gobies exposed to both visual and chemical stimuli from their natural predators, have proven that in some of the cases, additional visual predator exposures can sometimes strengthen anti-predator responses to chemical cues [143]. This confirms the importance of the interaction between both types of stimulants, as has been reported in other studies [63,84,85]. Thus, we believe that further studies involving visual cues and different concentrations of chemical stimuli and application methods should be prioritised to obtain a better overview of round goby feeding behaviour and prey–predator dynamics within local ecosystems.

Since fish, including round gobies, use olfaction not only to avoid predators but also to find food, orientate themselves, and reproduce [160–162], we anticipated that our tested fish would actively employ olfaction for predator recognition. The failure of our hypothesis underscores the complexity of predator–prey interactions and adaptation mechanisms. While many aquatic animals recognise predators via chemical cues and respond adaptively [163], in invasive or introduced species this ability seems to be reduced, which can provide an advantage when exploiting local resources. This can potentially also lead to competitive disadvantages in predator-rich environments, but if the visual cues permit a rapid response to a threat, the cost may not be so great. Therefore, understanding the perception and management of predation threats by invasive species deepens our insight into the factors driving the rapid invasion of unfamiliar habitats by some species, while others do not exhibit such behaviours.

5. Conclusions

In conclusion, our study sheds light on the behaviour of the invasive round goby concerning its response to chemical stimuli from predators and injured conspecifics. Contrary to expectations, round gobies did not exhibit discernible reactions to these cues, maintaining consistent feeding behaviour regardless of predator presence. This suggests species-specific thresholds in perceiving chemical cues, highlighting the complexity of predator–prey interactions and adaptation mechanisms. Further research involving visual cues and different concentrations of chemical stimuli should be prioritized to understand better round goby feeding efficiency and behaviour while including prey–predator dynamics in invaded ecosystems. Understanding the perception and management of predation threats by invasive species is crucial for elucidating factors driving their rapid invasion of new habitats, thus contributing to more effective conservation and management strategies.

Author Contributions: B.D. and P.F. research conceptualization; P.F. and B.D., sample design and methodology; N.Z.S., P.F. and V.M. investigation and data collection; M.L. and N.Z.S. data analysis and interpretation; M.B. funding provision; N.Z.S. writing—original draft; M.B., B.D., M.L. and N.Z.S. writing—review and editing. All authors have read and agreed to the published version of the manuscript.

Funding: Laboratory experiments were performed using the facilities of the CENAKVA research centre (LM 2023038) in accordance with the objectives of the European consortium DANUBIUS RI.

Institutional Review Board Statement: This study did not involve any endangered or protected species. All experimental manipulation was conducted according to the principles of the Institutional Animal Care and Use Committee (IACUC) of the University of South Bohemia, Faculty of Fisheries and Protection of Waters, Research Institute of Fish Culture and Hydrobiology, Vodňany, based on the EU harmonised animal welfare act of the Czech Republic. The above-mentioned Ethical Committee (IACUC) specifically approved this study (10/11/2022). The principles of laboratory animal care and national laws 246/1992 and regulations on animal welfare were followed at all times (Ref. number 22761/2009-17210).

Informed Consent Statement: Not applicable.

Data Availability Statement: The data presented in this study are available on request from the corresponding author.

Conflicts of Interest: The authors declare no conflicts of interest.

References

1. Pyšek, P.; Hulme, P.E.; Simberloff, D.; Bacher, S.; Blackburn, T.M.; Carlton, J.T.; Dawson, W.; Essl, F.; Foxcroft, L.C.; Genovesi, P.; et al. Scientists' warning on invasive alien species. *Biol. Rev.* **2020**, *95*, 1511–1534. [CrossRef] [PubMed]
2. Seebens, H.; Blackburn, T.M.; Dyer, E.E.; Genovesi, P.; Hulme, P.E.; Jeschke, J.M.; Pagad, S.; Essl, F. No saturation in the accumulation of alien species worldwide. *Nat. Commun.* **2017**, *8*, 14435. [CrossRef] [PubMed]
3. Seebens, H.; Bacher, S.; Blackburn, T.M.; Capinha, C.; Dawson, W.; Dullinger, S.; Pyšek, P.; Winter, M.; Arianoitsou, M.; Bacher, S.; et al. Projecting the continental accumulation of alien species through to 2050. *Glob. Chang. Biol.* **2021**, *27*, 970–982. [CrossRef] [PubMed]
4. Penk, M.R.; Jeschke, J.M.; Minchin, D.; Donohue, I. Warming can enhance invasion success through asymmetries in energetic performance. *J. Anim. Ecol.* **2016**, *85*, 419–426. [CrossRef] [PubMed]
5. Bellard, C.; Jeschke, J.M.; Leroy, B.; Mace, G.M. Insights from modeling studies on how climate change affects invasive alien species geography. *Ecol. Evol.* **2018**, *8*, 5688–5700. [CrossRef] [PubMed]
6. Diagne, C.; Leroy, B.; Vaissière, A.C.; Gozlan, R.E.; Roiz, D.; Jarić, I.; Salles, J.M.; Bradshaw, C.J.A.; Courchamp, F. High and rising economic costs of biological invasions worldwide. *Nature* **2021**, *592*, 571–576. [CrossRef] [PubMed]
7. Haubrock, P.J.; Cuthbert, R.N.; Ricciardi, A.; Diagne, C.; Courchamp, F. Economic costs of invasive bivalves in freshwater ecosystems. *Divers. Distrib.* **2022**, *28*, 1010–1021. [CrossRef]
8. Vilà, M.; Hulme, P.E. (Eds.) *Impact of Biological Invasions on Ecosystem Services*; Springer International Publishing: Cham, Switzerland, 2017; Volume 12.
9. Soto, I.; Cuthbert, R.N.; Ricciardi, A.; Ahmed, D.A.; Altermatt, F.; Schäfer, R.B.; Archambaud-Suard, G.; Bonada, N.; Cañedo-Argüelles, M.; Csabai, Z.; et al. The faunal Ponto-Caspianization of central and western European waterways. *Biol. Invasions* **2023**, *25*, 2613–2629. [CrossRef]
10. Neilson, M.E.; Stepien, C.A. Escape from the Ponto-Caspian: Evolution and biogeography of an endemic goby species flock (Benthophilinae: Gobiidae: Teleostei). *Mol. Phylogenetics Evol.* **2009**, *52*, 84–102. [CrossRef]
11. Copp, G.H.; Bianco, P.G.; Bogutskaya, N.G.; Erős, T.; Falka, I.; Ferreira, M.T.; Fox, M.G.; Freyhof, J.; Gozlan, R.E.; Grabowska, J.; et al. To be, or not to be, a non-native freshwater fish? *J. Appl. Ichthyol.* **2005**, *21*, 242–262. [CrossRef]
12. Corkum, L.D.; Sapota, M.R.; Skora, K.E. The round goby, *Neogobius melanostomus*, a fish invader on both sides of the Atlantic Ocean. *Biol. Invasions* **2004**, *6*, 173–181. [CrossRef]
13. Cerwenka, A.F.; Brandner, J.; Dashinov, D.; Geist, J. Small but mighty: The round goby (*Neogobius melanostomus*) as a model species of biological invasions. *Diversity* **2023**, *15*, 528. [CrossRef]

14. Sapota, M.R. The round goby (*Neogobius melanostomus*) in the Gulf of Gdańsk—A species introduction into the Baltic Sea. In *Biology of the Baltic Sea: Proceedings of the 17th BMB Symposium, Stockholm, Sweden, 25–29 November 2001*; Developments in Hydrobiology; Kautsky, H., Snoeijs, P., Eds.; Springer: Dordrecht, The Netherlands, 2004; Volume 176, pp. 219–224. [CrossRef]
15. Kornis, M.S.; Mercado-Silva, N.; Vander Zanden, M.J. Twenty years of invasion: A review of round goby *Neogobius melanostomus* biology, spread and ecological implications. *J. Fish Biol.* **2012**, *80*, 235–285. [CrossRef] [PubMed]
16. Masson, L.; Masson, G.; Beisel, J.N.; Gutowsky, L.F.G.; Fox, M.G. Consistent life history shifts along invasion routes? An examination of round goby populations invading on two continents. *Divers. Distrib.* **2018**, *24*, 841–852. [CrossRef]
17. Christensen, E.A.; Norin, T.; Tabak, I.; van Deurs, M.; Behrens, J.W. Effects of temperature on physiological performance and behavioral thermoregulation in an invasive fish, the round goby. *J. Exp. Biol.* **2021**, *224*, jeb237669. [CrossRef]
18. Diggins, T.P.; Kaur, J.; Chakraborti, R.K.; DePinto, J.V. Diet choice by the exotic round goby (*Neogobius melanostomus*) as influenced by prey motility and environmental complexity. *J. Great Lakes Res.* **2002**, *28*, 411–420. [CrossRef]
19. Copp, G.H.; Kováč, V.; Zweimüller, I.; Dias, A.; Nascimento, M.; Balážová, M. Preliminary study of dietary interactions between invading Ponto-Caspian gobies and some native fish species in the River Danube near Bratislava (Slovakia). *Aquat. Invasions* **2008**, *3*, 193–200. [CrossRef]
20. Penk, M.; Saul, W.C.; Dick, J.T.A.; Donohue, I.; Alexander, M.E.; Linzmaier, S.; Jeschke, J.M. A trophic interaction framework for identifying the invasive capacity of novel organisms. *Methods Ecol. Evol.* **2017**, *8*, 1786–1794. [CrossRef]
21. Saul, W.C.; Jeschke, J.M.; Heger, T. The role of eco- evolutionary experience in invasion success. *NeoBiota* **2013**, *17*, 57–74. [CrossRef]
22. Morkūnė, R.; Tomczak, M.T.; Bacevičius, E.; Gasiūnaitė, Z.R. Changes in the Baltic Sea coastal food web: A case study on the invasion of Round goby *Neogobius melanostomus* (Pallas, 1814). *Estuar. Coast. Shelf Sci.* **2024**, *296*, 108591. [CrossRef]
23. Mikl, L.; Adámek, Z.; Všetičková, L.; Janáč, M.; Roche, K.; Šlapanský, L.; Jurajda, P. Response of benthic macroinvertebrate assemblages to round (*Neogobius melanostomus*, Pallas 1814) and tubenose (*Proterorhinus semilunaris*, Heckel 1837) goby predation pressure. *Hydrobiologia* **2017**, *785*, 219–232. [CrossRef]
24. Oesterwind, D.; Bock, C.; Förster, A.; Gabel, M.; Henseler, C.; Kotterba, P.; Mengeb, M.; Myts, D.; Winkler, H.M. Predator and prey: The role of the round goby *Neogobius melanostomus* in the western Baltic. *Mar. Biol. Res.* **2017**, *13*, 188–197. [CrossRef]
25. Vašek, M.; Všetičková, L.; Roche, K.; Jurajda, P. Diet of two invading gobiid species (*Proterorhinus semilunaris* and *Neogobius melanostomus*) during the breeding and hatching season: No field evidence of extensive predation on fish eggs and fry. *Limnologica* **2014**, *46*, 31–36. [CrossRef]
26. Lutz, E.; Hirsch, P.E.; Bussmann, K.; Wiegleb, J.; Jermann, H.P.; Muller, R.; Burkhardt-Holm, P.; Adrian-Kalchhauser, I. Predation on native fish eggs by invasive round goby revealed by species-specific gut content DNA analyses. *Aquat. Conserv. Mar. Freshw. Ecosyst.* **2020**, *30*, 1566–1577. [CrossRef]
27. Dashinov, D.; Uzunova, E. Diet and feeding strategies of round goby, *Neogobius melanostomus* (Pallas, 1814) from the invasion front in the Danube River tributaries (Bulgaria): Ontogenetic shift and seasonal variation. *Limnologica* **2020**, *83*, 125796. [CrossRef]
28. Borcherding, J.; Dolina, M.; Heermann, L.; Knutzen, P.; Krüger, S.; Matern, S.; van Treeck, R.; Gertzen, S. Feeding and niche differentiation in three invasive gobies in the Lower Rhine, Germany. *Limnologica* **2013**, *43*, 49–58. [CrossRef]
29. Bhagat, Y.; Ruetz III, C.R.; Akins, A.L. Differential habitat use by the round goby (*Neogobius melanostomus*) and Dreissena spp. in coastal habitats of eastern Lake Michigan. *J. Great Lakes Res.* **2015**, *41*, 1087–1093. [CrossRef]
30. Števove, B.; Kováč, V. Ontogenetic variations in the diet of two invasive gobies, *Neogobius melanostomus* (Pallas, 1814) and *Ponticola kessleri* (Günther, 1861), from the middle Danube (Slovakia) with notice on their potential impact on benthic invertebrate communities. *Sci. Total Environ.* **2016**, *557*, 510–519. [CrossRef] [PubMed]
31. Barton, D.R.; Johnson, R.A.; Campbell, L.; Petruniak, J.; Patterson, M. Effects of round gobies (*Neogobius melanostomus*) on dreissenid mussels and other invertebrates in eastern Lake Erie, 2002–2004. *J. Great Lakes Res.* **2005**, *31*, 252–261. [CrossRef]
32. Lederer, A.M.; Janssen, J.; Reed, T.; Wolf, A. Impacts of the introduced round goby (*Apollonia melanostoma*) on dreissenids (*Dreissena polymorpha* and *Dreissena bugensis*) and on macroinvertebrate community between 2003 and 2006 in the littoral zone of Green Bay, Lake Michigan. *J. Great Lakes Res.* **2008**, *34*, 690–697. [CrossRef]
33. Henseler, C.; Oesterwind, D.; Kotterba, P.; Nordström, M.C.; Snickars, M.; Törnroos, A.; Bonsdorff, E. Impact of round goby on native invertebrate communities-An experimental field study. *J. Exp. Mar. Biol. Ecol.* **2021**, *541*, 151571. [CrossRef]
34. Ustups, D.; Bergström, U.; Florin, A.B.; Kruze, E.; Zilniece, D.; Elferts, D.; Knospina, E.; Uzars, D. Diet overlap between juvenile flatfish and the invasive round goby in the central Baltic Sea. *J. Sea Res.* **2016**, *107*, 121–129. [CrossRef]
35. Jůza, T.; Blabolil, P.; Baran, R.; Bartoň, D.; Čech, M.; Draštík, V.; Frouzová, J.; Holubová, M.; Ketelaars, H.A.M.; Kočvara, L.; et al. Collapse of the native ruffe (*Gymnocephalus cernua*) population in the Biesbosch lakes (The Netherlands) owing to round goby (*Neogobius melanostomus*) invasion. *Biol. Invasions* **2018**, *20*, 1523–1535. [CrossRef]
36. Ramler, D.; Keckeis, H. Occurrence of non-native fishes in the Danube east of Vienna (Austria) and potential interactions of invasive gobiids with native fishes. *J. Appl. Ichthyol.* **2019**, *35*, 850–862. [CrossRef]
37. Le Hen, G.; Balzani, P.; Haase, P.; Kouba, A.; Liu, C.; Nagelkerke, L.A.; Theissen, N.; Renault, D.; Soto, I.; Haubrock, P.J. Alien species and climate change drive shifts in a riverine fish community and trait compositions over 35 years. *Sci. Total Environ.* **2023**, *867*, 161486. [CrossRef] [PubMed]
38. Jacobs, P.; Hoedemakers, K. The round goby *Neogobius melanostomus* (Pallas, 1814) (Perciformes: Gobiidae), an invasive species in the Albert Canal (Belgium). *Belg. J. Zool.* **2013**, *143*, 148–153. [CrossRef]

39. Taraborelli, A.C.; Fox, M.G.; Johnson, T.B.; Schaner, T. Round goby (*Neogobius melanostomus*) population structure, biomass, prey consumption and mortality from predation in the Bay of Quinte, Lake Ontario. *J. Great Lakes Res.* **2010**, *36*, 625–632. [CrossRef]
40. Mikl, L.; Adámek, Z.; Roche, K.; Všetičková, L.; Šlapanský, L.; Jurajda, P. Invasive Ponto-Caspian gobies in the diet of piscivorous fish in a European lowland river. *Fundam. Appl. Limnol.* **2017**, *190*, 157–171. [CrossRef]
41. Bruestle, E.L.; Karboski, C.; Hussey, A.; Fisk, A.T.; Mehler, K.; Pennuto, C.; Gorsky, D. Novel trophic interaction between lake sturgeon (*Acipenser fulvescens*) and non-native species in an altered food web. *Can. J. Fish. Aquat. Sci.* **2019**, *76*, 6–14. [CrossRef]
42. Batabyal, A. Predator–prey systems as models for integrative research in biology: The value of a non-consumptive effects framework. *J. Exp. Biol.* **2023**, *226*, jeb245851. [CrossRef]
43. Lima, S.L. Nonlethal effects in the ecology of predator-prey interactions. *Bioscience* **1998**, *48*, 25–34. [CrossRef]
44. Mitchell, M.D.; Harborne, A.R. Non-consumptive effects in fish predator–prey interactions on coral reefs. *Coral Reefs* **2020**, *39*, 867–884. [CrossRef]
45. Kindinger, T.L.; Albins, M.A. Consumptive and non-consumptive effects of an invasive marine predator on native coral-reef herbivores. *Biol. Invasions* **2017**, *19*, 131–146. [CrossRef]
46. Creel, S.; Christianson, D. Relationships between direct predation and risk effects. *Trends Ecol. Evol.* **2008**, *23*, 194–201. [CrossRef] [PubMed]
47. Preisser, E.L.; Bolnick, D.I. The many faces of fear: Comparing the pathways and impacts of nonconsumptive predator effects on prey populations. *PLoS ONE* **2008**, *3*, e2465. [CrossRef] [PubMed]
48. Madin, E.M.; Gaines, S.D.; Warner, R.R. Field evidence for pervasive indirect effects of fishing on prey foraging behavior. *Ecology* **2010**, *91*, 3563–3571. [CrossRef] [PubMed]
49. Berejikian, B.A.; Tezak, E.P.; LaRae, A.L. Innate and enhanced predator recognition in hatchery-reared chinook salmon. *Environ. Biol. Fishes* **2003**, *67*, 241–251. [CrossRef]
50. Sih, A.; Bolnick, D.I.; Luttbeg, B.; Orrock, J.L.; Peacor, S.D.; Pintor, L.M.; Preisser, E.; Rehage, J.S.; Vonesh, J.R. Predator–prey naïveté, antipredator behavior, and the ecology of predator invasions. *Oikos* **2010**, *119*, 610–621. [CrossRef]
51. Dall, S.R. Managing risk: The perils of uncertainty. In *Evolutionary Behavioral Ecology*; Westneat, D.F., Fox, C.W., Eds.; Oxford University Press: Oxford, UK, 2010; pp. 194–206.
52. Smith, R.J.F. The evolution of chemical alarm signals in fishes. In *Chemical Signals in Vertebrates 4*; Springer: Boston, MA, USA, 1986; pp. 99–115.
53. Chivers, D.P.; Smith, R.J.F. Chemical alarm signalling in aquatic predator-prey systems: A review and prospectus. *Ecoscience* **1998**, *5*, 338–352. [CrossRef]
54. Kats, L.B.; Dill, L.M. The scent of death: Chemosensory assessment of predation risk by prey animals. *Ecoscience* **1998**, *5*, 361–394. [CrossRef]
55. Ferrari, M.C.O.; Wisenden, B.D.; Chivers, D.P. Chemical ecology of predator-prey interactions in aquatic ecosystems: A review and prospectus. *Can. J. Zool.* **2010**, *88*, 698–724. [CrossRef]
56. Alemadi, S.; Wisenden, B. Antipredator response to injury-released chemical alarm cues by convict cichlid young before and after independence from parental protection. *Behaviour* **2002**, *139*, 603–611. [CrossRef]
57. Hettyey, A.; Tóth, Z.; Thonhauser, K.E.; Frommen, J.G.; Penn, D.J.; Van Buskirk, J. The relative importance of prey-borne and predator-borne chemical cues for inducible antipredator responses in tadpoles. *Oecologia* **2015**, *179*, 699–710. [CrossRef] [PubMed]
58. Burks, R.L.; Lodge, D.M. Cued in: Advances and opportunities in freshwater chemical ecology. *J. Chem. Ecol.* **2002**, *28*, 1901–1917. [CrossRef] [PubMed]
59. Miyai, C.A.; Sanches, F.H.C.; Pinho-Neto, C.F.; Barreto, R.E. Effects of predator odour on antipredator responses of Nile tilapia. *Physiol. Behav.* **2016**, *165*, 22–27. [CrossRef] [PubMed]
60. Brown, G.E. Learning about danger: Chemical alarm cues and local risk assessment in prey fishes. *Fish Fish.* **2003**, *4*, 227–234. [CrossRef]
61. Wisenden, B.D. Chemically mediated strategies to counter predation. In *Sensory Processing in Aquatic Environments*; Collin, S.P., Marshall, N.J., Eds.; Springer: New York, NY, USA, 2003; pp. 236–251. [CrossRef]
62. Barcellos, L.J.G.; Volpato, G.L.; Barreto, R.E.; Coldebella, I.; Ferreira, D. Chemical communication of handling stress in fish. *Physiol. Behav.* **2011**, *103*, 372–375. [CrossRef] [PubMed]
63. Lehtiniemi, M. Swim or hide: Predator cues cause species specific reactions in young fish larvae. *J. Fish Biol.* **2005**, *66*, 1285–1299. [CrossRef]
64. Milano, D.; Lozada, M.; Zagarese, H.E. Predator-induced reaction patterns of landlocked *Galaxias maculatus* to visual and chemical cues. *Aquat. Ecol.* **2010**, *44*, 741–748. [CrossRef]
65. Sanches, F.H.C.; Miyai, C.A.; Pinho-Neto, C.F.; Barreto, R.E. Stress responses to chemical alarm cues in Nile tilapia. *Physiol. Behav.* **2015**, *149*, 8–13. [CrossRef]
66. Pereira, R.T.; Leutz, J.D.A.C.M.; Valença-Silva, G.; Barcellos, L.J.G.; Barreto, R.E. Ventilation responses to predator odors and conspecific chemical alarm cues in the frillfin goby. *Physiol. Behav.* **2017**, *179*, 319–323. [CrossRef] [PubMed]
67. Brown, G.E.; Rive, A.C.; Ferrari, M.C.; Chivers, D.P. The dynamic nature of antipredator behavior: Prey fish integrate threat-sensitive antipredator responses within background levels of predation risk. *Behav. Ecol. Sociobiol.* **2006**, *61*, 9–16. [CrossRef]
68. Jónsson, L. Chemical stimuli: Role in the behavior of fishes. In *Environmental Physiology of Fishes*; NATO Advanced Study Institutes Series; Ali, M.A., Ed.; Springer: Boston, MA, USA, 1980; Volume 35, pp. 353–367. [CrossRef]

69. Arvigo, A.L.; Miyai, C.A.; Sanches, F.H.; Barreto, R.E.; Costa, T.M. Combined effects of predator odor and alarm substance on behavioral and physiological responses of the pearl cichlid. *Physiol. Behav.* **2019**, *206*, 259–263. [CrossRef] [PubMed]
70. Mirza, R.S.; Chivers, D.P. Are chemical alarm cues conserved within salmonid fishes? *J. Chem. Ecol.* **2001**, *27*, 1641–1655. [CrossRef] [PubMed]
71. Hirsch, P.E.; N'Guyen, A.; Adrian-Kalchhauser, I.; Burkhardt-Holm, P. What do we really know about the impacts of one of the 100 worst invaders in Europe? A reality check. *Ambio* **2016**, *45*, 267–279. [CrossRef] [PubMed]
72. Bromley, P.J. The role of gastric evacuation experiments in quantifying the feeding rates of predatory fish. *Rev. Fish Biol. Fish.* **1994**, *4*, 36–66. [CrossRef]
73. Bajkov, A.D. How to estimate the daily food consumption of fish under natural conditions. *Trans. Am. Fish. Soc.* **1935**, *65*, 288–289. [CrossRef]
74. Elliott, J.M.; Persson, L. The estimation of daily rates of food consumption for fish. *J. Anim. Ecol.* **1978**, *47*, 977–991. [CrossRef]
75. Jobling, M. Mathematical models of gastric emptying and the estimation of daily rates of food consumption for fish. *J. Fish Biol.* **1981**, *19*, 245–257. [CrossRef]
76. Hedden, S.C.; Gido, K.B.; Hedden, C.K.; Pennock, C.A.; Duran, B.R.; Hines, B.A.; Gilbert, E.I.; McKinstry, M.C.; Durst, S.L.; Franssen, N.R. Quantifying consumption of native fishes by nonnative Channel Catfish in a desert river. *N. Am. J. Fish. Manag.* **2021**, *41*, 82–94. [CrossRef]
77. Houston, A.I.; McNamara, J.M.; Hutchinson, J.M. General results concerning the trade-off between gaining energy and avoiding predation. *Philos. Trans. R. Soc. Ser. B Biol. Sci.* **1993**, *341*, 375–397. [CrossRef]
78. Åbjörnsson, K.; Hansson, L.A.; Brönmark, C. Responses of prey from habitats with different predator regimes: Local adaptation and heritability. *Ecology* **2004**, *85*, 1859–1866. [CrossRef]
79. Pettersson, L.B.; Brönmark, C. Trading off safety against food: State dependent habitat choice and foraging in crucian carp. *Oecologia* **1993**, *95*, 353–357. [CrossRef]
80. Kristensen, E.A.; Closs, G.P. Anti-predator response of naïve and experienced common bully to chemical alarm cues. *J. Fish Biol.* **2004**, *64*, 643–652. [CrossRef]
81. McCormick, M.I.; Larson, J.K. Field verification of the use of chemical alarm cues in a coral reef fish. *Coral Reefs* **2007**, *26*, 571–576. [CrossRef]
82. McCormick, M.I.; Manassa, R. Predation risk assessment by olfactory and visual cues in a coral reef fish. *Coral Reefs* **2008**, *27*, 105–113. [CrossRef]
83. del Mar Palacios, M.; Warren, D.T.; McCormick, M.I. Sensory cues of a top-predator indirectly control a reef fish mesopredator. *Oikos* **2016**, *125*, 201–209. [CrossRef]
84. Palacios, M.D.M.; McCormick, M.I. Positive indirect effects of top-predators on the behaviour and survival of juvenile fishes. *Oikos* **2021**, *130*, 219–230. [CrossRef]
85. Foam, P.E.; Harvey, M.C.; Mirza, R.S.; Brown, G.E. Heads up: Juvenile convict cichlids switch to threat-sensitive foraging tactics based on chemosensory information. *Anim. Behav.* **2005**, *70*, 601–607. [CrossRef]
86. Holmes, T.H.; McCormick, M.I. Smell, learn and live: The role of chemical alarm cues in predator learning during early life history in a marine fish. *Behav. Process.* **2010**, *83*, 299–305. [CrossRef]
87. Buřič, M.; Bláha, M.; Kouba, A.; Drozd, B. Upstream expansion of round goby (*Neogobius melanostomus*)–first record in the upper reaches of the Elbe river. *Knowl. Manag. Aquat. Ecosyst.* **2015**, *416*, 32. [CrossRef]
88. Ray, W.J.; Corkum, L.D. Habitat and site affinity of the round goby. *J. Great Lakes Res.* **2001**, *27*, 329–334. [CrossRef]
89. Sadler, K. Effects of temperature on the growth and survival of the European eel, *Anguilla anguilla* L. *J. Fish Biol.* **1979**, *15*, 499–507. [CrossRef]
90. Seymour, E.A. Devising optimum feeding regimes and temperatures for the warmwater culture of eel, *Anguilla anguilla* L. *Aquac. Res.* **1989**, *20*, 311–324. [CrossRef]
91. Lee, V.A.; Johnson, T.B. Development of a bioenergetics model for the round goby (*Neogobius melanostomus*). *J. Great Lakes Res.* **2005**, *31*, 125–134. [CrossRef]
92. Golani, D.; Shefler, D.; Gelman, A. Aspects of growth and feeding habits of the adult European eel (*Anguilla anguilla*) in Lake Kinneret (Lake Tiberias), Israel. *Aquaculture* **1988**, *74*, 349–354. [CrossRef]
93. Emde, S.; Kochmann, J.; Kuhn, T.; Plath, M.; Klimpel, S. Getting what is served? Feeding ecology influencing parasite-host interactions in invasive round goby *Neogobius melanostomus*. *PLoS ONE* **2014**, *9*, e109971. [CrossRef] [PubMed]
94. Carman, S.M.; Janssen, J.; Jude, D.J.; Berg, M.B. Diel interactions between prey behaviour and feeding in an invasive fish, the round goby, in a North American river. *Freshw. Biol.* **2006**, *51*, 742–755. [CrossRef]
95. Didenko, A.; Volikov, Y.; Baranov, V.; Kruzhylina, S.; Gurbyk, A.; Bielikova, O. Chironomid diversity in the diets of Ponto-Caspian gobiids in a freshwater habitat: Implications for resource partitioning. *Limnologica* **2021**, *89*, 125890. [CrossRef]
96. Błońska, D.; Grabowska, J.; Kobak, J.; Rachalewski, M.; Bącela-Spychalska, K. Fish predation on sympatric and allopatric prey—A case study of Ponto-Caspian gobies, European bullhead and amphipods. *Limnologica* **2016**, *61*, 1–6. [CrossRef]
97. Verheggen, F.J.; Haubruge, E.; Mescher, M.C. Alarm pheromones—Chemical signaling in response to danger. *Vitam. Horm.* **2010**, *83*, 215–239. [CrossRef] [PubMed]

98. Richter, L.; Schwenkmezger, L.; Becker, J.; Winkelmann, C.; Hellmann, C.; Worischka, S. The very hungry amphipod: The invasive *Dikerogammarus villosus* shows high consumption rates for two food sources and independent of predator cues. *Biol. Invasions* **2018**, *20*, 1321–1335. [CrossRef]
99. Szydłowska, N.; Let, M.; Franta, P.; Buřič, M.; Worischka, S.; Richter, L.; Drozd, B. Gut evacuation rate as a tool for revealing feeding patterns in the invasive round goby (Neogobius melanostomus) under different feeding modes, food types and temperatures. *Aquat. Invasions* **2024**, *submitted*.
100. Hazlett, B. Responses to multiple chemical cues by the crayfish *Orconectes virilis*. *Behaviour* **1999**, *136*, 161–177. [CrossRef]
101. Manko, P. Stomach content analysis in freshwater fish feeding ecology. *University of Prešov* **2016**, *116*, 1–25.
102. Andersen, N.G. Depletion rates of gastrointestinal content in common goby (*Pomatoschistus microps* (Kr.)). Effects of temperature and fish size. *Dana* **1984**, *3*, 31–42.
103. Andersen, N.G. The effects of predator size, temperature, and prey characteristics on gastric evacuation in whiting. *J. Fish Biol.* **1999**, *54*, 287–301. [CrossRef]
104. Sveier, H.; Wathne, E.; Lied, E. Growth, feed and nutrient utilisation and gastrointestinal evacuation time in Atlantic salmon (*Salmo salar* L.): The effect of dietary fish meal particle size and protein concentration. *Aquaculture* **1999**, *180*, 265–282. [CrossRef]
105. Andersen, N.G. A gastric evacuation model for three predatory gadoids and implications of using pooled field data of stomach contents to estimate food rations. *J. Fish Biol.* **2001**, *59*, 1198–1217. [CrossRef]
106. Gillum, Z.D.; Facendola, J.J.; Scharf, F.S. Consumption and gastric evacuation in juvenile red drum *Sciaenops ocellatus* (Linnaeus): Estimation of prey type effects and validation of field-based daily ration estimates. *J. Exp. Mar. Biol. Ecol.* **2012**, *413*, 21–29. [CrossRef]
107. Seyhan, K.; Bascinar, N.S.; Bascinar, N.; Khan, U. Gastric evacuation in brook trout (*Salvelinus fontinalis*) fry: Effect of body size. *Pak. J. Zool.* **2020**, *52*, 1173. [CrossRef]
108. Persson, L. Rate of food evacuation in roach (*Rutilus rutilus*) in relation to temperature, and the application of evacuation rate estimates for studies on the rate of food consumption. *Freshw. Biol.* **1982**, *12*, 203–210. [CrossRef]
109. Temming, A.; Andersen, N.G. Modelling gastric evacuation in cod. *ICES CM* **1992**, *61*, 2–13.
110. Sweka, J.A.; Keith Cox, M.; Hartman, K.J. Gastric evacuation rates of brook trout. *Trans. Am. Fish. Soc.* **2004**, *133*, 204–210. [CrossRef]
111. Chabot, D.; Behrens, J.; Andersen, N.G. Digesting in hypoxia: Impact on gastric evacuation rate and postprandial metabolism (SDA) of Atlantic cod, *Gadus morhua*. In Proceedings of the ICES Annual Science Conference 2015, Copenhagen, Denmark, 21–25 September 2015.
112. Barreto, R.E.; Barbosa-Júnior, A.; Urbinati, E.C.; Hoffmann, A. Cortisol influences the antipredator behavior induced by chemical alarm cues in the Frillfin goby. *Horm. Behav.* **2014**, *65*, 394–400. [CrossRef] [PubMed]
113. Archard, G.A.; Earley, R.L.; Hanninen, A.F.; Braithwaite, V.A. Correlated behaviour and stress physiology in fish exposed to different levels of predation pressure. *Funct. Ecol.* **2012**, *26*, 637–645. [CrossRef]
114. Rehnberg, B.G.; Schreck, C.B. Chemosensory detection of predators by coho salmon (*Oncorhynchus kisutch*): Behavioural reaction and the physiological stress response. *Can. J. Zool.* **1987**, *65*, 481–485. [CrossRef]
115. Barreto, R.E.; Miyai, C.A.; Sanches, F.H.C.; Giaquinto, P.C.; Delicio, H.C.; Volpato, G.L. Blood cues induce antipredator behavior in *Nile tilapia* conspecifics. *PLoS ONE* **2013**, *8*, e54642. [CrossRef] [PubMed]
116. Giaquinto, P.C.; Volpato, G.L. Chemical cues related to conspecific size in pintado catfish, *Pseudoplatystoma coruscans*. *Acta Ethologica* **2005**, *8*, 65–69. [CrossRef]
117. Giaquinto, P.C.; Hoffmann, A. The scent of stress: Pintado catfish differentially respond to chemical cues from stressed conspecifics. *Behaviour* **2012**, *149*, 941–951. [CrossRef]
118. Fraser, D.F.; Huntingford, F.A. Feeding and avoiding predation hazard: The behavioral response of the prey. *Ethology* **1986**, *73*, 56–68. [CrossRef]
119. Chapple, D.G.; Simmonds, S.M.; Wong, B.B. Can behavioral and personality traits influence the success of unintentional species introductions? *Trends Ecol. Evol.* **2012**, *27*, 57–64. [CrossRef] [PubMed]
120. Ericsson, P.; Persson, A.; Behrens, J.W.; Brodin, T.; Hirsch, P.E.; Sundelin, A.; van Deurs, M.; von Friesen, L.W.; Nilsson, P.A. Personality-dependent inter-and intraspecific foraging competition in the invasive round goby, *Neogobius melanostomus*. *J. Fish Biol.* **2021**, *98*, 1234–1241. [CrossRef] [PubMed]
121. Galli, A.; Behrens, J.W.; Gesto, M.; Moran, N.P. Boldness and physiological variation in round goby populations along their Baltic Sea invasion front. *Physiol. Behav.* **2023**, *269*, 114261. [CrossRef] [PubMed]
122. Yeung, E.J.; Klemet-N'Guessan, S.; Hossie, T.J.; Fox, M.G. Boldness, movement and exploration tendency in round goby (*Neogobius melanostomus*) in Southern Ontario. *J. Fish Biol.* **2022**, *103*, 790–797. [CrossRef] [PubMed]
123. Sih, A.; Bell, A.; Johnson, J.C. Behavioral syndromes: An ecological and evolutionary overview. *Trends Ecol. Evol.* **2004**, *19*, 372–378. [CrossRef] [PubMed]
124. Mittelbach, G.G.; Ballew, N.G.; Kjelvik, M.K. Fish behavioral types and their ecological consequences. *Can. J. Fish. Aquat. Sci.* **2014**, *71*, 927–944. [CrossRef]
125. Ioannou, C.; Payne, M.; Krause, J. Ecological consequences of the bold-shy continuum: The effect of predator boldness on prey risk. *Oecologia* **2008**, *157*, 177–182. [CrossRef] [PubMed]

126. Thorlacius, M.; Brodin, T. Investigating large-scale invasion patterns using-small scale invasion successions—Phenotypic differentiation of the invasive round goby (*Neogobius melanostomus*) at invasion fronts. *Limnol. Oceanogr.* **2018**, *63*, 702–713. [CrossRef]
127. Behrens, J.W.; von Friesen, L.W.; Brodin, T.; Ericsson, P.; Hirsch, P.E.; Persson, A.; Sundelin, A.; van Deurs, M.; Nilsson, P.A. Personality-and size-related metabolic performance in invasive round goby (*Neogobius melanostomus*). *Physiol. Behav.* **2020**, *215*, 112777. [CrossRef]
128. Manassa, R.P.; McCormick, M.I. Social learning and acquired recognition of a predator by a marine fish. *Anim. Cogn.* **2012**, *15*, 559–565. [CrossRef]
129. Pūtys, Ž.; Ložys, L.; Būda, V. Respiratory response to the chemical cues of injured conspecifi cs and histology of skin in round goby, *Neogobius melanostomus* (Actinopterygii: Perciformes: Gobiidae). *Acta Ichthyol. Piscat.* **2015**, *45*, 411–415. [CrossRef]
130. Chivers, D.P.; Dixson, D.L.; White, J.R.; McCormick, M.I.; Ferrari, M.C. Degradation of chemical alarm cues and assessment of risk throughout the day. *Ecol. Evol.* **2013**, *3*, 3925–3934. [CrossRef] [PubMed]
131. Zhao, X.; Ferrar, M.C.; Chivers, D.P. Threat-sensitive learning of predator odours by a prey fish. *Behaviour* **2006**, *143*, 1103–1121. [CrossRef]
132. Zhao, X.; Chivers, D.P. Response of juvenile goldfish (*Carassius auratus*) to chemical alarm cues: Relationship between response intensity, response duration, and the level of predation risk. In *Chemical Signals in Vertebrates 10*; Mason, R.T., LeMaster, M.P., Müller-Schwarze, D., Eds.; Springer: Boston, MA, USA, 2005; pp. 334–341. [CrossRef]
133. Kłosiński, P.; Kobak, J.; Augustyniak, M.; Pawlak, R.; Jermacz, Ł.; Poznańska-Kakareko, M.; Kakareko, T. Behavioural responses to con-and heterospecific alarm cues by an alien and a coexisting native fish. *Hydrobiologia* **2022**, *849*, 985–1000. [CrossRef]
134. Wisenden, B.; Rugg, M.; Korpi, N.; Fuselier, L. Lab and field estimates of active time of chemical alarm cues of a cyprinid fish and an amphipod crustacean. *Behaviour* **2009**, *146*, 1423–1442.
135. Wagner, C.M.; Bals, J.D.; Byford, G.J.; Scott, A.M.; Feder, M.E. Olfactory sensitivity and threat-sensitive responses to alarm cue in an invasive fish. *Biol. Invasions* **2023**, *25*, 3083–3101. [CrossRef]
136. Atema, J. *Chemical Senses, Chemical Signals, and Feeding Behavior in Fishes*; Bardach, J.E., Magnuson, J.J., May, R.C., Reinhart, J.M., Eds.; FAO: Makati, Philippines, 1980; pp. 57–101.
137. Kasumyan, A.O. The olfactory system in fish: Structure, function, and role in behavior. *J. Ichthyol.* **2004**, *44*, 180.
138. Belanger, R.M.; Smith, C.M.; Corkum, L.D.; Zielinski, B.S. Morphology and histochemistry of the peripheral olfactory organ in the round goby, *Neogobius melanostomus* (Teleostei: Gobiidae). *J. Morphol.* **2003**, *257*, 62–71. [CrossRef]
139. Pfeiffer, W. The distribution of fright reaction and alarm substance cells in fishes. *Copeia* **1977**, 653–665. [CrossRef]
140. Smith, R.J.F. Alarm signals in fishes. *Rev. Fish Biol. Fish.* **1992**, *2*, 33–63. [CrossRef]
141. Marsh-Hunkin, K.E.; Gochfeld, D.J.; Slattery, M. Antipredator responses to invasive lionfish, Pterois volitans: Interspecific differences in cue utilization by two coral reef gobies. *Mar. Biol.* **2013**, *160*, 1029–1040. [CrossRef]
142. McLean, F.; Barbee, N.C.; Swearer, S.E. Avoidance of native versus non-native predator odours by migrating whitebait and juveniles of the common galaxiid, *Galaxias maculatus*. *N. Z. J. Mar. Freshw. Res.* **2007**, *41*, 175–184. [CrossRef]
143. Utne-Palm, A.C. Response of naïve two-spotted gobies *Gobiusculus flavescens* to visual and chemical stimuli of their natural predator, cod *Gadus morhua*. *Mar. Ecol. Prog. Ser.* **2001**, *218*, 267–274. [CrossRef]
144. Ferrari, M.C.; Messier, F.; Chivers, D.P. Degradation of chemical alarm cues under natural conditions: Risk assessment by larval woodfrogs. *Chemoecology* **2007**, *17*, 263–266. [CrossRef]
145. Mathis, A.; Smith, R.J.F. Chemical alarm signals increase the survival time of fathead minnows (*Pimephales promelas*) during encounters with northern pike (*Esox lucius*). *Behav. Ecol.* **1993**, *4*, 260–265. [CrossRef]
146. Wisenden, B.D.; Chivers, D.P. The role of public chemical information in antipredator behaviour. *Commun. Fishes* **2006**, *1*, 259–278.
147. Pollock, M.S.; Chivers, D.P.; Mirza, R.S.; Wisenden, B.D. Fathead minnows, *Pimephales promelas*, learn to recognize chemical alarm cues of introduced brook stickleback, *Culaea inconstans*. *Environ. Biol. Fishes* **2003**, *66*, 313–319. [CrossRef]
148. Dalesman, S.; Rundle, S.D.; Bilton, D.T.; Cotton, P.A. Phylogenetic relatedness and ecological interactions determine antipredator behavior. *Ecology* **2007**, *88*, 2462–2467. [CrossRef]
149. Vilhunen, S.; Hirvonen, H. Innate antipredator responses of Arctic charr (*Salvelinus alpinus*) depend on predator species and their diet. *Behav. Ecol. Sociobiol.* **2003**, *55*, 1–10. [CrossRef]
150. Mathis, A.; Smith, R.J.F. Chemical labeling of northern pike (*Esox lucius*) by the alarm pheromone of fathead minnows (*Pimephales promelas*). *J. Chem. Ecol.* **1993**, *19*, 1967–1979. [CrossRef] [PubMed]
151. Brown, G.E.; Godin, J.G.J. Who dares, learns: Chemical inspection behaviour and acquired predator recognition in a characin fish. *Anim. Behav.* **1999**, *57*, 475–481. [CrossRef] [PubMed]
152. Brown, G.E.; Godin, J.G.J. Anti-predator responses to conspecific and heterospecific skin extracts by threespine sticklebacks: Alarm pheromones revisited. *Behaviour* **1997**, *134*, 1123–1134. [CrossRef]
153. Pfeiffer, W. Alarm substances. *Experientia* **1963**, *19*, 113–123. [CrossRef] [PubMed]
154. Wisenden, B. Active space of chemical alarm cue in natural fish populations. *Behaviour* **2008**, *145*, 391–407. [CrossRef]
155. Olson, J.A.; Olson, J.M.; Walsh, R.E.; Wisenden, B.D. A method to train groups of predator-naive fish to recognize and respond to predators when released into the natural environment. *N. Am. J. Fish. Manag.* **2012**, *32*, 77–81. [CrossRef]

156. Rakauskas, V.; Pūtys, Ž.; Dainys, J.; Lesutienė, J.; Ložys, L.; Arbačiauskas, K. Increasing population of the invader round goby, *Neogobius melanostomus* (Actinopterygii: Perciformes: Gobiidae), and its trophic role in the Curonian Lagoon, SE Baltic Sea. *Acta Ichthyol. Piscat.* **2013**, *43*, 95–108. [CrossRef]
157. Jůza, T.; Blabolil, P.; Bartoň, D.; Čech, M.; Draštík, V.; Frouzova, J.; Holubová, M.; Ketelaars, H.A.M.; Kočvara, L.; Kubečka, J.; et al. Recovery of the ruffe (*Gymnocephalus cernua*) population after an invasion boom of round goby (*Neogobius melanostomus*) in De Gijster Lake (The Netherlands). *Aquat. Invasions* **2021**, *16*, 499–511. [CrossRef]
158. Ylönen, H.; Kortet, R.; Myntti, J.; Vainikka, A. Predator odor recognition and antipredatory response in fish: Does the prey know the predator diel rhythm? *Acta Oecologica* **2007**, *31*, 1–7. [CrossRef]
159. Utne, A.C.W.; Bacchi, B. The influence of visual and chemical stimuli from cod Gadus morhua on the distribution of two-spotted goby *Gobiusculus flavescens* (Fabricius). *Sarsia* **1997**, *82*, 129–135. [CrossRef]
160. Sreedharan, G.; Corkum, L.D.; Johnson, T.B. Response of the round goby, an invasive fish, to food odours. *Int. Ver. Theor. Angew. Limnol. Verhandlungen* **2009**, *30*, 1275–1278. [CrossRef]
161. Holmes, T.H.; McCormick, M.I. Response across a gradient: Behavioural reactions of newly settled fish to predation cues. *Anim. Behav.* **2011**, *81*, 543–550. [CrossRef]
162. Yavno, S.; Corkum, L.D. Round goby *Neogobius melanostomus* attraction to conspecific and heterospecific egg odours. *J. Fish Biol.* **2011**, *78*, 1944–1953. [CrossRef]
163. Hirvonen, H.; Holopainen, S.; Lempiäinen, N.; Selin, M.; Tulonen, J. Sniffing the trade-off: Effects of eel odours on nocturnal foraging activity of native and introduced crayfish juveniles. *Mar. Freshw. Behav. Physiol.* **2007**, *40*, 213–218. [CrossRef]

biology

MDPI

Article

The First Records of *Trissolcus japonicus* (Ashmead) and *Trissolcus mitsukurii* (Ashmead) (Hymenoptera, Scelionidae), Alien Egg Parasitoids of *Halyomorpha halys* (Stål) (Hemiptera, Pentatomidae) in Serbia

Aleksandra Konjević [1], Luciana Tavella [2] and Francesco Tortorici [2,*]

[1] Center of Excellence, Faculty of Agriculture, University of Novi Sad, Trg Dositeja Obradovica 8, 21000 Novi Sad, Serbia; sashak@polj.uns.ac.rs

[2] Dipartimento di Scienze Agrarie, Forestali e Alimentari (DISAFA), University of Torino, largo P. Braccini 2, 10095 Grugliasco, Italy; luciana.tavella@unito.it

* Correspondence: francesco.tortorici@unito.it

Simple Summary: In Serbia, there is a growing concern over the Brown Marmorated Stink Bug, which damages plants and creates aggregates in urban areas during the winter season. The bug population has increased, causing significant economic damage to crops. A study conducted in 2022 has discovered nine egg parasitoid species, including two foreign species, *Trissolcus japonicus* and *Tr. mitsukurii*, which were found in only one city in Serbia. This is the first time these species have been recorded in Serbia.

Abstract: Serbia has recently begun facing a serious problem with the Brown Marmorated Stink Bug, *Halyomorpha halys* (Stål), which was first recorded in October 2015. This species belongs to the Pentomidae family and is notorious for causing extensive damage to plants. During the winter, it tends to gather in urban areas, such as houses and different man-made facilities, which has raised concerns among producers and citizens. The population of this species has rapidly increased, causing significant economic damage to cultivated plants. However, despite the alarming situation no natural enemies have yet been identified in Serbia. Therefore, research in 2022 was focused on collecting stink bug eggs to investigate the presence of egg parasitoids. The study identified two foreign Hymenoptera species for the European region, *Trissolcus japonicus* (Ashmead) and *Tr. mitsukurii* (Ashmead) (Scelionidae), recorded for the first time in Serbia. Additionally, the list of egg parasitoid species belonging to the Hymenoptera order includes seven local species: *Anastatus bifasciatus* (Geoffroy), from the Eupelmidae family; *Ooencyrtus* sp., from the Encyrtidae family; and *Telenomus turesis* (Walker), *Tr. basalis* (Wollaston), *Tr. belenus* (Walker), *Tr. colemani* (Crawford), and *Tr. semistriatus* (Nees von Esenbeck), from the Scelionidae family. In total, nine egg parasitoid species were, for the first time, reported as parasitizing *H. halys* and related species in Serbia.

Keywords: brown marmorated stink bug; foreign insect species; first records; parasitoid list

1. Introduction

From its native range in eastern Asia, *Halyomorpha halys* (Stål) (Hemiptera, Pentatomidae) has spread throughout the Holarctic region and in a few countries in South America [1], to the point that it has become a worldwide invasive pest with more than 100 host plants reported [2].

Due to the severe damage caused by the stink bug and the resulting economic losses, the use of insecticides has increased as the control of this pest relies mainly on pesticides [3,4]. Because of the application of pesticides, including broad-spectrum insecticides, beneficial insects are also killed and integrated pest management programs are negatively impacted [5].

Among the natural enemies of *H. halys*, considerable emphasis has been placed on egg parasitoids. In its native range, the stink bug is regulated by several egg parasitoids. *Trissolcus japonicus* (Ashmead) and *Trissolcus mitsukurii* (Ashmead) are predominant and the most effective parasitoids in the native area [6–8].

In Europe, associated parasitoids belonging to Scelionidae (*Trissolcus* spp., *Telenomus* spp.), Eupelmidae (*Anastatus* spp.), and Encyrtidae (*Ooencyrtus* spp.) were found to parasitize *H. halys* eggs [9–16], but they were not fully effective in controlling *H. halys* populations. In particular, *Anastatus bifasciatus* (Geoffroy) and *Trissolcus kozlovi* Rjachovskij are the two native egg parasitoids in Europe that have shown some potential for the control of *H. halys* and have been considered for augmentative biological control strategies [10,13,17–20].

Several studies have been carried out to explore the ability of parasitoids to attack *H. halys* eggs in their native area as well as in the newly invaded areas worldwide [9,11,13,21]. Actually, only the two Asian species *Tr. japonicus* and *Tr. mitsukurii* appear to be promising candidates as biological control agents [8,22–25]. Since 2014, *Tr. japonicus* started to be detected in the USA [26], then in 2017 in Europe [18], and now the distribution area is widespread [27–29]. The first detection of *Tr. mitsukurii* outside the native area (i.e., in Italy) was in 2016 [30], and then the parasitoid spread to France [31].

In Serbia, *H. halys* was detected in October 2015 [32]. In the following years, research was conducted to assess the distribution of the species and the damage to local crops [33], when it was confirmed that the species had become established in the newly invaded areas, and at that time the list of endangered plant species was relatively short. Since 2018, monitoring using pyramid dead-in traps (AgBio Company) with aggregation pheromones (Tréce lures) has been carried out in agricultural and urban areas. The number of traps in the country increased from 20 to 47 in 2022 (Figure 1), and monitoring results showed a continuous spread throughout the country and an increase in the population. The list of endangered host plants has been significantly increased and includes both cultivated and ornamental plants [34]. Moderate to high-level damage has been reported in many cultivated plants since 2020, especially in hazelnuts, apples, cherries, peaches, and pears (Konjević, unpublished data). The protection of cultivated plants in Serbia relies almost entirely on chemical treatments, which have sometimes proven to be insufficiently effective.

Parasitoids of stink bugs, especially of exotic species, in Serbia were not monitored until 2022. In 2021, there were two accidental captures of three native parasitoid species, *Trissolcus basalis* (Wollaston), *Tr. belenus* (Walker), and *Telenomus* sp., which were morphologically identified at the Faculty of Biology, University "Al. I. Cuza" in Romania (Konjević and Popovici, unpublished data). Still, there was no evidence of the presence of exotic parasitoid species, or the autochthonous *A. bifasciatus*, to the eggs of *H. halys*.

Recent studies have estimated that in invaded areas, *Tr. japonicus* and *Tr. mitsukurii* are still spreading throughout much of the western Palearctic territory and have shown potential high habitat suitability in Serbia [28,29]. Therefore, the aim of this research was to monitor *H. halys* in urban and semi-urban environments and assess whether egg parasitoids are potentially present in such conditions.

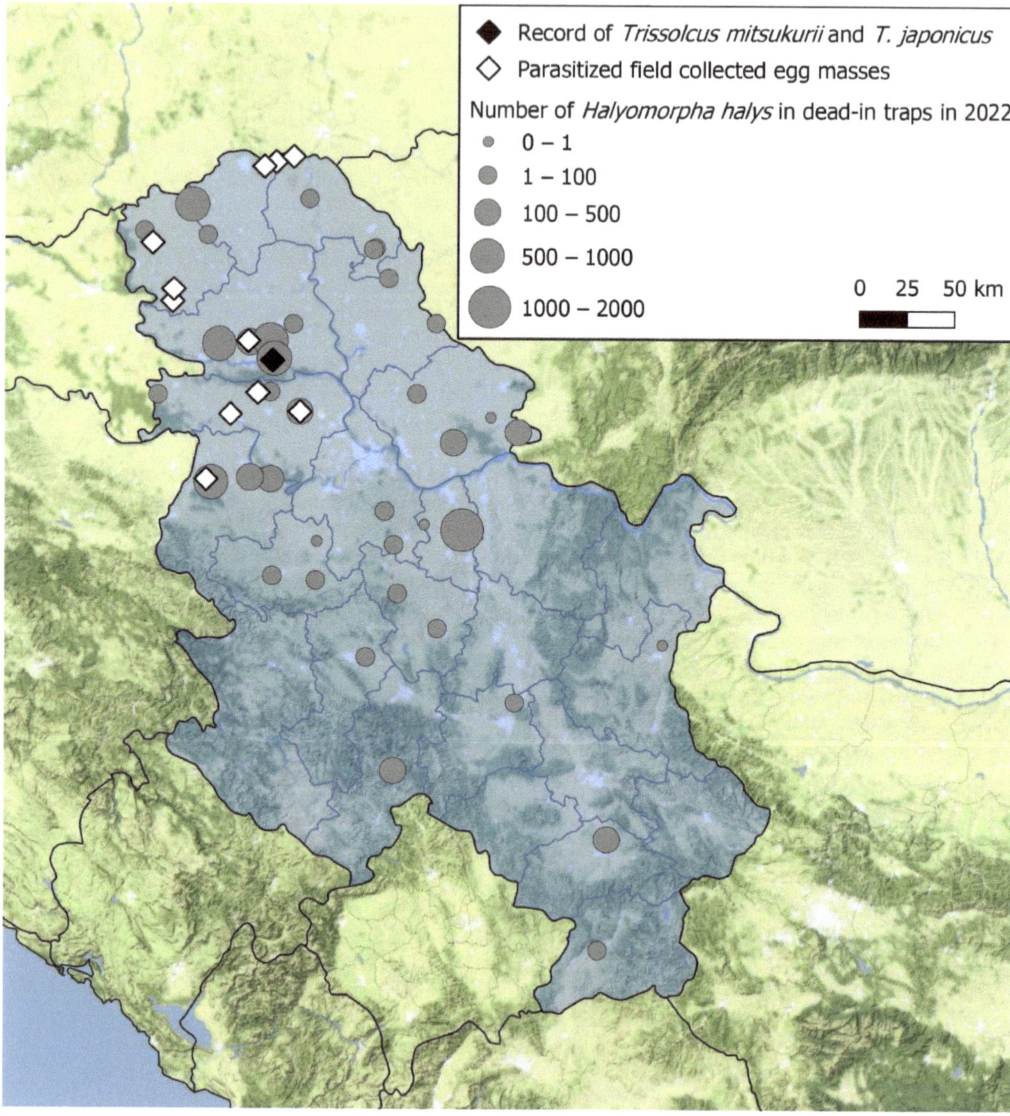

Figure 1. Sites with field-collected parasitized egg masses in 2022. Black rhombus: sites where *Tr. japonicus* and *Tr. mitsukurii* emerged from *H. halys* eggs. White rhombus: sites where only native parasitoid species were found. Gray circles: sites where dead-in traps were set in 2022 and the range of sampled *H. halys* specimens.

2. Materials and Methods

2.1. Field Collection and Laboratory Rearing of Stink Bug Egg Masses

The field survey was conducted from May to September 2022 at 27 different locations, mainly in the central and northern parts of Serbia, where the population of *H. halys* is regularly found in dead-in traps baited with aggregation pheromones. Firstly, the monitoring of egg masses started in the urban area of the second largest city in Serbia, Novi Sad, where three city parks were examined each week. Later, egg masses were sampled at the borders with Hungary, Croatia, and Romania in two visits during 2022, mainly urban and semi-urban environments. In addition, egg masses were occasionally collected in hazelnut, plum, and walnut orchards, and in a soybean field. All urban and semi-urban sites were characterized by a variety of plant species, but the search for eggs was focused on *Catalpa* sp., *Paulownia* sp., *Sambucus nigra* L., *Acer* sp., *Styphnolobium japonicum* (L.) Schott, and *Koelreuteria* sp., on which adults and nymphs of *H. halys* were observed. In many of these areas, there were no insecticide applications, except in orchards, where insecticide treatments were carried out regularly based on production needs.

Sampling in all areas was carried out in the morning hours, between 8 and 12 AM. The inspection time in each site was 1 h per person, looking at different parts of the same site. During the survey, woody and shrubby plant species were visually inspected by checking the leaves, especially the undersides where stink bugs usually lay their eggs. All field-collected stink bug eggs were brought to the laboratory and separated by sampling points and by insect species, comparing the morphology of the egg masses with images published on https://www.halyomorphahalys.com/wanzeneier-stink-bug-eggs.html (accessed on 7 February 2023) [35]. Eggs were then placed into petri dishes (10 or 15 cm in diameter) and stored under the room conditions for the emergence of either stink bug nymphs or adult parasitoids. All emerged parasitoids were stored in 96% ethanol at −4 °C, until their identification.

2.2. Species Identification

The parasitoid adults, previously stored in ethanol, were dried for morphological examination. A Wild M5 stereo microscope with up to 200× magnification was used for morphological diagnosis. Species of *Trissolcus* were determined using the characters and identification keys of Talamas et al. [36], Tortorici et al. [37], and Moraglio et al. [17]. Species of *Anastatus* were identified according to Peng et al. [38]. Species of *Telenomus* were determined using the keys of Kozlov and Kononova [39] and Johnson [40]. Images of the specimens were taken using a Canon 90D camera (Canon Inc., Tokyo, Japan) equipped with an extension tube and 10× and 20× LWD microscope lenses mounted on a macro-rail and illuminated with two speedlite flashes. The frames were merged with Zerene Stacker (PMax algorithm, Zerene Systems LLC, Richland, Washington, USA). All specimens used for morphological analysis were deposited in the Dipartimento di Scienze Agrarie, Forestali e Alimentari, University of Turin, Italy.

3. Results

A total of 568 naturally laid egg masses were collected, belonging to three pentatomid species: *H. halys*, *Nezara viridula* (L.), and an Asopinae (Hemiptera, Pentatomidae) (Table 1). Out of the 27 locations examined, parasitized egg masses were obtained from 12, reported in Table 1, while in the remaining locations either no eggs were recorded or only unparasitized eggs were found. Most of the eggs were collected on hazelnuts, in three orchards, while most of the parasitized eggs were sampled in the urban areas, on *St. japonicum* and *Sa. nigra*. Other host plants on which parasitized egg masses were collected are walnut, catalpa, plum, maple, paulownia, rose, and tomato. In total, parasitoid adults emerged successfully from 29 egg masses. Among the sampled eggs, there were eggs that were different in colour from the viable eggs, like dark yellow or brown, from which neither stink bug nymphs nor parasitoid adults emerged. Some eggs were already damaged by predators at the time of sampling.

Most of the parasitized eggs were found in areas where high numbers of *H. halys* specimens were monitored by dead-in traps, varying from 501 to 1000 per year/season (Figure 1).

Table 1. Sites in Serbia, where parasitized stink bug egg masses were collected in 2022. Hh, *Halyomorpha halys*; Nv, *Nezara viridula*; As, Asopinae. The asterisks (*) indicate cases of multiparasitisms.

ID	Site	Habitat	Coordinates	Altitude (m asl)	Total Number of Egg Masses	Parasitized Egg Masses
1	Backi Petrovac	Hazelnut orchard	45°19′38″ N 19°41′24″ E	80	1	Hh (1)
2	Backi Vinogradi	Semi-urban rural area, border with Hungary	46°7′26″ N 19°51′34″ E	95	2	Hh (2)
3	Horgos	Semi-urban rural area, border with Hungary	46°8′50″ N 19°58′11″ E	82	5	Hh (1)
4	Karavukovo	Rural area, border with Croatia	45°30′10″ N 19°12′22″ E	82	2	Hh (1 *)
5	Lacarak	Semi-urban rural area	44°59′49″ N 19°34′37″ E	81	3	Nv (2)
6	Ljukovo	Hazelnut orchard	45°0′32″ N 20°0′52″ E	101	306	Hh (1)
7	Mala Remeta	Agricultural area, orchards	45°5′40″ N 19°45′1″ E	205	10	Hh (1)
8	Novi Sad	Urban park	45°14′23″ N 19°50′36″ E	100	71	Hh (6 *), Nv (3 *), As (1)
9	Palic	Semi-urban rural area	46°6′14″ N 19°47′14″ E	98	1	Hh (1)
10	Ribari	Hazelnut orchard	44°42′9″ N 19°25′10″ E	93	150	Hh (3)
11	Sombor	Walnut orchard	45°45′53″ N 19°4′48″ E	84	12	Hh (2 *), Nv (3 *)
12	Srpski Miletic	Semi-urban rural area	45°33′27″ N 19°12′46″ E	83	5	Nv (1)

Anastatus bifasciatus was the most abundant species emerging from the eggs of *H. halys* and *N. viridula* (Table 2). *Ooencyrtus* sp. emerged from the eggs of both *H. halys* and *N. viridula*, while *Telenomus turesi* Walker emerged only from the eggs of *H. halys* (Table 2). Based on morphological analyses, *Trissolcus* specimens were identified as *Tr. basalis*, *Tr. belenus*, *Tr. colemani* (Crawford), and *Tr. semistriatus* (Nees von Esenbeck), and also as the exotic *Tr. japonicus* and *Tr. mitsukurii*. These two exotic parasitoid species emerged at the site of Novi Sad from egg masses of *H. halys* (Figure 1).

Table 2. Total number of individuals of *A. bifasciatus* (Ab), *Ooencyrtus* sp. (Oo), *Te. turesis* (Tt), *Tr. basalis* (Tba), *Tr. belenus* (Tbe), *Tr. colemani* (Tc), *Tr. japonicus* (Tj), *Tr. mitsukurii* (Tm), *Tr. semistriatus* (Ts) emerged from *H. halys*, *N. viridula*, and Asopinae egg masses.

Species	No. Egg Masses	No. and Species of Parasitoids Emerged in Laboratory
Halyomorpha halys	19	49 Tba, 32 Ab, 31 Tj, 19 Tt, 10 Oo, 3 Tc, 2 Tm
Nezara viridula	9	62 Ab, 37 Tba, 26 Ts, 1 Oo
Asopinae	1	4 Tbe

Trissolcus japonicus emerged from four egg masses and *Tr. mitsukurii* from one egg mass together with the first. The two species perfectly matched with the description in Talamas et al. [36] and morphological diagnosis in Sabbatini Peverieri et al. [41], and they did not show morphological anomalies that would complicate confirmation of species identity based on morphology alone. *Anastatus bifasciatus* emerged from nine egg masses of *H. halys* and four egg masses of *N. viridula*. Multiparasitism was recorded in four egg masses of *H. halys* and three egg masses of *N. viridula* with *Ooencyrtus* sp., *Te. turesis*, *Tr. basalis*, and *Tr. japonicus*. Furthermore, *Ooencyrtus* sp. shared the same egg masses together with *Tr. japonicus*.

Trissolcus japonicus and *Tr. mitsukurii* belong to the *flavipes*-group and the *basalis*-group, respectively [36]. The two species can be distinguished from other species in their respective groups by the following characteristics:

Trissolcus japonicus—Vertex between lateral ocelli with uniform and robust hyperoccipital carina (hoc) (Figure 2B,E). Microsculpture finely colliculate along the line between the dorsal margin of the hyperoccipital carina and the posterior margin of the anterior ocellus (haol) (Figure 2E). At intersection with malar sulcus, orbital furrow (of) expanded with medial margin well-defined (Figure 2D). Clypeus with four setae (cs) (Figure 2C). Antenna with five clavomeres (A7–A11 with basiconic sensilla) with moderately large clava (Figure 2A). Mesoscutum with notauli (not) and without median mesoscutal carina (mmc) (Figure 2B,E). The finely colliculated sculpture on the mesoscutum is uniformly extensive, reaching the posterior margin between the notauli (Figure 2B,E). Mesopleuron with episternal foveae (eps) forming a continuous line of cells from postacetabular sulcus (ats) to mesopleural pit (mpp) (Figure 2D). Sublateral seta absent on each side of tergite 1 (Figure 2B,E). Laterotergite 1 without a line of setae along dorsal margin (Figure 2A,D). Tergite 2 with striae present throughout anterior half of tergite (Figure 2B).

Figure 2. Female of *Tr. japonicus*. (**A**) head, mesosoma, metasoma, lateral view; (**B**) head, mesosoma, metasoma, dorsal view; (**C**) head, anterolateral view; (**D**) head, mesosoma, lateral view; (**E**) head, mesosoma, metasoma, dorsolateral view. A7–A11 = 5 clavomers; hoc = hyperoccipital carina; not = notauli; cs = clypeal setae; of = orbital furrow; mpp = mesopleural pit; eps = episternal foveae; ats = postacetabular sulcus; haol = line between the dorsal margin of hyperoccipital carina and posterior margin of anterior ocellus.

Trissolcus mitsukurii—Vertex between lateral ocelli without hyperoccipital carina (Figure 3B,E). Clypeus with six setae (cs) (Figure 3C). Antenna with five clavomeres

(A7–A11 with basiconic sensilla) with distinctly large clava (Figure 3A). Mesoscutum with notauli (not) (Figure 3B,E). Mesopleuron with two episternal foveae (eps) distant from postacetabular sulcus (ats) and mesopleural pit (mpp) (Figure 3D). Metapleuron without setae below metapleural sulcus (mtps) (Figure 3D). One or two sublateral seta (ss) present on each side of tergite 1 (Figure 3B,E). Laterotergite 1 with a line of setae along dorsal margin (slt1) (Figure 3C,D). Frons, from vertex to antennal scrobe, mesoscutum, and mesoscutellum coarsely rugose (Figure 3B,C,E).

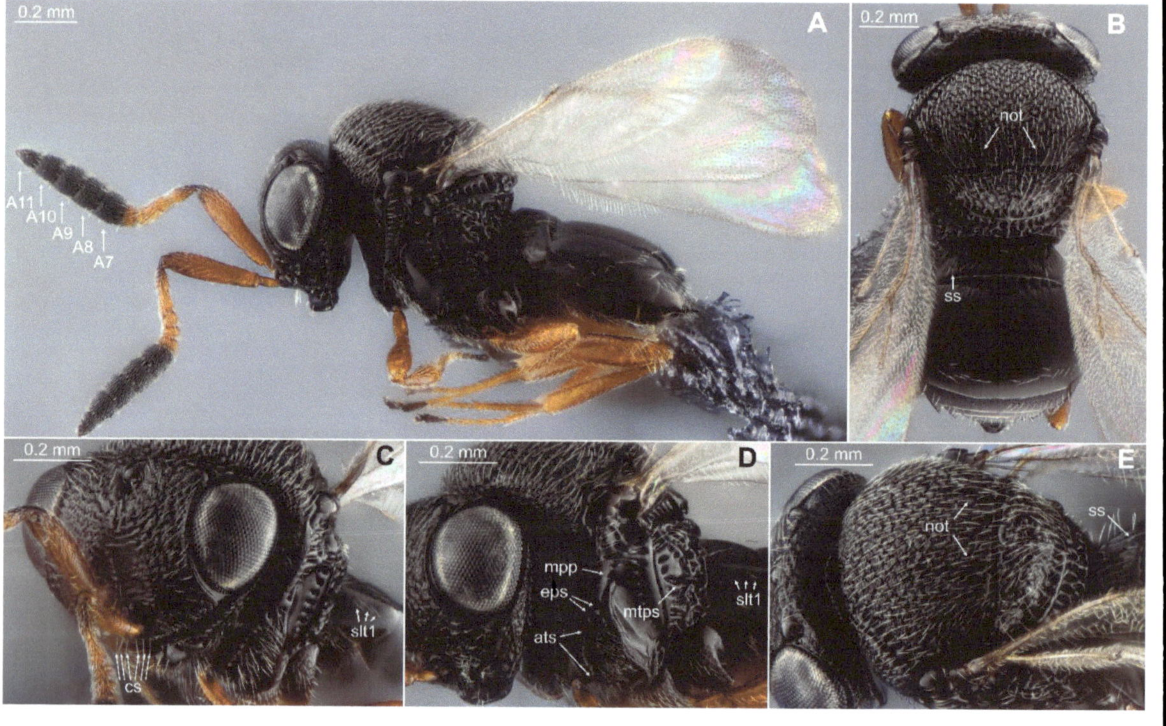

Figure 3. Female of *Tr. mitsukurii*. (**A**) head, mesosoma, metasoma, lateral view; (**B**) head, mesosoma, metasoma, dorsal view; (**C**) head, mesosoma, anterolateral view; (**D**) head, mesosoma, lateral view; (**E**) head, mesosoma, metasoma, dorsolateral view. A7–A11 = 5 clavomers; not = notauli; ss = sublateral seta; cs = clypeal setae; slt1 = setation of laterotergite 1; mpp = mesopleural pit; eps = episternal foveae; ats = postacetabular sulcus.

4. Discussion

The parasitized egg masses collected at Novi Sad represent the first record of the exotic *Tr. japonicus* and *Tr. mitsukurii* in Serbia and indicate the presence of an established adventive population of both parasitoids. This is confirmed by the fact that there have been neither deliberate releases of these species for classical biological control nor any known laboratory colonies in the area. Models of the potential interaction between *H. halys* and these two parasitoid species hypothesized that *Tr. mitsukurii* and *Tr. japonicus* exhibit high and medium-high habitat suitability in Serbia, respectively [29], and the record of the adventive populations of these two parasitoids in the same area confirms what was predicted. For the purpose of evaluating a biological control program, the model proposed by the same authors argues that *Tr. mitsukurii* shows better suitability than *Tr. japonicus* in northern Serbia, but the habitat suitability of both scelionids overlaps in much of the country. By contrast, in the southern-most mountainous areas of the country, the probability of the suitability of both parasitoids is very low.

Since the two exotic parasitoids are well characterized and distinguishable from the western Palearctic species [36], only morphological identification was performed here. Further molecular-based investigations could assess the similarity of these adventive populations to the haplotypes distributed in the newly invaded and/or native areas and in this way trace the origin of the populations found in Serbia.

The means of arrival of *Tr. japonicus* and *Tr. mitsukurii* in Serbia is unknown, but we can suppose that their arrival is connected with the transport of goods and passengers. As the most likely means of arrival of *H. halys* was as a stowaway in the transport of goods and passengers [42–44], it is very likely that the parasitoid species also arrived in the same way [45]. Namely, as of 2020 there are several foreign construction companies located in Novi Sad and its surroundings, which have the majority of necessary equipment arriving from abroad. A few of them are located at sites close to the point where both exotic parasitoid species were found in a park. It should be emphasized that the traffic between the headquarters, or the companies' place of business, and this park runs smoothly by the busy roads, leaving the possibility for their spread by vehicles. In contrast, in hazelnut orchards where the sampling was conducted in 2022 chemical treatments were carried out to control *H. halys*, which makes it difficult, although not impossible, to record and find potentially surviving parasitoid specimens. Also, hazelnut orchards are often located far from the main motorways; therefore, the spread of exotic parasitic species to those locations might need more time. These are the two main reasons why this survey was focused more on ornamental and not cultivated host plants in urban and semi urban areas.

The two species *Tr. japonicus* and *Tr. mitsukurii* can now be considered widespread throughout most European countries [18,27,30,31,41,46]. At present, in Serbia the two species were only found in one urban site and confirmed their preference for *H. halys* as the primary host. Since only a few specimens were found at one collection site, this may suggest that the spread of the adventive populations is in its early stage; therefore, the impact of the two parasitoids on the *H. halys* population cannot yet be evaluated. In subsequent years, the spread of both scelionids and their impact on the host population can be assessed by increasing the number of sites and expanding the collection season.

The Holarctic species *A. bifasciatus* is a generalist parasitoid [15,47], and its interspecific competition [48–50] makes it able to develop in different hosts and sharing a single egg mass together with other parasitoid species. Data obtained in the present study are consistent with those of other European studies and confirm the behaviour for this species. Very interesting are the data on other indigenous parasitoid species that emerged from *H. halys* eggs in a quantity similar to *A. bifasciatus*, particularly *Tr. basalis* and *Te. turesis*, which therefore deserve special attention to assess their possible adaptation and contribution to the control of the exotic pest. Due to the limited number of specimens of these native species found in the eggs of *H. halys*, it is difficult to predict if their presence has already affected or will potentially affect the decline of the alien and/or autochthonous stink bug species.

Given the economic importance of *H. halys* in Serbia, it will likely be beneficial to attract exotic parasitoids to agricultural fields and commercial fruit orchards, e.g., by increasing ecological infrastructures [51,52], and support them in a conservation biological control approach [53]. In this context, it may be warranted to further investigate the current status of the species in terms of spatial distribution and range of native hosts. Although laboratory studies were recently carried out [23,25,54], the interactions between populations of *Tr. japonicus* and *Tr. mitsukurii* in the field deserve close attention.

5. Conclusions

The brown marmorated stink bug, *H. halys*, is a serious threat to plant protection worldwide. Control strategies for this pest in Europe and America rely mainly on repeated application of chemical pesticides. From the perspective of an integrated control strategy with low pesticide use, attention has turned to natural enemies, such as egg parasitoids, which can be a valid strategy. To ensure the effectiveness of biological, a good knowledge of the pest and associated BCAs, including their biology, distribution, and diversity, is needed.

This study reports for the first time the presence of *Tr. japonicus* and *Tr. mitsukurii* in Serbia, thus providing additional valuable information on the geographic distribution of the two exotic parasitoids. Several countries are evaluating the introduction of these two parasitoids in classical biological control programs, and in some countries the release of either species is already underway. Knowing the spontaneous spread of both species is of paramount importance and can help better target and rationalize the control means, and guide scientists in implementing successful biological control strategies.

Author Contributions: Conceptualization, methodology, formal analysis, A.K., F.T. and L.T.; investigation, resources, data curation, A.K. and F.T.; writing—original draft preparation, review and editing, A.K., F.T. and L.T.; funding acquisition, A.K. and F.T. All authors have read and agreed to the published version of the manuscript.

Funding: The research was supported by the Plant Protection Directorate of the Ministry of Agriculture, Forestry, and Water Management of the Republic of Serbia, Agriserbia d.o.o., and Ferrero Trading Lux S.A. The study was also supported by the Ministry of Science, Technological Development, and Innovation of the Republic of Serbia, project number 451-03-47/2023-01/200117, and Agreement on the realization and co-financing of the scientific research work of the accredited Center of Exceptional Values in 2023—Agreement No. 451-03-1524/2023-04/16. This study was carried out within the Agritech National Research Center and received funding from the European Union Next-Generation EU (PIANO NAZIONALE DI RIPRESA E RESILIENZA (PNRR)—MISSIONE 4 COMPONENTE 2, INVESTIMENTO 1.4—D.D. 1032 17/06/2022, CN00000022). This manuscript reflects only the authors' views and opinions; neither the European Union nor the European Commission can be considered responsible for them.

Institutional Review Board Statement: Not applicable.

Informed Consent Statement: Not applicable.

Data Availability Statement: The raw data supporting the conclusions of this article will be made available by the authors on request.

Acknowledgments: A.K. would like to thank Verna Brdar, Milos Petrovic, and Nikola Lacarac from the Faculty of Agriculture, University of Novi Sad, Serbia, for their help in sampling.

Conflicts of Interest: The authors declare that the research was conducted in the absence of any commercial or financial relationships that could be construed as a potential conflict of interest.

References

1. EPPO. *Halyomorpha halys* (HALYHA) [World Distribution] EPPO Global Database. Available online: https://gd.eppo.int/taxon/HALYHA/distribution (accessed on 14 November 2022).
2. Bergmann, E.J.; Venugopal, P.D.; Martinson, H.M.; Raupp, M.J.; Shrewsbury, P.M. Host Plant Use by the Invasive *Halyomorpha halys* (Stål) on Woody Ornamental Trees and Shrubs. *PLoS ONE* **2016**, *11*, e0149975. [CrossRef] [PubMed]
3. Kuhar, T.P.; Kamminga, K. Review of the Chemical Control Research on *Halyomorpha halys* in the USA. *J. Pest Sci.* **2017**, *90*, 1021–1031. [CrossRef]
4. Leskey, T.C.; Nielsen, A.L. Impact of the Invasive Brown Marmorated Stink Bug in North America and Europe: History, Biology, Ecology, and Management. *Annu. Rev. Entomol.* **2018**, *63*, 599–618. [CrossRef] [PubMed]
5. Rice, K.B.; Bergh, C.J.; Bergmann, E.J.; Biddinger, D.J.; Dieckhoff, C.; Dively, G.; Fraser, H.; Gariepy, T.; Hamilton, G.; Haye, T.; et al. Biology, Ecology, and Management of Brown Marmorated Stink Bug (Hemiptera: Pentatomidae). *J. Integr. Pest Manag.* **2014**, *5*, A1–A13. [CrossRef]
6. Yang, Z.Q.; Yao, Y.X.; Qiu, L.F.; Li, Z.X. A New Species of *Trissolcus* (Hymenoptera: Scelionidae) Parasitizing Eggs of *Halyomorpha halys* (Heteroptera: Pentatomidae) in China with Comments on Its Biology. *Ann. Entomol. Soc. Am.* **2009**, *102*, 39–47. [CrossRef]
7. Lee, D.H. Current Status of Research Progress on the Biology and Management of *Halyomorpha halys* (Hemiptera: Pentatomidae) as an Invasive Species. *Appl. Entomol. Zool.* **2015**, *50*, 277–290. [CrossRef]
8. Zhang, J.P.; Zhang, F.; Gariepy, T.D.; Mason, P.G.; Gillespie, D.; Talamas, E.J.; Haye, T. Seasonal Parasitism and Host Specificity of *Trissolcus japonicus* in Northern China. *J. Pest Sci.* **2017**, *90*, 1127–1141. [CrossRef] [PubMed]
9. Abram, P.K.; Hoelmer, K.A.; Acebes-Doria, A.L.; Andrews, H.; Beers, E.H.; Bergh, J.C.; Bessin, R.; Biddinger, D.; Botch, P.; Buffington, M.L.; et al. Indigenous Arthropod Natural Enemies of the Invasive Brown Marmorated Stink Bug in North America and Europe. *J. Pest Sci.* **2017**, *90*, 1009–1020. [CrossRef]
10. Costi, E.; Haye, T.; Maistrello, L. Surveying Native Egg Parasitoids and Predators of the Invasive *Halyomorpha halys* in Northern Italy. *J. Appl. Entomol.* **2019**, *143*, 299–307. [CrossRef]

11. Dieckhoff, C.; Tatman, K.M.; Hoelmer, K.A. Natural Biological Control of *Halyomorpha halys* by Native Egg Parasitoids: A Multi-Year Survey in Northern Delaware. *J. Pest Sci.* **2017**, *90*, 1143–1158. [CrossRef]
12. Fusu, L.; Andreadis, S.S. *Ooencyrtus mirus* (Hymenoptera, Encyrtidae), Discovered in Europe Parasitizing Eggs of *Halyomorpha halys* (Hemiptera, Pentatomidae). *J. Hymenopt. Res.* **2023**, *96*, 1045–1060. [CrossRef]
13. Haye, T.; Fischer, S.; Zhang, J.P.; Gariepy, T.D. Can Native Egg Parasitoids Adopt the Invasive Brown Marmorated Stink Bug, *Halyomorpha halys* (Heteroptera: Pentatomidae), in Europe? *J. Pest Sci.* **2015**, *88*, 693–705. [CrossRef]
14. Rondoni, G.; Bertoldi, V.; Malek, R.; Foti, M.C.; Peri, E.; Maistrello, L.; Haye, T.; Conti, E. Native Egg Parasitoids Recorded from the Invasive *Halyomorpha halys* Successfully Exploit Volatiles Emitted by the Plant–Herbivore Complex. *J Pest Sci* **2017**, *90*, 1087–1095. [CrossRef]
15. Stahl, J.M.; Babendreier, D.; Haye, T. Using the Egg Parasitoid *Anastatus bifasciatus* against the Invasive Brown Marmorated Stink Bug in Europe: Can Non-Target Effects Be Ruled Out? *J. Pest Sci.* **2018**, *91*, 1005–1017. [CrossRef] [PubMed]
16. Tillman, G.; Toews, M.D.; Blaauw, B.R.; Sial, A.A.; Cottrell, T.E.; Talamas, E.J.; Buntin, D.G.; Joseph, S.V.; Balusu, R.R.; Fadamiro, H.; et al. Parasitism and Predation of Sentinel Eggs of the Invasive Brown Marmorated Stink Bug, *Halyomorpha halys* (Stål) (Hemiptera: Pentatomidae), in the Southeastern US. *Biol. Control* **2020**, *145*, 104247. [CrossRef]
17. Moraglio, S.T.; Tortorici, F.; Visentin, S.; Pansa, M.G.; Tavella, L. *Trissolcus kozlovi* in North Italy: Host Specificity and Augmentative Releases against *Halyomorpha halys* in Hazelnut Orchards. *Insects* **2021**, *12*, 464. [CrossRef] [PubMed]
18. Stahl, J.; Tortorici, F.; Pontini, M.; Bon, M.C.; Hoelmer, K.A.; Marazzi, C.; Tavella, L.; Haye, T. First Discovery of Adventive Populations of *Trissolcus japonicus* in Europe. *J. Pest Sci.* **2019**, *92*, 371–379. [CrossRef]
19. Stahl, J.; Babendreier, D.; Marazzi, C.; Caruso, S.; Costi, E.; Maistrello, L.; Haye, T. Can *Anastatus bifasciatus* Be Used for Augmentative Biological Control of the Brown Marmorated Stink Bug in Fruit Orchards? *Insects* **2019**, *10*, 108. [CrossRef] [PubMed]
20. Zapponi, L.; Tortorici, F.; Anfora, G.; Bardella, S.; Bariselli, M.; Benvenuto, L.; Bernardinelli, I.; Butturini, A.; Caruso, S.; Colla, R.; et al. Assessing the Distribution of Exotic Egg Parasitoids of *Halyomorpha halys* in Europe with a Large-Scale Monitoring Program. *Insects* **2021**, *12*, 316. [CrossRef]
21. Herlihy, M.V.; Talamas, E.J.; Weber, D.C. Attack and Success of Native and Exotic Parasitoids on Eggs of *Halyomorpha halys* in Three Maryland Habitats. *PLoS ONE* **2016**, *11*, e0150275. [CrossRef]
22. Sabbatini Peverieri, G.; Dieckhoff, C.; Giovannini, L.; Marianelli, L.; Roversi, P.F.; Hoelmer, K.A. Rearing *Trissolcus japonicus* and *Trissolcus mitsukurii* for Biological Control of *Halyomorpha halys*. *Insects* **2020**, *11*, 787. [CrossRef] [PubMed]
23. Costi, E.; Di Bella, E.; Iotti, D.; Maistrello, L. Biocontrol Implications of Multiparasitism by *Trissolcus mitsukurii* and *Trissolcus japonicus* on the Invasive Brown Marmorated Stink Bug. *Entomol. Exp. Appl.* **2022**, *170*, 772–781. [CrossRef]
24. Giovannini, L.; Sabbatini Peverieri, G.; Marianelli, L.; Rondoni, G.; Conti, E.; Roversi, P.F. Physiological Host Range of *Trissolcus mitsukurii*, a Candidate Biological Control Agent of *Halyomorpha halys* in Europe. *J. Pest Sci.* **2022**, *95*, 605–618. [CrossRef]
25. Giovannini, L.; Sabbatini-Peverieri, G.; Simoni, S.; Cervo, R.; Alan Hoelmer, K.; Federico Roversi, P. Interspecific Competition between *Trissolcus japonicus* and *Trissolcus mitsukurii*, Two Promising Candidates for Biocontrol of *Halyomorpha halys*. *Biol. Control* **2022**, *176*, 105068. [CrossRef]
26. Talamas, E.J.; Herlihy, M.V.; Dieckhoff, C.; Hoelmer, K.A.; Buffington, M.L.; Bon, M.C.; Weber, D.C. *Trissolcus japonicus* (Ashmead) (Hymenoptera, Scelionidae) Emerges in North America. *J. Hymenopt. Res.* **2015**, *43*, 119–128. [CrossRef]
27. Rot, M.; Maistrello, L.; Costi, E.; Bernardinelli, I.; Malossini, G.; Benvenuto, L.; Trdan, S. Native and Non-Native Egg Parasitoids Associated with Brown Marmorated Stink Bug (*Halyomorpha halys* [Stål, 1855]; Hemiptera: Pentatomidae) in Western Slovenia. *Insects* **2021**, *12*, 505. [CrossRef]
28. Yonow, T.; Kriticos, D.J.; Ota, N.; Avila, G.A.; Hoelmer, K.A.; Chen, H.; Caron, V. Modelling the Potential Geographic Distribution of Two *Trissolcus* Species for the Brown Marmorated Stink Bug, *Halyomorpha halys*. *Insects* **2021**, *12*, 491. [CrossRef]
29. Tortorici, F.; Bombi, P.; Loru, L.; Mele, A.; Moraglio, S.T.; Scaccini, D.; Pozzebon, A.; Pantaleoni, R.A.; Tavella, L. *Halyomorpha halys* and Its Egg Parasitoids *Trissolcus japonicus* and *T. mitsukurii*: The Geographic Dimension of the Interaction. *NeoBiota* **2023**, *85*, 197–221. [CrossRef]
30. Scaccini, D.; Falagiarda, M.; Tortorici, F.; Martinez-Sañudo, I.; Tirello, P.; Reyes-Domínguez, Y.; Gallmetzer, A.; Tavella, L.; Zandigiacomo, P.; Duso, C.; et al. An Insight into the Role of *Trissolcus mitsukurii* as Biological Control Agent of *Halyomorpha halys* in Northeastern Italy. *Insects* **2020**, *11*, 306. [CrossRef]
31. Bout, A.; Tortorici, F.; Hamidi, R.; Warot, S.; Tavella, L.; Thomas, M. First Detection of the Adventive Egg Parasitoid of *Halyomorpha halys* (Stål) (Hemiptera: Pentatomidae) *Trissolcus mitsukurii* (Ashmead) (Hymenoptera: Scelionidae) in France. *Insects* **2021**, *12*, 761. [CrossRef]
32. Šeat, J. *Halyomorpha halys* (Stål, 1855) (Heteroptera: Pentatomidae) a New Invasive Species in Serbia. *Acta Entomol. Serbic* **2015**, *20*, 167–171. [CrossRef]
33. Musolin, D.L.; Konjević, A.; Karpun, N.N.; Protsenko, V.Y.; Ayba, L.Y.; Saulich, A.K. Invasive Brown Marmorated Stink Bug *Halyomorpha halys* (Stål) (Heteroptera: Pentatomidae) in Russia, Abkhazia, and Serbia: History of Invasion, Range Expansion, Early Stages of Establishment, and First Records of Damage to Local Crops. *Arthropod-Plant Interact.* **2018**, *12*, 517–529. [CrossRef]
34. Konjević, A. True Bugs (Heteroptera) as Pests in Ornamentals. In *Sustainable Practices in Horticulture and Landscape Architecture*; Ostojić, J., Cig, A., Eds.; Iksad: Ankara, Turkey, 2022; pp. 123–144.

35. Haye, T.; Wyniger, D. Stink Bug Eggs—The Marbled Stink Bug, *Halyomorpha halys*. Available online: https://www.halyomorphahalys.com/wanzeneier-stink-bug-eggs.html (accessed on 7 February 2023).
36. Talamas, E.J.; Buffington, M.L.; Hoelmer, K.A. Revision of Palearctic *Trissolcus* Ashmead (Hymenoptera, Scelionidae). *J. Hymenopt. Res.* **2017**, *56*, 3–185. [CrossRef]
37. Tortorici, F.; Talamas, E.J.; Moraglio, S.T.; Pansa, M.G.; Asadi-Farfar, M.; Tavella, L.; Caleca, V. A Morphological, Biological and Molecular Approach Reveals Four Cryptic Species of *Trissolcus* Ashmead (Hymenoptera, Scelionidae), Egg Parasitoids of Pentatomidae (Hemiptera). *J. Hymenopt. Res.* **2019**, *93*, 153–200. [CrossRef]
38. Peng, L.; Gibson, G.A.P.; Tang, L.; Xiang, J. Review of the Species of *Anastatus* (Hymenoptera: Eupelmidae) Known from China, with Description of Two New Species with Brachypterous Females. *Zootaxa* **2020**, *4767*, 351–401. [CrossRef]
39. Kozlov, M.A.; Kononova, S.V. *Telenominae of the Fauna of the USSR*; Nauka: Leningrad, Russia, 1983.
40. Johnson, N.F. Systematics of Nearctic *Telenomus*: Classification and Revisions of the *podisi* and *phymatae* Species Groups (Hymenoptera: Scelionidae). *Bull. Ohio Biol. Surv.* **1984**, *6*, 1–113. [CrossRef]
41. Sabbatini Peverieri, G.; Talamas, E.J.; Bon, M.C.; Marianelli, L.; Bernardinelli, I.; Malossini, G.; Benvenuto, L.; Roversi, P.F.; Hoelmer, K.A. Two Asian Egg Parasitoids of *Halyomorpha halys* (Stål) (Hemiptera, Pentatomidae) Emerge in Northern Italy: *Trissolcus mitsukurii* (Ashmead) and *Trissolcus japonicus* (Ashmead) (Hymenoptera, Scelionidae). *J. Hymenopt. Res.* **2018**, *67*, 37–53. [CrossRef]
42. Haye, T.; Wyniger, D.; Gariepy, T. Recent Range Expansion of Brown Marmorated Stink Bug in Europe. In Proceedings of the 8th International Conference on Urban Pests, Zurich, Switzerland, 20–23 July 2014; Müller, G., Pospischil, R., Robinson, W.H., Eds.; Executive Committee of the International Conference on Urban Pests: Zurich, Switzerland, 2014; pp. 309–314.
43. Cesari, M.; Maistrello, L.; Ganzerli, F.; Dioli, P.; Rebecchi, L.; Guidetti, R. A Pest Alien Invasion in Progress: Potential Pathways of Origin of the Brown Marmorated Stink Bug *Halyomorpha halys* Populations in Italy. *J. Pest Sci.* **2015**, *88*, 1–7. [CrossRef]
44. Konjević, A. First Records of the Brown Marmorated Stink Bug *Halyomorpha halys* (Stål, 1855) (Hemiptera: Pentatomidae) in Republic of North Macedonia. *Acta Zool. Bulg.* **2020**, *72*, 687–690.
45. Buffington, M.L.; Talamas, E.J.; Hoelmer, K.A. Team *Trissolcus*: Integrating Taxonomy and Biological Control to Combat the Brown Marmorated Stink Bug. *Am. Entomol.* **2018**, *64*, 224–232. [CrossRef]
46. Dieckhoff, C.; Wenz, S.; Renninger, M.; Reißig, A.; Rauleder, H.; Zebitz, C.P.W.; Reetz, J.; Zimmermann, O. Add Germany to the List—Adventive Population of *Trissolcus japonicus* (Ashmead) (Hymenoptera: Scelionidae) Emerges in Germany. *Insects* **2021**, *12*, 414. [CrossRef] [PubMed]
47. Noyes, J. Universal Chalcidoidea Database—World Wide Web Electronic Publication. Available online: https://www.nhm.ac.uk/our-science/data/chalcidoids/database/ (accessed on 27 March 2023).
48. Konopka, J.K.; Haye, T.; Gariepy, T.D.; McNeil, J.N. Possible Coexistence of Native and Exotic Parasitoids and Their Impact on Control of *Halyomorpha halys*. *J. Pest Sci.* **2017**, *90*, 1119–1125. [CrossRef]
49. Stahl, J.M.; Babendreier, D.; Haye, T. Life History of *Anastatus bifasciatus*, a Potential Biological Control Agent of the Brown Marmorated Stink Bug in Europe. *Biol. Control* **2019**, *129*, 178–186. [CrossRef]
50. Mi, Q.; Zhang, J.P.; Haye, T.; Zhang, B.; Zhao, C.; Lei, Y.; Li, D.; Zhang, F. Fitness and Interspecific Competition of *Trissolcus japonicus* and *Anastatus japonicus*, Egg Parasitoids of *Halyomorpha halys*. *Biol. Control* **2021**, *152*, 104461. [CrossRef]
51. Olson, D.M.; Wäckers, F.L. Management of Field Margins to Maximize Multiple Ecological Services. *J. Appl. Ecol.* **2007**, *44*, 13–21. [CrossRef]
52. Winkler, K.; Wäckers, F.; Pinto, D.M. Nectar-Providing Plants Enhance the Energetic State of Herbivores as Well as Their Parasitoids under Field Conditions. *Ecol. Entomol.* **2009**, *34*, 221–227. [CrossRef]
53. Abram, P.K.; Mills, N.J.; Beers, E.H. Review: Classical Biological Control of Invasive Stink Bugs with Egg Parasitoids—What Does Success Look Like? *Pest Manag. Sci.* **2020**, *76*, 1980–1992. [CrossRef]
54. Scala, M.; Fouani, J.M.; Zapponi, L.; Mazzoni, V.; Wells, K.E.; Biondi, A.; Baser, N.; Verrastro, V.; Anfora, G. Attraction of Egg Parasitoids *Trissolcus mitsukurii* and *Trissolcus japonicus* to the Chemical Cues of *Halyomorpha halys* and *Nezara viridula*. *Insects* **2022**, *13*, 439. [CrossRef]

biology

MDPI

Opinion

Biological Invasion Data Gaps in China: Examples of Distribution, Inventories, and Impact

Jiayao He [1], Ke Chen [1], Peishan Sun [2], Han Xu [1] and Xubin Pan [1,*]

1 Institute of Plant Inspection and Quarantine, Chinese Academy of Inspection and Quarantine, Beijing 100176, China
2 Research Center of GACC for International Inspection and Quarantine Standards and Technical Regulations, Beijing 100013, China
* Correspondence: xubin.hu.pan@gmail.com

Simple Summary: The growing impact of invasive alien species highlights the need for effective management. However, the lack of baseline data on invasive alien species often hinder the efficient allocation of resources. Data requirements for invasive alien species remain under-recognized, leading to insufficient evaluation of the gaps. Our study identifies critical data gaps and proposes recommendations for improving the collection, curation, and quality control of invasive alien species data in China.

Abstract: The impact of invasive alien species (IAS) on nature and society is increasing globally. It is crucial to utilize information systems for evidence-based management, enabling the assessment and supporting survey and control actions. However, the lack of accessible and comprehensive baseline IAS data often impedes the ability to prioritize and allocate resources efficiently. Despite the increased public awareness of biological invasions in China over the past decades, the critical importance of data requirements has not been fully recognized, leading to gaps in available data. Here, we outline the key data demands for the management of biological invasions and highlight the current lack of high-quality data for invasion management in China, and critically assess data gaps in IAS distribution, inventory, and impact. Additionally, we propose a conceptual framework to illustrate the data requirements throughout the invasion management process, along with indicators to assess data quality within three dimensions: relevance, resolution, and reliability.

Keywords: biosecurity; data demands; data gap analysis; proactive management

1. Introduction

The growing threats to biodiversity, ecosystem services, and society sustainability posed by invasive alien species (IAS) underline the need to integrate information and develop assessment approaches on IAS [1–4]. Accurate and sufficient information on IAS facilitates the management of biological invasions and the ability to respond rapidly. In addition, reliable data provide a basis for discerning invasion patterns and, thus, affect policy-making and management actions related to reducing IAS threats directly [5,6]. These data can be used to prioritize IAS and allocate management resources more objectively [7]. However, comprehensive standardized baseline data have not been available, and there are multiple data gaps despite a substantial increase in invasion research on IAS.

China encompasses many endemic species and biological resources that are critically vulnerable to a wide range of IAS [8]. Despite many efforts, biological invasions have been a major cause of the accelerated loss of biodiversity. Previous studies have demonstrated more than 660 invasive alien species, according to a government report (2019 Report on the State of National Ecology and Environment). However, this report lacks detailed information on the specific composition of these invasive organisms. In addition, annual economic losses caused by IAS in China have been estimated at more than $18.9 billion, with invasive alien insects accounting for a large proportion of these losses [9]. However, several studies have shown that the number of documented invasive alien species and the economic losses caused by them seem to be underestimated [10–13]. As global trade has intensified the movement of many species across natural dispersal barriers, non-native species have become naturalized and even invasive [14]. Importantly, trade volume and human footprint are important indicators of biological invasion, and IAS has become a driver of biodiversity threats, particularly in developing countries [15]. China is now suffering more from these threats than at any other time in the past few decades [16].

It is crucial to understand the specific data required at each stage of the biological invasion process to manage invasive alien species, further contributing to sustainable development goals. IAS data that is available, accessible, and of sufficient quality is essential for supporting evidence-based management. However, existing information on IAS data in China, which is fundamentally important to both scientific and political agendas, is currently fragmented. Our study aimed to make a data gap analysis within IAS distribution, inventories, and impact in China, indicating the specific role of data in invasion research and management and then providing insights into the baseline information needed to underpin policy-making and management settings for IAS.

2. IAS Data Demands and Gaps

To determine which alien species are "invasive" and to predict the risk of a species being introduced, establishing, spreading, and causing consequences, relevant information regarding its taxonomy, pathways, habitat types, abiotic and biotic factors, and impacts associated with these species, should be required [17]. The availability of the above information can help with evidence-based management and play a vital role in data-driven research in invasion biology [18,19]. In addition, they can provide opportunities for improvements in related analytical approaches and can contribute to an in-depth understanding of invasion mechanisms in the global context [1]. Accordingly, inventories with established numbers and distribution range of IAS allow for an enriched knowledge in invasion patterns [20], while incomplete and unstructured data of biological invasions would get in the way of management efforts [21]. Here, we propose a conceptual framework to illustrate the data requirements throughout the biological invasion process and management actions (Figure 1).

2.1. Distribution

Data on spatial and temporal distributions is an essential part of bio-invasion knowledge, which is necessary to estimate existing/potential threats and manage IAS [21]. Interestingly, identifying species introductions from a historical perspective—geographic origins and year of first record—was found even more effective than traits in explaining invasion success along multiple dimensions of invasiveness [22,23]. Historical and current data on species distributions are needed to track changes in biodiversity [24]; however, the reliability of inferences from these records is highly impacted by data quality. Notably, occurrence records are important for examining species' responses or adaptation to environmental variables. Integrating distribution data and spatial variables can inform assessments, surveys, and control. For example, the year and sites of an IAS first record can infer the primary introduction pathway of this species, which maybe link to accidental escapes from ornamental and horticultural activities, thereby assisting the regulation of IAS gateway areas and improving pathway management [25].

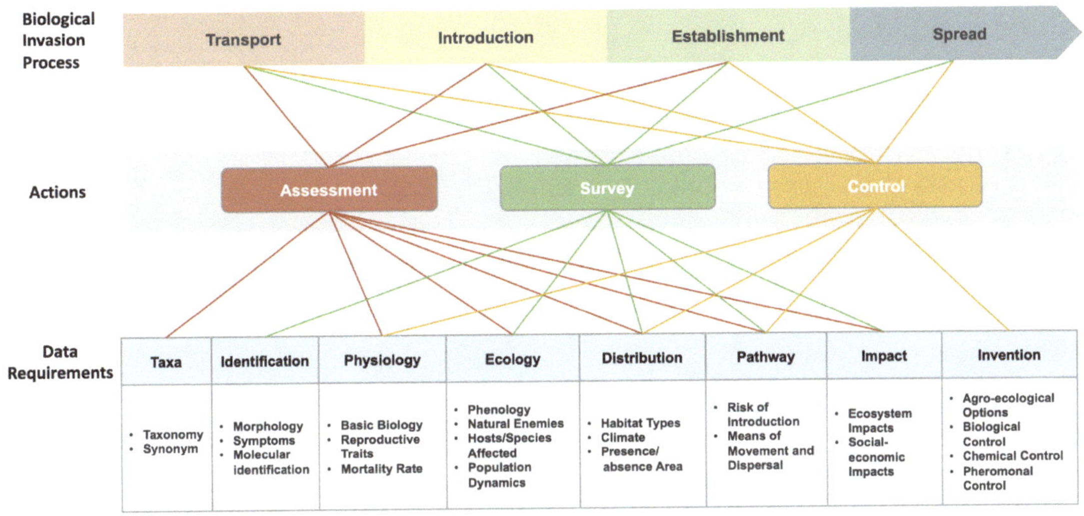

Figure 1. General outline of data required for the stages of invasion and generation of key management actions. Each stage has particular data requirements to ensure effective responses. These data are essential for decision-making, prioritizing resource allocations, and managing biological invasions.

Many government professionals and researchers have regarded "evaluating the quality of distribution data" as the most difficult part of usage for species occurrence information systems [26]. The geographical distribution of the IAS can be compiled at sub-regional administrative divisions. However, various factors such as environmental variables and vegetation cover, that could account for biological invasion are not consistent among these artificially zoned sites. Thus, their applications are difficult owning to such discrepancies. As such, finer distribution data rather than summaries at coarser resolutions are required, and monitoring of invasion trends requires more than a single snapshot of the distribution [24].

High-resolution distribution data are often lacking or only exist at a small spatial or short temporal scale in China, expecially in the remote areas such as marine islands or high elevation mountainous area (Table 1). In several studies, species occurrence has been used to analyze biological traits related to invasion success [27], estimate the spatial extent of invasive patterns [28], and identify climate, land use, and socioeconomic factors that could affect IAS distribution patterns [29–32]. However, occurrence data have mostly been complied at the county level [33]. Moreover, the occurrences of some high-profile IAS have only been collated at the provincial and county levels [34,35]. In addition, early catalogue that has not been updated for a long time which has often been used in recent research and is likely to undermine the credibility of the results. For example, basic distribution data used in a 2020 study to analyze spatiotemporal patterns were recorded in Ma (2014) [33,36]; however, the introduction and spread of IAS has been rapidly growing in recent years. Despite the effectiveness of technologies and tools such as species distribution models (SDMs) in predicting the potential geographical distribution of IAS, which is useful to inform proactive management strategies, the lack of baseline data at finer distribution resolutions remains one of the key gaps that need to be addressed.

Table 1. Key data gaps regarding the distribution, inventories, and impacts of invasive alien species in China are identified and collated. These gaps are expected to be addressed through nationally coordinated research and development as public awareness increases over time.

Data Type	Current Issues	Gaps	Ongoing Efforts
Distribution	Lack of georeferenced data at detailed temporal/spatial scales.	- Inaccuracy - Inadequate data	Systematic surveys and ongoing monitoring are currently underway.
Inventories	Insufficient documentation of introduced species.	- Outdated - Incomplete or underestimated	Joint efforts are underway to update the inventory.
Impact	Evidence of impacts is fragmentary.	- Sporadic research across species - Limited estimations on impact sectors - Underestimated	We are working on the consolidation of impact data.

2.2. Inventories

IAS inventories from distribution are necessary to provide historical baselines and inform management strategies. Therefore, it is essential to compile inventories of exotic species and develop effective tools for storing IAS data. Ensuring reliability and completeness is a major challenge when facing enormous volumes of data. Additionally, the depth and range of surveillance activities can directly affect the resolution of IAS data.

Previous studies have shown a positive correlation between the number of documented invasive alien species and land area [37,38]. In Europe, more than 12,000 species of alien plants and animals have been documented [39]. The United States, with a surface area comparable to China, hosts 3,885 invasive species out of 12,981 introduced species records [40]. Moreover, China and US have similar ecogeographic regions; both are major importers and exporters of goods. Notably, a widely-used comprehensive national inventory for invasive species was provided by Xu et al. [13], and the reported number (488) may be a significant underestimate despite the time lag between the introduction and subsequent establishment of non-native species. A verified national checklist was presented on the The Global Register of Introduced and Invasive Species (GRIIS), including 917 records of introduced (alien) and invasive species known to occur in China. [41]. Despite significant efforts and the use of effective prioritization tools to identify potential IAS targets and update quarantine pest lists, the lack of integration for established IAS/quarantine pest lists remains a critical issue (Table 1).

Globally, at least 13,939 plant taxa have become naturalized in regions outside their native distributions [42]. There are 950 plant species that have been naturalized in China, and approximately 400 species are considered invasive [43]. Intensive studies have focused on checklists of alien plants in China [36,44–47]. However, inventories of other taxonomic groups are not well documented. Additionally, the alien status and introduction dates of many species have rarely been stated and clarified; only a limited number of studies [9,13,34] have annotated the year when an invasive alien species was first detected. Thus, one crucial requirement is the establishment of an information integration system that includes national IAS inventories, promoting the information consolidation of invasive species.

2.3. Impact

The impact data of IAS can provide intuitive and specific conclusions and are crucial for supporting invasion-related policies, and critical information regarding the risks or impacts of alien species can inform invasion management and control actions. For example, identifying extremely harmful alien species will guide agencies in taking actions to prevent introductions and contain their spread [48]. However, the gravity of biological invasions has not been generally acknowledged by the public and is undervalued by decision-makers, which is a consequence of the limited knowledge and societal awareness of IAS impacts [49]

Several studies assessing the impact of alien species have emerged in recent years. A recent study showed that the biological invasion of insects costs a minimum of US$76.0 billion per year [2]. As such, the total reported costs of global bio-invasion reached US$1.288 trillion over the past few decades (1970–2017), and the annual cost has steadily increased over time, with no signs of slowing down [50]. Importantly, some effects may be underestimated, as many assessments do not obtain full coverage of the impacted sectors [1], along with biased estimated values due to geographically uneven data. Accordingly, IAS impacts in high-income areas are easily available and simultaneously support decisions regarding the control of invasive species [50,51]. A previous study predicted that China would suffer the highest economic loss in agriculture due to alien invasive pests [52]. The number of alien species for which effects have been reported represents only 3.3% of all introduced species in Asia. Similarly, in China, the number of introduced species is five times higher than those with cost estimates [11], highlighting the absence of impact assessments for a substantial portion of alien species. In particular, given the high GDP, significant trade volume, and the vast numbers of IASs with unreported costs in China, comprehensive and collective assessments of their monetary costs across socioeconomic sectors, taxonomic groups, and geographic regions are still lacking (Table 1).

We extracted records of economic cost estimations from China between 1995 and 2017 based on an established dataset [11], which compiled the actual economic costs of invasions from the InvaCost database [53], incorporating both Chinese and English data. Consistent with our expectations, invasive species, such as *Eichhornia crassipes* (water hyacinth), *Hyphantria cunea* (fall webworms), and *Bursaphelenchus xylophilus* (pine wood nematodes), accounted for a large proportion of the reported analyses. Some analyses have examined how invasions affect ecosystem services and genetic resources, particularly from agricultural, forestry, and fishery perspectives [54]. However, diversifying impacts such as social welfare, urban vegetation, and infrastructure are rarely addressed or quantified. Relevant studies on invasion impacts in China remain insufficient, and existing analyses cover only a small fraction of invasive species, which might lead to taxonomic biases. Since the widely cited analyses [9,55] have been outdated and less representative owing to rapidly escalating biological invasions over decades, quantifying invasion impacts is required to fill the gaps.

3. IAS Data Evaluation

Considering the quality of IAS data is crucial during the collection and storage process, we have identified three key indicators: relevance, resolution, and reliability (Figure 2). These indicators can be assessed across multiple dimensions, such as taxonomic, spatial, and temporal aspects. "Relevance" refers to the extent to which the data aligns with the specific needs and objectives. "Resolution" refers to the level of detail and precision within the data, which may range from the resolution of spatial data, the frequency and timeliness of temporal data, to the specification of taxonomic data. "Reliability" refers to the trustworthiness and consistency of the data, ensuring that it is from credible sources and can be confidently used in research, management, and policy-making. Furthermore, the report titled "The Thematic Assessment Report on Invasive Alien Species and Their Control' from the Intergovernmental Science-Policy Platform on Biodiversity and Ecosystem Services (IPBES)" provides a good example of evaluating gaps in invasive species data and knowledge [56].

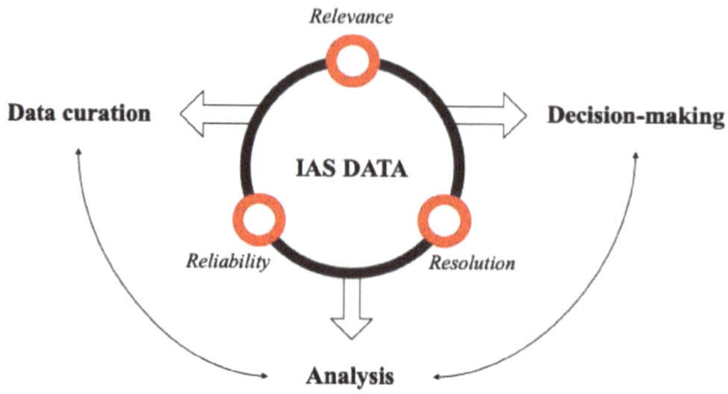

Figure 2. Framework for data quality evaluation of invasive alien species. The processes of data curation (organizing and managing a collection of relevant datasets) and analyses may be essential for evidence-based decision-making. Before that, data should be accessible and analytical. The framework emphasizes three key indicators of IAS data gap analysis that focus on relevance, resolution, and reliability.

4. Future Development and Recommendations for China's IAS Data

4.1. Biosecurity Law and Nationwide General Survey of IAS

The Biosecurity Law of the People's Republic of China has been in effect since 2021, aiming to enhance the management level of biosecurity and establish a national biosecurity coordination mechanism. This law emphasizes national multiagency coordination to enhance the identification, survey, and management of IAS. Additionally, a nationwide general survey of IAS has been conducted in China, providing up-to-date data on biological invasions. On this basis, subsequent research should prioritize data integration and assess the status of alien species, classifying them as invasive where appropriate. In addition to IAS prioritization, efforts should focus on population management and increasing the availability of data in the poluplation dynamics of invasive alien species to ensure effective management strategies. These efforts are expected to bridge significant gaps in IAS distribution, inventories, and impact, delivering reliable information to policymakers and researchers, thereby enabling actions to mitigate the detrimental impacts of IAS.

4.2. Balance of IAS Data Security and Sharing

Accurate, detailed, and timely information on biological invasions is necessary; however, certain restrictions on the sharing of phytosanitary data currently hinder the development of cyberinfrastructure [57]. Given the significance and sensitivity of biosecurity data to national security and economic benefits, the misuse or unauthorized disclosure of such data can have serious implications for stakeholders, national security, and international trade. Promoting biosecurity data sharing requires the involvement of various entities and strikes a balance between data security and sharing, while the FAIR principles (i.e., findable, accessible, interoperable, and reusable) provide general guidance for promoting the reuse and integration of scientific data [58,59]. The CARE principles (i.e., collective benefit, authority to control, responsibility, and ethics) are centered around people and purpose, recognizing the pivotal role of data in advancing innovation, governance, and self-determination [60]. Thus, IAS data sharing following FAIR and CARE principles must not harm national biosecurity and abide by the law.

4.3. National Integrated System for IAS

Harmonizing and consolidating reliable data is crucial for converting scattered information into valuable knowledge. China has made significant investments in data collection for IAS; nevertheless, there is still a lack of integrated network information platforms. It is imperative to urgently aggregate and integrate existing data on IAS in China through collaborative efforts. In addition, ensuring quality control is essential to providing accessible data that is reliable and credible, and establishing links to the corresponding information sources for easy tracking is recommended.

5. Conclusions

The emerging trends of IAS have been increasing at unprecedented rates. Deficiency of IAS data may lead to costly and ineffective investment in invasion management. Bridging gaps in regard to data availability and quality to aid assessment, survey and control actions of IAS is an urgent but challenging task in the long term. This study proposes a conceptual framework for the required IAS data in order to trigger more concern about data collection, storage, and curation. Main biological invasion data gaps of distribution, inventories, and impact in China is illustrated. Collaboration between the scientific and management sectors is essential for delivering more fruitful outcomes and ensuring that the management resources are allocated towards effective objectives.

Author Contributions: Conceptualization, X.P. and J.H.; methodology, X.P. and J.H.; validation, H.X., P.S. and K.C.; investigation, K.C.; data curation, J.H.; writing—original draft preparation, J.H.; writing—review and editing, X.P.; visualization, X.P. and J.H.; supervision, K.C.; project administration, K.C.; funding acquisition, X.P. All authors have read and agreed to the published version of the manuscript.

Funding: This work was funded by the Basic Scientific Research Foundation of the Chinese Academy of Inspection and Quarantine (Grant No. 2021JK007) and the National Key R&D Program of China (Grant No. 2023YFC2605200).

Institutional Review Board Statement: Not applicable.

Informed Consent Statement: Not applicable.

Data Availability Statement: No new data were created or analyzed in this study.

Conflicts of Interest: The authors declare no conflict of interest.

References

1. Bartz, R.; Kowarik, I. Assessing the environmental impacts of invasive alien plants: A review of assessment approaches. *NeoBiota* **2019**, *43*, 69–99. [CrossRef]
2. Bradshaw, C.J.A.; Leroy, B.; Bellard, C.; Roiz, D.; Albert, C.; Fournier, A.; Barbet-Massin, M.; Salles, J.-M.; Simard, F.; Courchamp, F. Massive yet grossly underestimated global costs of invasive insects. *Nat. Commun.* **2016**, *7*, 12986. [CrossRef] [PubMed]
3. David, P.; Thébault, E.; Anneville, O.; Duyck, P.F.; Chapuis, E.; Loeuille, N. Impacts of Invasive Species on Food Webs: A Review of Empirical Data. In *Advances in Ecological Research*; David, A.B., Alex, J.D., François, M., Eds.; Academic Press: Massachusetts, MA, USA, 2017; Volume 56, pp. 1–60.
4. Van Rees, C.B.; Hand, B.K.; Carter, S.C.; Bargeron, C.; Cline, T.J.; Daniel, W.; Ferrante, J.A.; Gaddis, K.; Hunter, M.E.; Jarnevich, C.S.; et al. A framework to integrate innovations in invasion science for proactive management. *Biol. Rev.* **2022**, *97*, 1712–1735. [CrossRef] [PubMed]
5. Meyerson, L.; Mooney, H. Invasive alien species in an era of globalization. *Front. Ecol. Environ.* **2007**, *5*, 199–208. [CrossRef]
6. Pyšek, P.; Richardson, D.; Rejmanek, M.; Webster, G.; Williamson, M.; Kirschner, J. Alien Plants in Checklists and Floras: Towards Better Communication between Taxonomists and Ecologists. *Taxon* **2004**, *53*, 131–143. [CrossRef]
7. McGeoch, M.A.; Genovesi, P.; Bellingham, P.J.; Costello, M.J.; McGrannachan, C.; Sheppard, A. Prioritizing species, pathways, and sites to achieve conservation targets for biological invasion. *Biol. Invasions* **2016**, *18*, 299–314. [CrossRef]
8. Weber, E.; Li, B. Plant Invasions in China: What Is to Be Expected in the Wake of Economic Development? *BioScience* **2008**, *58*, 437–444. [CrossRef]
9. Wan, F.H.; Yang, N.W. Invasion and Management of Agricultural Alien Insects in China. *Annu. Rev. Entomol.* **2016**, *61*, 77–98. [CrossRef]
10. Axmacher, J.; Sang, W. Plant Invasions in China—Challenges and Chances. *PLoS ONE* **2013**, *8*, e64173. [CrossRef]
11. Liu, C.; Diagne, C.; Angulo, E.; Banerjee, A.; Chen, Y.; Cuthbert, R.; Haubrock, P.; Kirichenko, N.; Pattison, Z.; Watari, Y. Economic costs of biological invasions in Asia. *NeoBiota* **2021**, *67*, 53–78. [CrossRef]
12. Wu, S.-H.; Sun, H.-T.; Teng, Y.-C.; Rejmanek, M.; Chaw, S.-M.; Yang, T.-Y.; Hsieh, C.-F. Patterns of plant invasions in China: Taxonomic.; biogeographic, climatic approaches and anthropogenic effects. *Biol. Invasions* **2010**, *12*, 2179–2206. [CrossRef]
13. Xu, H.; Qiang, S.; Genovesi, P.; Ding, H.; Wu, J.; Meng, L.; Zheng, M.; Miao, J.; Hu, B.; Guo, J. An inventory of invasive alien species in China. *Neobiota.* **2012**, *15*, 1–26. [CrossRef]

14. Ni, M.; Deane, D.C. Annual first record rate of naturalised non-native plants in China driven by intentional introductions. *Biol. Invasions* **2022**, *24*, 603–606. [CrossRef]
15. Lenzen, M.; Moran, D.; Kanemoto, K.; Foran, B.; Lobefaro, L.; Geschke, A. International trade drives biodiversity threats in developing nations. *Nature* **2012**, *486*, 109–112. [CrossRef]
16. Liu, X.; Blackburn, T.; Song, T.; Li, X.; Huang, C.; Yiming, L. Risks of Biological Invasion on the Belt and Road. *Curr. Biol.* **2019**, *29*, 499–505.e4. [CrossRef]
17. Richardson, D.M.; Pyšek, P. Naturalization of introduced plants: Ecological drivers of biogeographical patterns. *New Phytol.* **2012**, *196*, 383–396. [CrossRef] [PubMed]
18. Joppa, L.; O'Connor, B.; Visconti, P.; Smith, C.; Geldmann, J.; Hoffmann, M.; Watson, J.; Butchart, S.; Virah-Sawmy, M.; Halpern, B. Filling in biodiversity threat gaps. *Science* **2016**, *352*, 416–418. [CrossRef]
19. Pyšek, P.; Pergl, J.; Essl, F.; Lenzner, B.; Dawson, W.; Kreft, H.; Weigelt, P.; Winter, M.; Kartesz, J.; Nishino, M. Naturalized alien flora of the world: Species diversity, taxonomic and phylogenetic patterns, geographic distribution and global hotspots of plant invasion. *Preslia* **2017**, *89*, 203–274. [CrossRef]
20. Pyšek, P.; Meyerson, L.A.; Simberloff, D. Introducing "Alien Floras and Faunas", a new series in Biological Invasions. *Biol. Invasions* **2018**, *20*, 1375–1376. [CrossRef]
21. Latombe, G.; Pyšek, P.; Jeschke, J.M.; Blackburn, T.M.; Bacher, S.; Capinha, C.; Costello, M.J.; Fernández, M.; Gregory, R.D.; Hobern, D.; et al. A vision for global monitoring of biological invasions. *Biol. Conserv.* **2017**, *213*, 295–308. [CrossRef]
22. Fristoe, T.S.; Chytrý, M.; Dawson, W.; Essl, F.; Heleno, R.; Kreft, H.; Maurel, N.; Pergl, J.; Pyšek, P.; Seebens, H. Dimensions of invasiveness: Links between local abundance, geographic range size, and habitat breadth in Europe's alien and native floras. *Proc. Natl. Acad. Sci. USA* **2021**, *118*, e2021173118. [CrossRef] [PubMed]
23. Pyšek, P.; Manceur, A.M.; Alba, C.; McGregor, K.F.; Pergl, J.; Štajerová, K.; Chytrý, M.; Danihelka, J.; Kartesz, J.; Klimešová, J. Naturalization of central European plants in North America: Species traits, habitats, propagule pressure, residence time. *Ecology* **2015**, *96*, 762–774. [CrossRef] [PubMed]
24. Boakes, E.; McGowan, P.; Fuller, R.; Chang-qing, D.; Clark, N.; O'Connor, K. Distorted Views of Biodiversity: Spatial and Temporal Bias in Species Occurrence Data. *PLoS Biol.* **2010**, *8*, e1000385. [CrossRef] [PubMed]
25. Arianoutsou, M.; Bazos, I.; Christopoulou, A.; Kokkoris, I.; Zikos, A.; Zervou, S.; Delipetrou, P.; Cardoso, A.; Deriu, I.; Gervasini, E. Alien plants of Europe: Introduction pathways, gateways and time trends. *PeerJ* **2021**, *9*, e11270. [CrossRef] [PubMed]
26. Martín-Mora, E.; Ellis, S.; Page, L. Use of web-based species occurrence information systems by academics and government professionals. *PLoS ONE* **2020**, *15*, e0236556. [CrossRef]
27. Ni, M.; Deane, D.C.; Li, S.; Wu, Y.; Sui, X.; Xu, H.; Chu, C.; He, F.; Fang, S. Invasion success and impacts depend on different characteristics in non-native plants. *Divers. Distrib.* **2021**, *27*, 1194–1207. [CrossRef]
28. Chen, J.; Ma, F.; Zhang, Y.; Wang, C.; Xu, H. Spatial distribution patterns of invasive alien species in China. *Glob. Ecol. Conserv.* **2021**, *26*, e01432. [CrossRef]
29. Chen, X.-L.; Ning, D.-D.; Xiao, Q.; Jiang, Q.-Y.; Lu, Y.-Y.; Xu, Y.-J. Factors affecting the geographical distribution of invasive species in China. *J. Integr. Agric.* **2022**, *21*, 1116–1125. [CrossRef]
30. Pan, X.; Zhang, J.; Xu, H.; Zhang, X.; Zhang, W.; Song, H.; Zhu, S. Spatial similarity in the distribution of invasive alien plants and animals in China. *Nat. Hazards* **2015**, *77*, 1751–1764. [CrossRef]
31. Xu, G.; Li, H.; Mo, X.; Meng, W. Composition and spatial-temporal distribution of Chinese naturalized plants. *Chin. J. Plant Ecol.* **2019**, *43*, 601–610. [CrossRef]
32. Zhou, Q.; Wang, Y.; Li, X.; Liu, Z.; Wu, J.; Musa, A.; Ma, Q.; Yu, H.; Cui, X.; Wang, L. Geographical distribution and determining factors of different invasive ranks of alien species across China. *Sci. Total Environ.* **2020**, *722*, 137929. [CrossRef] [PubMed]
33. Li, Y.; Shen, Z. Roles of Dispersal Limit and Environmental Filtering in Shaping the Spatiotemporal Patterns of Invasive Alien Plant Diversity in China. *Front. Ecol. Evol.* **2020**, *8*, 544670. [CrossRef]
34. Chen, C.; Wang, Q.-H.; Wu, J.-Y.; Huang, D.; Zhang, W.-H.; Zhao, N.; Li, X.-F.; Wang, L.-X. Historical introduction, geographical distribution, and biological characteristics of alien plants in China. *Biodivers. Conserv.* **2017**, *26*, 353–381. [CrossRef]
35. Yan, X.; Liu, Q.; Shou, H.; Zeng, X.; Zhang, Y.; Chen, L.; Liu, Y.; Ma, H.; Qi, S.; Ma, J. The categorization and analysis on the geographic distribution patterns of Chinese alien invasive plants. *Biodivers. Sci.* **2014**, *22*, 667–676.
36. Ma, J. *The Survey Reports on Chinese Alien Invasive Plants*; Part I & II; Higher Education Press: Beijing, China, 2014.
37. Hulme, P.E. Contrasting alien and native plant specie-area relationships: The importance of spatial grain and extent. *Glob. Ecol. Biogeogr.* **2008**, *17*, 641–647. [CrossRef]
38. McGeoch, M.; Butchart, S.; Spear, D.; Marais, E.; Kleynhans, E.; Symes, A.; Chanson, J.; Hoffmann, M. Global indicators of biological invasion: Species numbers, biodiversity impact and policy responses. *Divers. Distrib.* **2010**, *16*, 95–108. [CrossRef]
39. Vilà, M.; Basnou, C.; Pyšek, P.; Josefsson, M.; Genovesi, P.; Gollasch, S.; Nentwig, W.; Olenin, S.; Roques, A.; Roy, D.; et al. DAISIE partners How well do we understand the impacts of alien species on ecosystem services? A pan-European, cross-taxa assessment. *Front. Ecol. Environ.* **2010**, *8*, 135–144. [CrossRef]
40. Simpson, A.; Turner, R.; Blake, R.; Liebhold, A.; Dorado, M. United States Register of Introduced and Invasive Species: U.S. Geological Survey Data Release. Available online: https://www.sciencebase.gov/catalog/item/6357fcfed34ebe4425031fb6 (accessed on 26 May 2024).

41. Zhao, C.; Liu, Q.; Li, F.; Wong, L.J.; Pagad, S. Global Register of Introduced and Invasive Species—China. Version 1.3. Invasive Species Specialist Group ISSG. Checklist Dataset. 2020. Available online: https://www.gbif.org/dataset/6d11211b-caa0-4e63-b99c-e944099d5017 (accessed on 26 May 2024).
42. Van Kleunen, M.; Pyšek, P.; Dawson, W.; Essl, F.; Kreft, H.; Pergl, J.; Weigelt, P.; Stein, A.; Dullinger, S.; König, C. The Global Naturalized Alien Flora (GloNAF) database. *Ecology* **2019**, *100*, e02542. [CrossRef] [PubMed]
43. Lin, Q.; Yu, S.; Tang, S.; Cui, X.; Gao, X.; Wang, H.; Liu, Q.; Ma, J. Current status of naturalized alien species in China and its relative problem. *Chin. J. Plant Ecol.* **2021**, *45*, 1275–1280. [CrossRef]
44. Jiang, H.; Fan, Q.; Li, J.-T.; Shi, S.; Li, S.-P.; Liao, W.-B.; Shu, W.-S. Naturalization of alien plants in China. *Biodivers. Conserv.* **2011**, *20*, 1545–1556. [CrossRef]
45. Liu, J.; Dong, M.; Miao, S.L.; Li, Z.Y.; Song, M.H.; Wang, R.Q. Invasive alien plants in China: Role of clonality and geographical origin. *Biol. Invasions* **2006**, *8*, 1461–1470. [CrossRef]
46. Ma, J.; Li, H. *The Checklist of the Alien Invasive Plants of China*; Higher Education Press: Beijing, China, 2018.
47. Weber, E.; Shi, G.S.; Li, B. Invasive alien plants in China: Diversity and ecological insights. *Biol. Invasions* **2008**, *10*, 1411–1429. [CrossRef]
48. Blackburn, T.M.; Essl, F.; Evans, T.; Hulme, P.E.; Jeschke, J.M.; Kühn, I.; Kumschick, S.; Marková, Z.; Mrugała, A.; Nentwig, W.; et al. A Unified Classification of Alien Species Based on the Magnitude of their Environmental Impacts. *PLoS Biol.* **2014**, *12*, e1001850. [CrossRef]
49. Courchamp, F.; Fournier, A.; Bellard, C.; Bertelsmeier, C.; Bonnaud, E.; Jeschke, J.M.; Russell, J.C. Invasion Biology: Specific Problems and Possible Solutions. *Trends Ecol. Evol.* **2017**, *32*, 13–22. [CrossRef]
50. Diagne, C.; Leroy, B.; Vaissière, A.-C.; Gozlan, R.E.; Roiz, D.; Jarić, I.; Salles, J.-M.; Bradshaw, C.J.A.; Courchamp, F. High and rising economic costs of biological invasions worldwide. *Nature* **2021**, *592*, 571–576. [CrossRef] [PubMed]
51. Early, R.; Bradley, B.A.; Dukes, J.S.; Lawler, J.J.; Olden, J.D.; Blumenthal, D.M.; Gonzalez, P.; Grosholz, E.D.; Ibañez, I.; Miller, L.P. Global threats from invasive alien species in the twenty-first century and national response capacities. *Nat. Commun.* **2016**, *7*, 12485. [CrossRef]
52. Paini, D.R.; Sheppard, A.W.; Cook, D.C.; De Barro, P.J.; Worner, S.P.; Thomas, M.B. Global threat to agriculture from invasive species. *Proc. Natl. Acad. Sci. USA* **2016**, *113*, 7575–7579. [CrossRef] [PubMed]
53. Diagne, C.; Leroy, B.; Gozlan, R.; Vaissière, A.-C.; Assailly, C.; Nuninger, L.; Roiz, D.; Jourdain, F.; Jarić, I.; Courchamp, F. InvaCost—A public database of the economic costs of biological invasions worldwide. *Sci. Data* **2020**, *7*, 277. [CrossRef]
54. Liu, N.; Li, Y.; Zhang, R. Invasion of Colorado potato beetle, Leptinotarsa decemlineata, in China: Dispersal, occurrence, and economic impact. *Entomol. Exp. Appl.* **2012**, *143*, 207–217. [CrossRef]
55. Xu, H.; Ding, H.; Li, M.; Qiang, S.; Guo, J.; Han, Z.; Huang, Z.; Sun, H.; He, S.; Wu, H. The distribution and economic losses of alien species invasion to China. *Biol. Invasions* **2006**, *8*, 1495–1500. [CrossRef]
56. IPBES. *Thematic Assessment Report on Invasive Alien Species and their Control of the Intergovernmental Science-Policy Platform on Biodiversity and Ecosystem Services*; Roy, H.E., Pauchard, A., Stoett, P., Renard Truong, T., Eds.; IPBES Secretariat: Bonn, Germany, 2023. [CrossRef]
57. Magarey, R.; Colunga-Garcia, M.; Fieselmann, D. Plant Biosecurity in the United States: Roles, Responsibilities, and Information Needs. *BioScience* **2009**, *59*, 875–884. [CrossRef]
58. Carroll, S.R.; Herczog, E.; Hudson, M.; Russell, K.; Stall, S. Operationalizing the CARE and FAIR Principles for Indigenous data futures. *Sci. Data* **2021**, *8*, 108. [CrossRef] [PubMed]
59. Wilkinson, M.D.; Dumontier, M.; Aalbersberg, I.J.; Appleton, G.; Axton, M.; Baak, A.; Blomberg, N.; Boiten, J.-W.; da Silva Santos, L.B.; Bourne, P.E. The FAIR Guiding Principles for scientific data management and stewardship. *Sci. Data* **2016**, *3*, 160018. [CrossRef] [PubMed]
60. Carroll, S.; Garba, I.; Figueroa-Rodríguez, O.; Holbrook, J.; Lovett, R.; Materechera, S.; Parsons, M.; Raseroka, K.; Rodriguez-Lonebear, D.; Rowe, R. The CARE Principles for Indigenous Data Governance. *Data Sci. J.* **2020**, *19*, 43. [CrossRef]

MDPI AG
Grosspeteranlage 5
4052 Basel
Switzerland
Tel.: +41 61 683 77 34

Biology Editorial Office
E-mail: biology@mdpi.com
www.mdpi.com/journal/biology

www.ingramcontent.com/pod-product-compliance
Lightning Source LLC
Chambersburg PA
CBHW041411210926
51850CB00013B/3236
* 9 7 8 3 7 2 5 8 4 4 9 7 5 *